Harmonic Superspace

This is the first pedagogical introduction to the harmonic superspace method in extended ($N > 1$) supersymmetry. Inspired by recent exciting developments in superstring theory, it provides a systematic treatment of the quantum field theories with $N = 2$ and $N = 3$ extended supersymmetry in harmonic superspace. This new kind of superspace, proposed and developed by the authors, is now generally recognized as the most natural geometric framework for such theories.

The authors present the harmonic superspace approach as a means of providing a concise and covariant off-shell description of all of the $N = 2$ supersymmetric theories, at both the classical and quantum levels. Furthermore, they show how it offers a unique way to construct an off-shell formulation of a theory with higher supersymmetry, namely, the $N = 3$ supersymmetric Yang–Mills theory. On shell it is equivalent to the famous $N = 4$ super-Yang–Mills theory which was the first example of an ultraviolet-finite supersymmetric model. Harmonic superspace makes manifest many remarkable geometric properties of the $N = 2$ theories, for example, the one-to-one correspondence between $N = 2$ supersymmetric matter, and hyper-Kähler and quaternionic manifolds. Moreover, this methodology offers new and important insights into the geometry of these complex Riemannian manifolds.

This book will be of great interest to researchers and graduate students working in the areas of supersymmetric quantum field theory, string theory and complex geometries.

ALEXANDER GALPERIN, EVGENY IVANOV, VICTOR OGIEVETSKY (deceased) and EMERY SOKATCHEV invented harmonic superspace whilst working together at the Joint Institute for Nuclear Research at Dubna, Russia, during the 1980s. They are world experts on the subject and developed the formalism to the present state of the art. Since that time, the authors have dispersed across the globe.

After working at Imperial College, London, and the Johns Hopkins University, Baltimore, Galperin is now employed by a finance company in Washington DC, but remains interested in the field. Ivanov is a professor at the Bogoliubov Laboratory of Theoretical Physics, Joint Institute for Nuclear Research, Dubna, where he heads the research group 'Problems of Supersymmetry'. He succeeded Professor Ogievetsky who was the head of this research group for many years, until passing away in 1996. Ogievetsky was a pioneer in the theory of supersymmetry and during his lifetime was honoured with the I. E. Tamm Gold Medal of the Academy of Sciences of the USSR and the von Humboldt Foundation Award. Since leaving Dubna, Sokatchev has worked at CERN, the Bulgarian Academy of Sciences, and the University of Bonn, and is now a professor at the Laboratoire d'Annecy-le-Vieux de Physique Théorique (CNRS) and at the Université de Savoie, France – where he teaches.

Between them, Galperin, Ivanov, Ogievetsky and Sokatchev have authored several hundred papers in the fields of supersymmetry, gauge theories, string theory, supergravity and integrable systems.

CAMBRIDGE MONOGRAPHS ON
MATHEMATICAL PHYSICS

J. Ambjørn, B. Durhuus and T. Jonsson *Quantum Geometry: A Statistical Field Theory Approach*
A. M. Anile *Relativistic Fluids and Magneto-Fluids*
J. A. de Azcárraga and J. M. Izquierdo *Lie Groups, Lie Algebras, Cohomology and Some Applications in Physics*†
V. Belinski and E. Verdaguer *Gravitational Solitons*
J. Bernstein *Kinetic Theory in the Early Universe*
G. F. Bertsch and R. A. Broglia *Oscillations in Finite Quantum Systems*
N. D. Birrell and P. C. W. Davies *Quantum Fields in Curved Space*†
S. Carlip *Quantum Gravity in 2 + 1 Dimensions*
J. C. Collins *Renormalization*†
M. Creutz *Quarks, Gluons and Lattices*†
P. D. D'Eath *Supersymmetric Quantum Cosmology*
F. de Felice and C. J. S. Clarke *Relativity on Curved Manifolds*†
P. G. O. Freund *Introduction to Supersymmetry*†
J. Fuchs *Affine Lie Algebras and Quantum Groups*†
J. Fuchs and C. Schweigert *Symmetries, Lie Algebras and Representations: A Graduate Course for Physicists*
A. S. Galperin, E. A. Ivanov, V. I. Ogievetsky and E. S. Sokatchev *Harmonic Superspace*
R. Gambini and J. Pullin *Loops, Knots, Gauge Theories and Quantum Gravity*†
M. Göckeler and T. Schücker *Differential Geometry, Gauge Theories and Gravity*†
C. Gómez, M. Ruiz Altaba and G. Sierra *Quantum Groups in Two-dimensional Physics*
M. B. Green, J. H. Schwarz and E. Witten *Superstring Theory, volume 1: Introduction*†
M. B. Green, J. H. Schwarz and E. Witten *Superstring Theory, volume 2: Loop Amplitudes, Anomalies and Phenomenology*†
S. W. Hawking and G. F. R. Ellis *The Large-Scale Structure of Space-Time*†
F. Iachello and A. Aruna *The Interacting Boson Model*
F. Iachello and P. van Isacker *The Interacting Boson–Fermion Model*
C. Itzykson and J.-M. Drouffe *Statistical Field Theory, volume 1: From Brownian Motion to Renormalization and Lattice Gauge Theory*†
C. Itzykson and J.-M. Drouffe *Statistical Field Theory, volume 2: Strong Coupling, Monte Carlo Methods, Conformal Field Theory, and Random Systems*†
J. I. Kapusta *Finite-Temperature Field Theory*†
V. E. Korepin, A. G. Izergin and N. M. Boguliubov *The Quantum Inverse Scattering Method and Correlation Functions*†
M. Le Bellac *Thermal Field Theory*†
N. H. March *Liquid Metals: Concepts and Theory*
I. M. Montvay and G. Münster *Quantum Fields on a Lattice*†
A. Ozorio de Almeida *Hamiltonian Systems: Chaos and Quantization*†
R. Penrose and W. Rindler *Spinors and Space-Time, volume 1: Two-Spinor Calculus and Relativistic Fields*†
R. Penrose and W. Rindler *Spinors and Space-Time, volume 2: Spinor and Twistor Methods in Space-Time Geometry*†
S. Pokorski *Gauge Field Theories*, 2nd edition
J. Polchinski *String Theory, volume 1: An Introduction to the Bosonic String*
J. Polchinski *String Theory, volume 2: Superstring Theory and Beyond*
V. N. Popov *Functional Integrals and Collective Excitations*†
R. G. Roberts *The Structure of the Proton*†
J. M. Stewart *Advanced General Relativity*†
A. Vilenkin and E. P. S. Shellard *Cosmic Strings and Other Topological Defects*†
R. S. Ward and R. O. Wells Jr *Twistor Geometry and Field Theories*†

† Issued as a paperback

The list of errors found in the first edition of the book "Harmonic Superspace"

1. p. 31, Eq. (2.22): should be the factor i before commutators in the l.h.s.;
2. p. 33, the sentence continuing Eq. (2.32): should be "is the three-dimensional space projection of the matrix part of L^{ab}" just instead of "is the angular momentum operator";
3. p. 35, the very end of 4th line after Eqs. (2.39): should be "2^{2N}";
4. p. 42, Eq. (2.66): "bars" above ψ and κ should be shifted to the second positions;
5. p. 76, the sentence just preceding Eq. (5.9): should be "Weyl" instead of "Dirac";
6. p. 93, Eq. (5.101): "du " under the integral is misleading and should be removed;
7. p. 95, Eq. (5.111): the same;
8. p. 114, Eq. (6.37): the r.h.s. of 3d item should read "$-\delta_i^j$";
9. p. 157, Eq. (8.61): in the r.h.s. there should be $(D_{\bar{1}}^{--})^2$;
10. p. 248, Eq. (11.210): the signs of both items in the r.h.s. should be altered;
11. p. 260, Eq. (11.284): the sign of the 1st item in the r.h.s. should be altered;
12. p. 276, the beginning of 2nd line after Eq. (12.68): should be "$D^{(2,-1)}, D^{(-2,1)}$";
13. p. 294, Ref. [G13]: should be "912 - 916" instead of "155 - 158";
14. p. 297, Ref. [H9]: should be "(1982)" instead of "(1984)";
15. p. 301, Ref. [S8]: should be "275 - 277" instead of "278 - 289".

Harmonic Superspace

A. S. Galperin, E. A. Ivanov, V. I. Ogievetsky, E. S. Sokatchev

CAMBRIDGE UNIVERSITY PRESS
Cambridge, New York, Melbourne, Madrid, Cape Town, Singapore, São Paulo

Cambridge University Press
The Edinburgh Building, Cambridge CB2 2RU, UK

Published in the United States of America by Cambridge University Press, New York

www.cambridge.org
Information on this title: www.cambridge.org/9780521801645

First published 2001
This digitally printed first paperback version (with corrections) 2007

A catalogue record for this publication is available from the British Library

Library of Congress Cataloguing in Publication data

Harmonic superspace / A.S. Galperin . . . [et al.].
p. cm.
Includes bibliographical references and index.
ISBN 0 521 80164 8
1. Supersymmetry. I. Galperin, A. S. (Alexander Samoilovich), 1954–
QC174.17.S9 H37 2001
539.7′25–dc21 2001025481

ISBN-13 978-0-521-80164-5 hardback
ISBN-10 0-521-80164-8 hardback

ISBN-13 978-0-521-02042-8 paperback
ISBN-10 0-521-02042-5 paperback

Contents

Preface

This book is a pedagogical introduction to the harmonic superspace method in extended supersymmetry. There exist quite a few monographs, textbooks and reviews devoted to simple (or $N = 1$) supersymmetry, including detailed presentations of the $N = 1$ superfield techniques, the natural language for $N = 1$ supersymmetry. However, until now there has been no systematic treatment of the analogous issues in extended ($N \geq 2$) supersymmetry. In view of the growing interest in extended supersymmetries, mainly inspired by the impressive developments in superstring theory during the last decade, the need for such a presentation is becoming urgent. The present book is intended to partly fill this gap, mainly with regard to the simplest extended supersymmetry, the $N = 2$ one. We hope to convince the reader that the natural framework for dealing with $N = 2$ supersymmetry is harmonic superspace, an extension of the ordinary superspace of Salam and Strathdee by the bosonic coordinates of the supersymmetry automorphism (or R symmetry) group. The harmonic superspace approach provides a concise, manifestly covariant and off-shell description of all of the $N = 2$ supersymmetric theories, at both the classical and quantum levels. It also offers the unique way to construct an off-shell formulation of a theory with higher supersymmetry, namely, the $N = 3$ supersymmetric Yang–Mills theory. On shell it is equivalent to the famous $N = 4$ super-Yang–Mills theory which was the first example of an ultraviolet-finite supersymmetric model. Harmonic superspace also makes manifest many remarkable geometric properties of the $N = 2$ theories; for instance, the one-to-one correspondence between $N = 2$ supersymmetric matter and hyper-Kähler and quaternionic manifolds. Moreover, the harmonic superspace point of view offers new insights into the geometry of these complex Riemannian manifolds.

While writing this book, we benefitted a lot from discussions and exchanges of ideas with many of our colleagues and friends. We wish to express our warmest gratitude to them. Special thanks are due to Paul Townsend who encouraged us at the final stages of the writing, as well as to François Delduc, Paul Howe, Oleg Ogievetsky and Boris Zupnik whose comments (sometimes critical) and suggestions allowed us (we hope) to make the presentation more comprehensive and readable. We apologize for the unavoidable incompleteness of the list of references; we have made an effort to refer to those papers which are of conceptual value and are most relevant to the specific subjects covered by the book.

The work on the manuscript was completed during multiple visits of E.I. to LAPTH (Annecy-le-Vieux). He would like to thank the Director of LAPTH Professor Paul Sorba for kind hospitality in Annecy. His work was supported by the PICS Project No. 593, NATO Grant No. PST. GLG 974874, RFBR-CNRS grant No. 9802-22034 and RFBR grant No. 99-02-18417.

The first draft of this book was written some years ago, when all the four authors were working together at the Bogoliubov Laboratory of Theoretical Physics at JINR, Dubna, the place where harmonic superspace was born. Soon after that, life scattered us all over the world. Destiny decided that we had to finish the book without our teacher, colleague and dear friend Victor Isaakovich Ogievetsky who passed away in March of 1996.

The three other authors (A.G., E.I. and E.S.) hope that this book will be a tribute to the bright memory of V. I. Ogievetsky.

1
Introductory overview

We start by overviewing the origins, motivations, basic ideas and results of the harmonic superspace (and space) approach. Our major aim here is to give the reader a preliminary impression of the subject before immersion into the main body of the book.

1.1 Brief motivations

It is hardly possible to overestimate the rôle of symmetries in the development of physics. The place they occupy is becoming more and more important every year. The very family of symmetries is getting richer all the time: Besides the old symmetries based on Lie algebras we are now exploiting new kinds of symmetries. These include supersymmetries which mix bosons with fermions and are based on superalgebras, symmetries associated with non-linear algebras of Zamolodchikov's type, symmetries connected to quantum groups, etc. To date, the supersymmetric models have been studied in most detail. They turn out to have quite remarkable features. They open a new era in the search for a unified theory of all interactions including gravity. They help to solve the hierarchy problem in the grand unification theories. For the first time in the history of quantum field theory, supersymmetry has led to the discovery of a class of ultraviolet-finite local four-dimensional field theories. In these finite theories the ultraviolet divergences in the boson and fermion loops 'miraculously' cancel against each other. Supersymmetries underlie the superstring theories, which provide the first consistent scheme for quantization of gravity. The research programs of the leading accelerator laboratories include searches for supersymmetric partners of the known particles (predicted by supersymmetry but not yet discovered).

In view of this impressive development, it is imperative to be able to formulate the supersymmetric theories in a systematic, consistent and clear way. There already exist several reviews [F2, F3, F4, F7, N2, O4, S13, V1] and textbooks [B17, G26, W7, W12] devoted to the simplest kind of supersymmetry, $N = 1$ (i.e., containing one spinor generator in its superalgebra). It was this supersymmetry that was first discovered in the pioneering articles [G38, G39, V4, V5, V6, W8]. The superfield approach appropriate to this case was developed in the 1970s. However, extended supersymmetries (i.e., those containing more than one spinor generator) turned out much more difficult. Each new step in understanding them requires new notions and approaches. Even in the simplest extended $N = 2$ supersymmetry, until 1984 no way to formulate all such theories off shell, in a manifestly supersymmetric form and in terms of unconstrained superfields, was known. Such formulations are preferable not only because of their intrinsic beauty, but also since they provide an efficient technique, in particular, in quantum calculations or in the proof of finiteness. The invention of a new, harmonic superspace [G4, G13] made it possible to develop off-shell unconstrained formulations of all the $N = 2$ supersymmetric theories (matter, Yang–Mills and supergravity) and of $N = 3$ Yang–Mills theory.

$N = 2$ harmonic superspace is standard superspace augmented by the two-dimensional sphere $S^2 \sim SU(2)/U(1)$. In such an enlarged superspace it is possible to introduce a new kind of analyticity, Grassmann harmonic [G4, G13]. This proved to be the key to the adequate off-shell unconstrained formulations, just like chirality [F13], the simplest kind of Grassmann analyticity [G8], is a keystone in $N = 1$ supersymmetry. This new analyticity amounts to the existence of an analytic subspace of harmonic superspace whose odd dimension is half of that of the full superspace. All $N = 2$ theories mentioned above are naturally described by *Grassmann analytic* superfields, i.e., the unconstrained superfields in this subspace. A similar kind of analyticity underlies the $N = 3$ gauge theory [G5, G6].

A most unusual and novel feature of the analytic superfields is the unavoidable presence of infinite sets of auxiliary and/or gauge degrees of freedom in their component expansions. They naturally emerge from the harmonic expansions on the two-sphere S^2 with respect to a new sort of bosonic coordinates, the harmonic variables, which describe S^2 in a parametrization-independent way. These infinite sets, instead of being a handicap, proved to be very helpful indeed. It is due to their presence in the analytic superfield describing the $N = 2$ scalar multiplet (hypermultiplet) [F1, S12] off shell that one can circumvent the so-called 'no-go' theorem [H18, S21] claiming that such a formulation is not possible. In fact, the no-go theorems always implicitly assume the existence of a *finite* set of auxiliary fields.

The Grassmann analytic superfields with their infinite towers of components can be handled in much the same way as ordinary superfields, using a set of simple rules and tools. In [G14, G15] we worked out the quantization scheme for the $N = 2$ matter and gauge theories in harmonic superspace. The crucial importance of formulating quantum perturbation theory in supersymmetric models in terms of *unconstrained* off-shell superfields has repeatedly been pointed out in the literature (see, e.g., [H16]). Such formulations allow one to understand the origin of many remarkable properties of quantum supersymmetric theories which seem miraculous in the context of the component or constrained superfield formulations. Above all, this concerns the cancellation of ultraviolet divergences. Harmonic superspace is the only known approach which provides unconstrained off-shell formulations of both the matter and gauge $N = 2$ multiplets and as such it is indispensable for quantum calculations in the theories involving these multiplets. A particular representative of this class of theories is $N = 4$ super-Yang–Mills theory which, from the $N = 2$ perspective, is just the minimal coupling of the hypermultiplet in the adjoint representation of the gauge group to the $N = 2$ super-Yang–Mills multiplet.

It is worthwhile to emphasize that the harmonic superspace approach is very close to the twistor one which is an effective tool for solving the self-dual Yang–Mills and Einstein equations. In fact, harmonic superspace could be regarded as an isotwistor superspace. However, even when applied to the purely bosonic self-duality problems, the harmonic space approach has some advantages, one of them being as follows. We use harmonics (the fundamental isospin 1/2 spherical functions) as abstract global coordinates spanning the whole two-sphere. This is in contrast with, e.g., polar or stereographic coordinates which require two charts on the sphere. So, if one succeeds in solving a self-duality equation in terms of harmonics, there will be no need to attack the famous Riemann–Hilbert problem which is central in conventional twistor approaches. We also wish to stress that the harmonic (super)space formalism heavily uses the Cartan coset technique, transparent and familiar to many physicists.

A surge of interest in the harmonic superspace methods and, above all, in the methods for off-shell quantum calculations was mainly motivated by two remarkable developments in our understanding of supersymmetric field theories during the 1990s.

The first one stems from the seminal paper by Seiberg and Witten [S5] where it was suggested that $N = 2$ gauge theories are exactly solvable at the full quantum level under some reasonable hypotheses like S duality intimately related to extended supersymmetry [W17]. The study of the structure of the quantum low-energy effective actions of $N = 2$ gauge theories, in both the perturbative and non-perturbative sectors, is of great importance in this respect. The quantum harmonic superspace methods were successfully applied for this purpose, in particular for computing the holomorphic and non-holomorphic contributions to the effective action (see [B14, B15, B16, I8] and references therein).

The second source of interest is the famous Maldacena AdS/CFT conjecture [G42, M1, W16]. This is the idea that the quantum $N = 4$ super Yang–Mills theory in the limit of large number of colors and strong coupling is dual to the type IIB superstring on $AdS_5 \times S^5$ and contains the corresponding supergravity as a sub-sector of its Hilbert space. This conjecture greatly stimulated thorough analysis of the structure of this exceptional gauge theory from different points of view using different calculational means. The harmonic superspace methods, as was shown in several recent papers [E1, E3, E4, E5, H14], can drastically simplify the calculations and allow one to make far reaching predictions in $N = 4$ super-Yang–Mills theory.

All this justifies the need for a comprehensive introduction to the harmonic superspace approach. We hope that the present book will meet, at least partly, this quest. Here we do not discuss the latest developments but prefer to concentrate on the basics of the harmonic superspace method. Some developments are briefly addressed in the Conclusions. When reading this book one may find it helpful to consult the reviews and books mentioned above. We also point out that there are a few papers devoted to the mathematical aspects of harmonic superspace and, in particular, to a more rigorous definition of it, e.g., [H3, H10, H12, R6, S4]. We do not address these special issues in our rather elementary exposition.

1.2 Brief summary

The present book has been conceived as a pedagogical review of all the extended supersymmetric $N = 2$ theories and of $N = 3$ Yang–Mills theory in the framework of harmonic superspace. The details of these theories are discussed, as well as some applications. A special emphasis is put on their geometrical origin and on the relationship with hyper-Kähler and quaternionic complex manifolds which appear as the target manifolds of $N = 2$ supersymmetric sigma models in a flat background and in the presence of supergravity, respectively. The Cartan coset techniques are used systematically with emphasis on their power and simplicity. The self-duality Yang–Mills and Einstein equations are treated in this language with stress on their deep affinity with $N = 2$ supersymmetric theories and on comparing the harmonic space approach with the twistor one.

A detailed outline of the content of this book is given at the end of Chapter 1. In order to help the reader, we preface the main body of the book with an overview of the basic ideas, notions and origins. We begin with a discussion of spaces and superspaces for the realization of symmetries and supersymmetries, emphasizing the importance of making the right choice: *The same symmetry can be realized in different ways, one of them being much more appropriate for a given problem than the others.*

1.3 Spaces and superspaces

Manifestly invariant formulations of field theories make use of some space (or superspace) where a given symmetry (or supersymmetry) is realized geometrically by coordinate transformations. Two examples are well known:

(i) In Minkowski space $\mathbb{M}^4 = (x^a)$ the Poincaré group transformations have the form

$$x'^a = \Lambda^a_b x^b + c^a \,. \tag{1.1}$$

In classical and quantum field theories the action principle and the equations of motion are manifestly invariant under (1.1), the form of the corresponding field transformations being completely fixed by (1.1) and the tensor properties of the field, e.g.,

$$f'(x') = f(x) \tag{1.2}$$

for a scalar field $f(x)$. It is important that this transformation law does not depend on the model under consideration.

(ii) One usually attempts to formulate manifestly invariant N-extended supersymmetric theories in the standard superspace [S1]

$$\mathbb{R}^{4|4N} = (x^a, \theta^\alpha_i, \bar{\theta}^{\dot{\alpha}i}) \,, \qquad i = 1, 2, \ldots, N \tag{1.3}$$

involving the spinor *anticommuting* coordinates $\theta^\alpha_i, \bar{\theta}^{\dot{\alpha}i}$ in addition to x^a. Their transformation rules under the Poincaré group are evident, while the transformations under supersymmetry (supertranslations with anticommuting parameters $\epsilon^\alpha_i, \bar{\epsilon}^{\dot{\alpha}i}$) are given by

$$\delta x^a = i(\epsilon^i \sigma^a \bar{\theta}_i - \theta^i \sigma^a \bar{\epsilon}_i) \,, \qquad \delta\theta^\alpha_i = \epsilon^\alpha_i \,, \qquad \delta\bar{\theta}^{\dot{\alpha}i} = \bar{\epsilon}^{\dot{\alpha}i} \,. \tag{1.4}$$

Superfields $\Phi(x, \theta, \bar{\theta})$ are defined as functions on this superspace and their transformation law is completely determined by (1.2). For example, for a scalar superfield

$$\Phi'(x', \theta', \bar{\theta}') = \Phi(x, \theta, \bar{\theta}) \,. \tag{1.5}$$

Of course, this law is model-independent. Expanding a superfield $\Phi(x, \theta, \bar{\theta})$ in powers of the spinor (anticommuting, hence nilpotent) variables $\theta, \bar{\theta}$ yields a finite set of usual fields $f(x)$, $\psi^\alpha(x), \ldots$, called components of the superfield.

As an alternative to $\mathbb{R}^{4|4N}$, N-extended supersymmetry can also be realized in the so-called chiral superspace $\mathbb{C}^{4|2N}$ which is *complex* and contains only *half* of the spinor coordinates [F13]:

$$\delta x^a_L = -2i\theta^i_L \sigma^a \bar{\epsilon}_i \,, \qquad \delta\theta^\alpha_{L\,i} = \epsilon^\alpha_i \,. \tag{1.6}$$

In fact, the real superspace $\mathbb{R}^{4|4N}$ can be viewed as a real hypersurface in the complex superspace $\mathbb{C}^{4|2N}$:

$$x^a_L = x^a + i\theta^i \sigma^a \bar{\theta}_i \,, \qquad \theta^\alpha_{L\,i} = \theta^\alpha_i \,, \qquad \overline{\theta^\alpha_{L\,i}} = \bar{\theta}^{\dot{\alpha}\,i} \,. \tag{1.7}$$

1.4 Chirality as a kind of Grassmann analyticity

The superfields $\Phi(x_L, \theta_L) = \Phi(x + i\theta\sigma\bar{\theta}, \theta)$ defined in $\mathbb{C}^{4|2N}$ can be treated as Grassmann analytic superfields. Indeed, they obey the constraint

$$\bar{D}_{\dot{\alpha}i}\Phi = \left(-\frac{\partial}{\partial\bar{\theta}^{\dot{\alpha}i}} - i(\theta_i\sigma^a)_{\dot{\alpha}}\frac{\partial}{\partial x^a}\right)\Phi = 0\,, \tag{1.8}$$

where $\bar{D}^i_{\dot{\alpha}}$ is the covariant (i.e., commuting with the supersymmetry transformations) spinor derivative. In the basis $(x_L, \theta_L, \bar{\theta})$ this derivative simplifies to $\bar{D}^i_{\dot{\alpha}} = -\partial/\partial\bar{\theta}^{\dot{\alpha}i}$. Then the constraint (1.8) takes the form of a Cauchy–Riemann condition,

$$\frac{\partial\,\Phi}{\partial\,\bar{\theta}^{\dot{\alpha}}_i} = 0\,, \tag{1.9}$$

which means that Φ is a function of θ_L but is independent of $\bar{\theta}$ (cf. the standard theory of analytic functions where the Cauchy–Riemann condition $\partial f(z)/\partial\bar{z} = 0$ means that the function depends on the variable z and is independent of its conjugate $\bar{z}$). The notion of Grassmann analyticity [G8] in this simplest form is most useful in $N = 1$ supersymmetry. In this book the reader will see that there exist non-trivial generalizations of this concept which underlie the $N = 2$ and $N = 3$ supersymmetric theories.

It should by emphasized that finding the adequate superspace for a given theory is, as a rule, a non-trivial problem. The above superspaces $\mathbb{R}^{4|4N}$ and $\mathbb{C}^{4|2N}$ prove to be appropriate for off-shell formulations only in the simplest case of $N = 1$ supersymmetry. These 'standard' superspaces cease to be so useful in the extended ($N > 1$) supersymmetric theories. Finding and using the adequate superspaces for $N = 2,\ 3$ is the main subject of this book.

Now, before approaching the main problem, we recall in a few words some key points in $N = 1$ supersymmetry.

1.5 $N = 1$ chiral superfields

As already said, $N = 1$ supersymmetric theories can be formulated in the superspaces $\mathbb{R}^{4|4}$ or $\mathbb{C}^{4|2}$. Consider, for example, the simplest $N = 1$ supermultiplet, the matter one. On shell it contains a spin 1/2 field ψ_α and a complex scalar field $A(x)$. In $\mathbb{R}^{4|4}$ it can be described by a scalar superfield $\Phi(x, \theta, \bar{\theta})$. However, the latter involves too many fields in its θ expansion: Four real scalars, two Majorana spinors and a vector of various dimensions. To eliminate the extra fields, it is necessary to impose a *constraint* on the superfield, which turns out to be just the chirality (Grassmann analyticity) condition

$$\bar{D}_{\dot{\alpha}}\Phi = 0\,. \tag{1.10}$$

As explained above, this constraint means that Φ is an analytic superfield. In the $N = 1$ case the expansion of such a superfield (written down in the chiral basis) is very short [F13]:

$$\Phi(x, \theta, \bar{\theta}) = \Phi(x_L, \theta_L) = \phi(x_L) + \theta_L^\alpha \psi_\alpha(x_L) + \theta_L^\alpha \theta_{L\alpha} F(x_L) . \qquad (1.11)$$

The fields ϕ, ψ_α, F form the *off-shell* $N = 1$ matter supermultiplet.

The chiral ($N = 1$ analytic) superspace $\mathbb{C}^{4|2}$ is the cornerstone of all the $N = 1$ theories: They are either formulated in terms of chiral superfields (matter and its self-couplings) or are based on gauge principles which respect chirality (Yang–Mills and supergravity and their couplings to matter). The reader will see that for $N = 2$ and $N = 3$ the suitably modified concept of Grassmann analyticity will also be crucial.

1.6 Auxiliary fields

Besides the physical fields $\phi(x_L)$, $\psi_\alpha(x_L)$, the superfield $\Phi(x_L, \theta_L)$ also contains an *auxiliary* complex scalar field $F(x_L)$ of non-physical dimension 2. As a consequence, this field can only appear in an action without derivatives and thus can be eliminated by its equation of motion. In the presence of auxiliary fields the supersymmetry transformations are *model-independent* and so have the same form off and on shell. They form a *closed* supersymmetry algebra. For example, in the case of the chiral scalar superfield above one obtains from (1.5), (1.6) and (1.11)

$$\begin{aligned} \delta\phi(x) &= -\epsilon^\alpha \psi_\alpha(x) , \\ \delta\psi_\alpha(x) &= -2i\sigma^a_{\alpha\dot{\alpha}} \bar{\epsilon}^{\dot{\alpha}}\, \partial_a \phi(x) - 2\epsilon_\alpha F(x) , \\ \delta F(x) &= -i\bar{\epsilon}^{\dot{\alpha}} \sigma^a_{\alpha\dot{\alpha}}\, \partial_a \psi^\alpha(x) . \end{aligned} \qquad (1.12)$$

The commutator of two such supertranslations yields an ordinary translation with a parameter composed in accordance with the supersymmetry algebra. (See Chapter 2 for more details on the realization of supersymmetry in terms of fields.)

Of course, one can find a realization of supersymmetry on the physical fields only, with the auxiliary fields eliminated by the equations of motion of a given model. In fact, the first known realizations of supersymmetry were of just such a kind, and it was to some extent an 'art' to simultaneously find the invariant action and the supersymmetry transformations leaving it invariant. In contrast with the transformations in the presence of auxiliary fields, now one has:

(i) Supersymmetry transformations depending on the choice of the specific field model. They are in general non-linear and the structure of this non-linearity varies from one action to another.

(ii) The algebra of these transformations closes only modulo the equations of motion, i.e., on shell. Such algebras are referred to as *open* or *soft*.

These complications cause difficulties when trying to exploit the consequences of supersymmetry, in particular, in studying the ultraviolet behavior. Working in a manifestly invariant manner, in terms of the appropriate superfields, has undeniable advantages for such purposes. Note that some people prefer to avoid the use of superfields and instead work directly with the off-shell supermultiplets of fields including the auxiliary ones (e.g., $\phi(x)$, $\psi_\alpha(x)$, $F(x)$ in our $N = 1$ example). Then one needs a set of rules for handling such multiplets, known as *tensor calculus*. The superfield approach automatically reproduces all such rules in a nice geometrical way. This concerns the composition rule for supermultiplets (it amounts to multiplication of superfields), the building of invariant actions, etc.

The reader should realize that the notion of auxiliary fields is not peculiar to supersymmetry, it also appears in the usual non-supersymmetric theories. For instance, the Coulomb field is auxiliary in quantum electrodynamics.

The auxiliary fields play an extremely important rôle in the theories with extended supersymmetry, their number there may even become infinite. The reader will learn from the present book that this is due to a new feature of the harmonic superspace: It involves auxiliary bosonic coordinates. This superspace of a new kind is the only one that provides us with a systematic tool for off-shell realizations of all the $N = 2$ extended supersymmetries and the $N = 3$ Yang–Mills theory.

1.7 Why standard superspace is not adequate for $N = 2$ supersymmetry

'Not adequate' means that in the framework of the standard superspaces $\mathbb{R}^{4|8}$ and $\mathbb{C}^{4|4}$ it is impossible to find off-shell actions for an unconstrained description of all the $N = 2$ supersymmetric theories. We illustrate this on the example of the Fayet–Sohnius matter hypermultiplet [F1, S12]. On shell this supermultiplet contains four scalar fields forming an $SU(2)$ doublet $f^i(x)$ and two isosinglet spinor fields $\psi^\alpha(x)$, $\bar\kappa^{\dot\alpha}(x)$. To incorporate them as components of a standard superfield one has to use [S12] an isodoublet superfield $q^i(x, \theta, \bar\theta)$ defined in $\mathbb{R}^{4|8}$. Due to the large number of spinor variables this superfield contains a lot of redundant field components in addition to the physical ones listed above. The extra fields are eliminated by imposing the constraint [S12]

$$D^{(i}_\alpha q^{j)} = \bar D^{(i}_{\dot\alpha} q^{j)} = 0\,, \tag{1.13}$$

where (ij) means symmetrization and D^i_α, $\bar D^i_{\dot\alpha}$ are the supercovariant spinor derivatives obeying the algebra

$$\{D^i_\alpha, \bar D_{\dot\alpha j}\} = -2i\delta^i_j \sigma^a_{\alpha\dot\alpha} \frac{\partial}{\partial x^a} \tag{1.14}$$

(for their precise definition see Chapter 3). These constraints eliminate the extra field components of q^i, leaving only the above physical fields (and their derivatives in the higher terms of the θ expansion):

$$q^i(x, \theta, \bar{\theta}) = f^i(x) + \theta^{i\alpha}\psi_\alpha(x) + \bar{\theta}^i_{\dot{\alpha}}\bar{\kappa}^{\dot{\alpha}}(x) + \text{ derivative terms.} \tag{1.15}$$

However, at the same time the above constraints put all the physical fields on the free mass shell:

$$\Box f^i(x) = (\sigma^a)^{\alpha\dot{\alpha}}\frac{\partial}{\partial x^a}\psi_\alpha(x) = \sigma^a_{\alpha\dot{\alpha}}\frac{\partial}{\partial x^a}\bar{\kappa}^{\dot{\alpha}}(x) = 0. \tag{1.16}$$

The reason for this is that the constraints (1.13) are not integrable off shell: The supercovariant spinor derivatives do not anticommute. Equations (1.16) follow from the constraints (1.13) and the algebra (1.14), taking into account the definitions

$$f^i(x) = q^i|_{\theta=\bar{\theta}=0}, \qquad \psi_\alpha(x) = \frac{1}{2}D^i_\alpha q_i|_{\theta=\bar{\theta}=0}, \qquad \kappa^{\dot{\alpha}(x)} = \frac{1}{2}\bar{D}^{\dot{\alpha}}_i q^i|_{\theta=\bar{\theta}=0}. \tag{1.17}$$

In order to extend this theory off shell and to introduce interactions it has been proposed to relax, in one way or another, the constraints (1.13) [H15, Y1]. However, according to the general no-go theorem [H18, S21] (see Chapter 2), this is impossible in the framework of the standard $N = 2$ superspaces $\mathbb{R}^{4|8}$ or $\mathbb{C}^{4|4}$ using a *finite number* of auxiliary fields (or, equivalently, a *finite number* of standard $N = 2$ superfields). A natural way out was to look for other superspaces.

1.8 Search for conceivable superspaces (spaces)

Above we saw that it is helpful to consider different superspaces even in the simplest case $N = 1$. For any (super)symmetry there exists a number of admissible (super)spaces. The inadequacy of the standard superspaces $\mathbb{R}^{4|8}$ and $\mathbb{C}^{4|4}$ for off-shell realizations of $N = 2$ supersymmetry suggested to start searching through the list of other available superspaces. This list is provided by the standard coset construction due to E. Cartan [C4]* that allows one to classify the different (super)spaces of some (super)group G and to handle them effectively. One has to examine the conceivable quotients (we prefer the term 'coset') G/H of the group G over some of its subgroups H. For instance, Minkowski space is the coset $\mathbb{M}^4 = \mathcal{P}/\mathcal{L} = (x^a)$ of the Poincaré group $\mathcal{P}$ over its Lorentz subgroup $\mathcal{L}$. As we shall see later, the Poincaré group for the Euclidean space $\mathbb{R}^4$ can also be realized in another way, using the coset space $\mathcal{P}/SU(2) \times U(1)$, with $SU(2) \times U(1)$ being a subgroup of the rotation group $SO(4) = SU(2) \times SU(2)$. This space is closely related to the so-called *twistor space* (more precisely, the traditional twistor space is related by a similar procedure to the Poincaré group of the complexified Minkowski space $\mathbb{M}^4$).

* Subsequently rediscovered by physicists [C11, O3, V3].

Analogously, the standard real superspaces $\mathbb{R}^{4|4N}$ are the coset spaces

$$\mathbb{R}^{4|4N} = \frac{SuP_N}{\mathcal{L}} = (x^a, \theta_i^\alpha, \bar{\theta}^{\dot{\alpha}i}), \qquad (1.18)$$

where SuP_N is the N-extended super-Poincaré group involving the generators of the Poincaré group and the spinor supersymmetry generators Q_α^i, $\bar{Q}_{\dot{\alpha}i}$. In the same way, the chiral superspaces are the following coset spaces

$$\mathbb{C}^{4|2N} = \frac{SuP_N}{\{\mathcal{L}, \bar{Q}_{\dot{\alpha}i}\}} = (x^a, \theta_i^\alpha). \qquad (1.19)$$

Note the important difference between (1.18) and (1.19). In the latter the stability *supergroup* contains half of the spinor generators in addition to the Lorentz group ones. In Chapter 3 the coset techniques [C4, C11, O3, V3] are presented in detail. These techniques provide simple rules on how to find explicit transformation laws, how to construct invariants making use of covariant derivatives (obtained from the appropriate Cartan forms), etc.

1.9 $N = 2$ harmonic superspace

Certainly, $\mathbb{R}^{4|4N}$ and $\mathbb{C}^{4|2N}$ do not exhaust the list of possible superspaces for realizations of N-extended supersymmetry. Let us briefly outline some general features of $N = 2$ harmonic superspace, our main topic of interest in this book.

The $N = 2$ superalgebra

$$\left\{Q_\alpha^i, \bar{Q}_{\dot{\alpha}j}\right\} = 2\delta_j^i (\sigma^a)_{\alpha\dot{\alpha}} P_a\,, \qquad i, j = 1, 2 \qquad (1.20)$$

possesses an $SU(2)$ group of automorphisms, Q_α^i, $\bar{Q}_{\dot{\alpha},j}$ being $SU(2)$ doublets (indices i, j) and P_a being a singlet. In the standard case of eqs. (1.18) and (1.19) (with $N = 2$) this $SU(2)$ can be viewed as present both in the numerator and the denominator, thus effectively dropping out. To obtain the harmonic superspace, one has to keep only the $U(1)$ subgroup of $SU(2)$ in the denominator instead of the whole $SU(2)$:

$$\mathbb{H}^{4+2|8} = \frac{SuP_2}{\mathcal{L}} \times \frac{SU(2)}{U(1)}\,. \qquad (1.21)$$

In other words, one has to enlarge the $N = 2$ supersymmetry group by its automorphisms group $SU(2)$ realized in the coset space $SU(2)/U(1)$. The latter is a two-dimensional space known to have the topology of the two-sphere S^2. So, harmonic superspace is a tensor product of $\mathbb{R}^{4|8}$ and a two-sphere S^2.

1.10 Dealing with the sphere S^2

Before discussing the harmonic superspace as a whole it is instructive to study its much more familiar part $SU(2)/U(1)$. Of course, one could choose polar

(θ, ϕ) or stereographic $(t, \bar{t})$ coordinates on this sphere. However, it turns out much more convenient to coordinatize it by some 'zweibeins' $u^{+i}, u_i^- = \overline{u^{+i}}$ having $SU(2)$ indices i and $U(1)$ charges $\pm$. After imposing the constraint

$$u^{+i} u_i^- = 1 \,, \tag{1.22}$$

the matrix

$$\| u \| = \begin{pmatrix} u_1^+ & u_1^- \\ u_2^+ & u_2^- \end{pmatrix} = \frac{1}{\sqrt{1 + t\bar{t}}} \begin{pmatrix} e^{i\psi} & -\bar{t}e^{-i\psi} \\ te^{i\psi} & e^{-i\psi} \end{pmatrix}, \qquad 0 \le \psi < 2\pi \tag{1.23}$$

represents the group $SU(2)$ in the familiar stereographic parametrization. We are interested in its coset space $SU(2)/U(1)$. This means that the zweibeins have to be defined up to a $U(1)$ phase corresponding to a transformation of the $U(1)$ group in the coset denominator:

$$u_i^{+'} = e^{i\alpha} u_i^+ \,, \qquad u_i^{-'} = e^{-i\alpha} u_i^- \tag{1.24}$$

(this transformation can be realized as right multiplications of the matrix (1.23) with the Pauli matrix τ^3 as the generator). So the phase ψ in the parametrization (1.23) is inessential and one effectively deals only with the complex coordinates $t, \bar{t}$. In order for the phase not to show up, the 'functions' on the sphere must have a *definite* $U(1)$ charge q and, as a consequence, all the terms in their harmonic expansion must contain only products of zweibeins u^+, u^- of the given charge q. For instance, for $q = +1$

$$f^+(u) = f^i u_i^+ + f^{(ijk)} u_i^+ u_j^+ u_k^- + \cdots \,. \tag{1.25}$$

Such quantities undergo homogeneous $U(1)$ phase transformations, according to their overall charge. This requirement on the harmonic functions can be called $U(1)$ charge preservation. In each term in (1.25) complete symmetrization in the indices $i, j, k, \ldots$ is assumed, otherwise the term can be reduced to the preceding ones by eq. (1.22).

In fact, the zweibeins u_i^+, u_i^- are the fundamental spin 1/2 spherical harmonics familiar from quantum mechanics, and (1.25) is an example of a harmonic decomposition on S^2. This is why we call u_i^+, u_i^- *harmonic variables* (or simply 'harmonics').

1.10.1 Comparison with the standard harmonic analysis

We would like to point out the following important features of the harmonic space approach that differ from the standard ones in textbooks and reviews on harmonic analysis [B9, C2, C3, G31, G37, H5, H6, V2, W13]:

(i) We use the harmonics themselves as coordinates of the sphere. This amounts to refraining from using any explicit parametrization like the stereographic one (1.23). Instead, we assume the defining constraint (1.22) together with the requirement of $U(1)$ charge preservation.

(ii) We deal with symmetrized products of harmonics instead of sets of special functions, like the Jacobi polynomials or the spherical functions familiar from the harmonic analysis on the two-dimensional sphere.

These formal modifications turn out very convenient for the following main reasons:

(i) The coefficients in the harmonic expansions (like f^i, $f^{(ijk)}$, ... in (1.25)) transform as irreducible representations of the $SU(2)$ group of the coset numerator. This is of special value in $N = 2$ supersymmetry because the $N = 2$ supermultiplets are classified, in particular, according to the $SU(2)$ automorphism group.

(ii) Working with local coordinates one is confronted with the Riemann–Hilbert problem: Two maps are needed to cover the two-sphere or the extended complex plane. So, given a function which is well defined in the northern hemisphere, one has to worry about defining it consistently in the southern hemisphere. Remarkably enough, this problem does not appear if one exploits the harmonics u_i^+, u_i^- as 'global' coordinates on S^2. If one has succeeded in solving some equation in terms of harmonics, then the solution obtained is well defined on the entire sphere, after substitution of the parametrization (1.23) (or any other local one).

The latter statement can be illustrated by the following simple example. On the sphere S^2 one may introduce two covariant derivatives consistent with the constraint (1.22) and having $U(1)$ charges +2 and −2:

$$D^{++} = u^{+i}\frac{\partial}{\partial u^{-i}} \quad \text{and} \quad D^{--} = u^{-i}\frac{\partial}{\partial u^{+i}} . \tag{1.26}$$

They will be heavily used in what follows and referred to as harmonic derivatives. In the twistor literature [I2, H2, H22, K8] these derivatives are known as the *edth* and *antiedth* operators and their expressions in terms of polar or stereographic coordinates are used, e.g.,

$$D^{++} f^{(q)}(u) = -e^{(q+2)i\psi}\left[(1+t\bar{t})\frac{\partial\phi^{(q)}}{\partial\bar{t}} + \frac{qt}{2}\phi^{(q)}\right]. \tag{1.27}$$

Here $f^{(q)}(u) = e^{iq\psi}\phi^{(q)}(t,\bar{t})$ is a harmonic function of $U(1)$ charge q. As explained above, it depends on the coordinate ψ associated with the $U(1)$ charge through a simple phase factor. In contrast to the harmonic form (1.26), the edth operator D^{++} (1.27) explicitly involves the $U(1)$ charge q.

From the definition (1.26) follow the obvious rules for the action of D^{++} on the harmonics

$$D^{++}u_i^+ = 0\,, \qquad D^{++}u_i^- = u_i^+\,. \tag{1.28}$$

Let us consider the simple harmonic differential equation

$$D^{++}f^+ = 0\,. \tag{1.29}$$

In harmonics its solution is immediately obtained from (1.25):

$$f^+ = f^i u_i^+\,, \tag{1.30}$$

where f^i are arbitrary constants. Indeed, f^+ has this form because all other terms in its harmonic expansion include u^-.

Now it is instructive to compare this 'harmonic' procedure with solving the same equation as a partial differential equation with respect to the complex coordinates $t, \bar{t}$. It is easy to find the general solution of this equation for $q = 1$ in the form

$$f^+(t, \bar{t}, \psi) = e^{i\psi}(1 + t\bar{t})^{-\frac{1}{2}}F(t)\,, \tag{1.31}$$

where $F(t)$ is an arbitrary holomorphic function. However, we are interested in solutions well-behaved on the whole two-sphere (we wish to solve the Riemann–Hilbert problem). This requirement restricts the function $F(t)$ to the form of a polynomial of degree 1: $F(t) = f^1 + f^2 t$, where f^1, f^2 are arbitrary constants. In this way one obtains the same solution (1.30) in the particular parametrization (1.23). Note, however, that the solution (1.30) is *manifestly* $SU(2)$ *covariant* (the constants f^i form a doublet) whereas in a particular parametrization $SU(2)$ is realized as a non-linear coordinate transformation.

Finally, a word about integration on the two-sphere. In the harmonic approach it is defined by the following formal rules:

$$\int du\ 1 = 1\,, \qquad \int du\ u^+_{(i_1}..u^+_{i_k}u^-_{i_{k+1}}..u^-_{i_{k+l})} = 0\,. \tag{1.32}$$

This definition means the vanishing of the integrals of any spherical function with spin (represented by symmetrized products of harmonics). Of course, it admits integration by parts, etc. These rules can be justified by the use of some specific parametrization for the harmonics, e.g., (1.23). However, the abstract form (1.32) is most convenient in field theory, as the reader will have a number opportunities to see.

1.11 Why harmonic superspace helps

We now return to the harmonic superspace $\mathbb{H}^{4+2|8} = \mathbb{R}^{4|8} \times SU(2)/U(1)$ with the coordinates $\{x^a, \theta^\alpha_i, \bar{\theta}^{\dot{\alpha}i}, u^\pm_i\}$. We explain, on the example of the Fayet–Sohnius hypermultiplet, why adding the two-sphere is so crucial for the off-shell formulation of $N = 2$ supersymmetric field theories.

With the help of the harmonics $u_i^\pm$ we can give the constraints (1.13) another, more suggestive form. Let us multiply them by u_i^+, u_j^+. Introducing the notation

$$D_\alpha^+ = u_i^+ D_\alpha^i, \qquad \bar{D}_{\dot\alpha}^+ = u_i^+ \bar{D}_{\dot\alpha}^i \tag{1.33}$$

and

$$q^+ = u_i^+ q^i \tag{1.34}$$

one rewrites (1.13) as

$$D_\alpha^+ q^+ = 0\,, \qquad \bar{D}_{\dot\alpha}^+ q^+ = 0\,. \tag{1.35}$$

Equation (1.34) simply means that q^+ depends linearly on the harmonics u_i^+ (in this basis). It can be recast in an equivalent form using the harmonic derivative D^{++}:

$$D^{++} q^+ = 0 \tag{1.36}$$

(cf. (1.29)). Equations (1.35) and (1.36) together are clearly equivalent to the constraints (1.13). However, these equations turn out to have a deeper meaning than (1.13). First of all, we remark that the derivatives entering the modified constraints (1.35), (1.36) mutually (anti)commute,

$$\{D_\alpha^+, \bar{D}_{\dot\alpha}^+\} = [D^{++}, D_\alpha^+] = [D^{++}, \bar{D}_{\dot\alpha}^+] = 0\,, \tag{1.37}$$

in contrast to the derivatives entering the original constraints (1.13). This property is of great significance. Owing to it one can consider equations (1.35) as the *generalized Cauchy–Riemann condition of Grassmann analyticity*. To reveal its meaning one should choose an adequate basis in superspace, the so-called *analytic basis* (an analog of the chiral basis (1.7)):

$$x_A^a = x^a - 2i\theta^{(i}\sigma^a\bar{\theta}^{j)}u_i^+ u_j^-\,, \qquad \theta_{A\alpha}^\pm = u_i^\pm \theta_\alpha^i\,, \qquad \bar{\theta}_{A\dot\alpha}^\pm = u_i^\pm \bar{\theta}_{\dot\alpha}^i\,. \tag{1.38}$$

In this basis the spinor derivatives D_α^+ and $\bar{D}_{\dot\alpha}^+$ become simple partial derivatives and the constraints (1.35) take the form

$$D_\alpha^+ q^+ = \frac{\partial}{\partial\theta^{-\alpha}} q^+ = 0\,, \qquad \bar{D}_{\dot\alpha}^+ q^+ = \frac{\partial}{\partial\bar{\theta}^{-\dot\alpha}} q^+ = 0\,. \tag{1.39}$$

Like the Cauchy–Riemann condition of ordinary analyticity or that of $N = 1$ Grassmann analyticity (chirality), equations (1.39) express the fact that q^+ is independent of half of the relevant variables, this time of the spinor coordinates $\theta^{-\alpha}, \bar{\theta}^{-\dot\alpha}$. Their solution is

$$q^+ = q^+(x_A, \theta^+, \bar{\theta}^+, u^\pm)\,. \tag{1.40}$$

The same condition can be imposed on harmonic superfields with $U(1)$ charges different from $+1$. We refer to conditions like (1.35) or (1.39) as *Grassmann analyticity* conditions and to the subspace

$$\mathbb{HR}^{4+2|8} = (x_A, \theta^+, \bar{\theta}^+, u^\pm) = (\zeta, u^\pm) \tag{1.41}$$

as *Grassmann analytic superspace*. It contains only half of the original spinor coordinates (those having $U(1)$ charge equal to $+1$) and yet it is closed under the full $N = 2$ supersymmetry transformations. We can state that all the $N = 2$ supersymmetric theories (matter, Yang–Mills and supergravity) are most adequately formulated in its framework.

In the analytic basis (1.38) the harmonic derivative takes the form

$$D^{++} = u^{+i}\frac{\partial}{\partial u^{-i}} - 2i\theta^{+}\sigma^{a}\bar{\theta}^{+}\frac{\partial}{\partial x_A^a}\,, \tag{1.42}$$

where one sees space-time derivatives. As a consequence, eq. (1.36) becomes dynamical and yields the free equations of motion for all physical matter fields. So, we have succeeded in reformulating the original constraints (1.13) (whose rôle was to eliminate the extra fields and simultaneously to put the remaining physical fields on shell) into the analyticity constraints (1.39) (having the evident solution (1.40)) and the (free) equation of motion (1.36). This was achieved due to the presence of the harmonics $u_i^{\pm}$. Now we can go a step further and introduce general self-interactions. This simply amounts to inserting a general source J^{+++} of $U(1)$ charge $+3$ in the right-hand side of eq. (1.36):

$$D^{++}q^{+} = J^{+++}(q^{+}, u^{\pm})\,. \tag{1.43}$$

One has to realize that the harmonic expansion of the analytic superfield q^+ contains an *infinite number* of auxiliary fields. This is how harmonic superspace gets around the no-go theorem [H18, S21] asserting that it is not possible to describe the above complex hypermultiplet off shell with a finite number of auxiliary fields.

1.12 $N = 2$ supersymmetric theories

Now we are ready to very briefly overview the $N = 2$ matter, Yang–Mills and supergravity theories formulated in harmonic superspace in order to give some guidelines to the main text where the reader will find all the details and, the authors hope, a deeper insight.

1.12.1 $N = 2$ matter hypermultiplet

The general action for $N = 2$ supersymmetric matter is written down as an analytic superspace integral:

$$S = -\int du\, d\zeta^{(-4)}\, [\tilde{q}^{+}D^{++}q^{+} - L^{+4}(q^{+}, \tilde{q}^{+}, u)]\,. \tag{1.44}$$

Here L^{+4} is an arbitrary function of its arguments carrying $U(1)$ charge $+4$. It gives rise to the source term in the equation of motion (1.43), $J^{+++} =$

$\partial L^{+4}/\partial \tilde{q}^+$. The operation $\tilde{}$ is a special involution preserving the analytic harmonic superspace (1.41) (it is reduced to ordinary complex conjugation for the u-independent quantities), and the integration measure of the analytic superspace is defined as

$$d\zeta^{(-4)} = d^4x\, d^2\theta^+\, d^2\bar{\theta}^+ . \tag{1.45}$$

This measure carries negative $U(1)$ charge because Grassmann integration is equivalent to differentiation [B7] with respect to the odd coordinates of the analytic superspace $\theta^+_\alpha, \bar{\theta}^+_{\dot{\alpha}}$. These formulas look simple. However, in order to be able to effectively work with them one needs precise definitions and details, especially of the harmonic calculus on S^2. In particular, one needs to know how to solve differential equations on S^2 to which the auxiliary field equations of motion following from (1.44) are reduced. All this will be explained in the main body of the book.

Here we make a few comments only. The off-shell action (1.44) corresponds to the general $N = 2$ supersymmetric sigma model. The target spaces of such sigma models are known to belong to a remarkable class of $4n$-dimensional complex manifolds: According to the theorem of ref. [A2] they are the so-called hyper-Kähler manifolds. This means that they admit a triplet of covariantly constant complex structures forming the algebra of quaternionic units or, equivalently, that their holonomy group lies in $Sp(n)$. The essentially new point in the harmonic approach is that the interaction Lagrangian $L^{+4}(q, \tilde{q}, u)$ appears as the hyper-Kähler potential which encodes the complete information about the local properties of a given manifold. For example, $L^{+4} = \lambda(q^+)^2(\tilde{q}^+)^2$ describes the well-known Taub–NUT hyper-Kähler manifold. It is worthwhile mentioning that the four-dimensional hyper-Kähler manifolds (corresponding to a single hypermultiplet action) represent solutions of the self-dual Einstein equations, among them the gravitational instantons.

1.12.2 $N = 2$ *Yang–Mills theory*

$N = 2$ supersymmetric Yang–Mills theory is similar to ordinary ($N = 0$) Yang–Mills theory. It is based on making an internal symmetry group local in the analytic harmonic superspace $(x_A, \theta^+, \bar{\theta}^+, u^\pm) = (\zeta, u^\pm)$ (1.41) (instead of just Minkowski space in the $N = 0$ case):

$$\delta q^+_r = i\lambda^k (t_k)_{rs} q^+_s \quad \Rightarrow \quad \delta q^+_r(\zeta, u^\pm) = i\lambda^k(\zeta, u^\pm)(t_k)_{rs} q^+_s(\zeta, u^\pm) , \tag{1.46}$$

where t_k are the generators of the internal symmetry group and λ^k are the corresponding parameters. As usual, one should covariantize the derivatives entering the action. In our case, it is the harmonic one:

$$D^{++} \quad \Rightarrow \quad \mathcal{D}^{++} = D^{++} + iV^{++}(\zeta, u) . \tag{1.47}$$

The gauge connection $V^{++}(\zeta, u) = V^{++k}(x_A, \theta^+, \bar{\theta}^+, u^\pm)t_k$ is a Lie algebra-valued analytic harmonic superfield. It transforms under the gauge group according to the standard rule

$$\delta V^{++}(\zeta, u) = -\mathcal{D}^{++}\lambda(\zeta, u), \qquad \lambda(\zeta, u) \equiv \lambda^k(\zeta, u)t_k. \tag{1.48}$$

This superfield describes just the off-shell $N = 2$ Yang–Mills supermultiplet, as the reader will see in Chapter 7. This multiplet consists of a gauge vector field $A_a(x)$, a doublet of Weyl spinors $\psi^i_\alpha(x)$, a complex scalar field $\phi(x)$ and a triplet of auxiliary fields $D^{(ij)}(x)$. It should be pointed out that, as opposed to the matter hypermultiplet $q^+(\zeta, u)$, the gauge superfield $V^{++}(\zeta, u)$ contains a *finite* number of auxiliary fields. Instead, it has an infinite number of pure gauge degrees of freedom which are gauged away by the transformations (1.48). The harmonic superspace formulation reveals the close similarity between $N = 2$ super-Yang–Mills theory and the ordinary bosonic ($N = 0$) Yang–Mills theory.

Having defined the covariant derivative (1.47), one can immediately introduce the minimal Yang–Mills–matter coupling by simply covariantizing the action (1.44). The details of how to construct an invariant action for the Yang–Mills superfield itself will be given in Chapter 7. The general class of $N = 2$ Yang–Mills field theories in interaction with hypermultiplets is known to contain a subclass of four-dimensional *ultraviolet finite quantum field theories* (in particular, $N = 4$ Yang–Mills theory). They also reveal remarkable properties of duality [S5]. Harmonic superspace considerably simplifies many aspects and makes manifest many features of these theories, e.g., the proof of non-renormalization theorems, finding out the full structure of the quantum effective actions, etc.

The harmonic approach is also convenient for the description of general non-minimal self-couplings of vector $N = 2$ supermultiplets. These theories are unique because they are the only $N = 2$ supersymmetric field-theoretical models that admit a natural chiral structure of interactions. For this reason they may be useful in the phenomenological context as a possible basis of $N = 2$ GUT models. Sigma models inherent to these couplings are of interest in their own right. Their tangent manifolds are of some special Kähler type [C7, C10] and have been discussed in connection with the so-called $c*$-map [C8, C9].

A historical comment is due here. Unlike $N = 2$ matter, the $N = 2$ Yang–Mills theory can be formulated in terms of standard *unconstrained* $\mathbb{R}^{4|8}$ superfields (since it only involves a finite set of auxiliary fields). Such a more 'traditional' formulation of $N = 2$ Maxwell theory was first given in [M3] and its non-Abelian version in [G28]. The main drawback of this approach is the lack of geometric meaning of the Yang–Mills prepotential and gauge group, which makes quantization particularly cumbersome. In Chapter 7 we shall show that these $\mathbb{R}^{4|8}$ objects can be derived from the harmonic superspace ones by a special choice of gauge with respect to the transformations (1.46), (1.48).

1.12.3 $N = 2$ supergravity

Now we make a few comments on the $N = 2$ supergravity theory. It is an extension of Einstein's theory of gravity describing the metric field $g_{mn}(x)$ (graviton) and its $N = 2$ superpartners: an $SU(2)$ doublet of Rarita–Schwinger fields $\psi^i_{m\alpha}(x)$, $\bar{\psi}^i_{m\dot{\alpha}}(x)$ ('gravitini') and a vector gauge field $A_m(x)$ ('graviphoton'). The underlying principle is gauge invariance under some supergroup containing the diffeomorphism group of four-dimensional space-time as a subgroup. To formulate $N = 2$ supergravity theory one has to answer the following questions:

(i) What kind of superspace is appropriate?
(ii) What is the gauge supergroup needed?
(iii) What are the unconstrained prepotentials?
(iv) How to construct the invariant action?
(v) How many versions of the theory do exist and what are the differences between them?

The answers to the above questions given in this book are as follows:

(i) The superspace for $N = 2$ supergravity is harmonic superspace.
(ii) The appropriate $N = 2$ (conformal) supergravity gauge supergroup is the superdiffeomorphism group of the harmonic analytic superspace (ζ, u):

$$\begin{aligned} \delta x^m &= \lambda^m(\zeta, u)\,, \\ \delta\theta^{\mu+} &= \lambda^{\mu+}(\zeta, u)\,, \qquad \delta\bar{\theta}^{\dot{\mu}+} = \bar{\lambda}^{\dot{\mu}+}(\zeta, u)\,, \\ \delta u^+_i &= \lambda^{++}(\zeta, u)u^-_i\,, \qquad \delta u^-_i = 0\,, \end{aligned} \tag{1.49}$$

where the local parameters λ are arbitrary analytic harmonic functions. Note that only the harmonics u^+ but not u^- transform, a peculiarity due to the special realization of the $N = 2$ superconformal group (see Chapter 9).

(iii) As in $N = 2$ Yang–Mills theory, the $N = 2$ supergravity prepotentials appear in the covariantized harmonic derivative

$$\mathcal{D}^{++} = u^+_i \frac{\partial}{\partial u^-_i} + H^{++++} u^-_i \frac{\partial}{\partial u^+_i} + H^{++m} \frac{\partial}{\partial x^m_A} + H^{++\hat{\mu}+} \frac{\partial}{\partial \theta^{\hat{\mu}+}} \tag{1.50}$$

(here $\hat{\mu} = \mu, \dot{\mu}$). Covariantization is achieved by adding to the flat harmonic derivative $D^{++} = u^+_i \partial/\partial u^-_i$ appropriate vielbein terms with analytic vielbeins $H^{++++}(\zeta, u)$, $H^{++m}(\zeta, u)$, $H^{++\hat{\mu}+}(\zeta, u)$ (the counterparts of the gauge connection $V^{++}(\zeta, u)$ in the Yang–Mills case). These vielbeins are the unconstrained prepotentials of $N = 2$ supergravity. Their gauge transformation laws follow from (1.49).

In fact, the supergroup (1.49) and the prepotentials (1.50) are relevant to the so-called conformal (or Weyl) supergravity. This kind of supergravity possesses a somewhat bigger gauge symmetry than Einstein $N = 2$ supergravity. The extra gauge transformations have to be compensated by coupling conformal supergravity to some matter supermultiplets called *compensators*. This procedure is widely used in gravity and supergravity theories [D17, D18, D19, D20, D21, D22, F14, G26, G27].

(iv) The action for $N = 2$ Einstein supergravity is written down as the action for the $N = 2$ compensators in the background of $N = 2$ conformal supergravity. Two such compensating supermultiplets are needed. One of them is always an Abelian vector supermultiplet, but there exist several alternative choices for the second one.

(v) It should be stressed that different sets of compensators lead to different off-shell versions of $N = 2$ Einstein supergravity having different sets of auxiliary fields. In the harmonic superspace approach one can reproduce all the versions previously found in the component field approach [D15, D16, D17, D18, D19, F15]. Naturally, the latter always contains a *finite* set of auxiliary fields. Consequently, the corresponding compensators are described by *constrained* Grassmann analytic superfields (e.g., by the so-called tensor or non-linear multiplets). The presence of such constraints restricts the possible form of matter couplings, the latter have to be consistent with the former. For instance, the matter hypermultiplets self-couplings must possess some isometries. However, the harmonic superspace approach provides a new, 'principal' version of $N = 2$ Einstein supergravity with an *unconstrained* hypermultiplet q^+ as a compensator. This version admits the most general matter couplings. At the same time, it naturally contains an infinite number of auxiliary fields and thus could not be discovered by traditional methods. The bosonic target manifolds of the corresponding $N = 2$ sigma models are *quaternionic* [B1, B4] in contrast to the hyper-Kähler ones in the flat $N = 2$ case. The harmonic superspace approach clearly exhibits this important property [G7, G19] and offers an efficient tool for the explicit calculation of quaternionic metrics [G7].

1.13 $N = 3$ Yang–Mills theory

The harmonic superspace concept is not limited to $N = 2$ supersymmetry only. However, going to $N > 2$ requires some major changes. At present, $N = 3$ Yang–Mills theory is fully understood [G5, G6, G12]. Here are some of the basic ideas. $N = 3$ harmonic superspace is a tensor product of the standard real superspace $\mathbb{R}^{4|12}$ and the six-dimensional coset space $SU(3)/U(1) \times U(1)$, where $SU(3)$ is the automorphism group of $N = 3$ supersymmetry. So, instead of just one we now deal with *two* $U(1)$ charges. The corresponding harmonics

$$u_i^I = (u_i^{(1,0)}, u_i^{(0,-1)}, u_i^{(-1,1)}) \; ; \qquad u_I^i = \overline{u_i^I} \; ; \qquad i = 1, 2, 3 \qquad (1.51)$$

are subject to the defining conditions

$$u_I^i u_i^J = \delta_I^J \; ; \qquad u_I^i u_j^I = \delta_j^i \; ; \qquad \det u = 1 \, .$$

The spinor variables $\theta_{i\alpha}, \bar{\theta}^i_{\dot{\alpha}}$ form the representations 3 and $\underline{3}$ of $SU(3)$. With the help of the harmonics (1.51) they are projected onto six independent variables:

$$\theta_\alpha^{(-1,0)} \, , \quad \theta_\alpha^{(0,1)} \, , \quad \theta_\alpha^{(1,-1)} \, , \quad \bar{\theta}_{\dot{\alpha}}^{(1,0)} \, , \quad \bar{\theta}_{\dot{\alpha}}^{(0,-1)} \, , \quad \bar{\theta}_{\dot{\alpha}}^{(-1,1)} \, .$$

The analytic $N=3$ superspace contains only four of them, $\theta_\alpha^{(1,-1)}$, $\theta_\alpha^{(0,1)}$, $\bar{\theta}_{\dot{\alpha}}^{(1,0)}$, $\bar{\theta}_{\dot{\alpha}}^{(-1,1)}$ (and not half, as was the case in $N = 2$). The analytic Yang–Mills prepotentials $V^{(1,1)}$, $V^{(2,-1)}$, $V^{(-1,2)}$ are introduced as the gauge connections for the harmonic derivatives $\mathcal{D}^{(1,1)}$, $\mathcal{D}^{(2,-1)}$, $\mathcal{D}^{(-1,2)}$. They have the usual transformation law $\delta V^{(a,b)} = \mathcal{D}^{(a,b)}\lambda$, where λ is a chargeless analytic superfield parameter. The action is very unusual, it is written down as a Chern–Simons term:

$$\begin{aligned} S^{N=3}_{\rm SYM} \;\; = & \int du \, d\zeta_A^{(-2,-2)} {\rm Tr} \left(V^{(2,-1)} F^{(0,3)} + V^{(-1,2)} F^{(3,0)} + V^{(1,1)} F^{(1,1)} \right. \\ & \left. - i V^{(1,1)} [V^{(2,-1)}, V^{(-1,2)}] \right) , \end{aligned} \qquad (1.52)$$

where the three F's are the field strengths, e.g., $F^{(3,0)} = -i[\mathcal{D}^{(1,1)}, \mathcal{D}^{(2,-1)}]$. Note that the Chern–Simons-type action (1.52) was proposed as early as 1985 and it describes a very non-trivial dynamics. Nowadays Chern–Simons actions are becoming popular in connection with string field theory and topological field theory [W15].

There remain a lot of important problems in supersymmetric theories that one can hope to solve within the harmonic superspace approach. These techniques have already been employed to approach $N = 4$ supersymmetric Yang–Mills theory [S18], ten-dimensional Yang–Mills and supergravity theories in the context of superparticle and superstring models [G3, N3, N4, N5, S16, S17], etc.

1.14 Harmonics and twistors. Self-duality equations

The harmonic superspace approach has a close relationship to the famous twistor theory [P2, P3, P4]. Common for both is an extension of space-time (in twistor theory) and superspace (in the harmonic superspace approach) by adding some two-dimensional sphere S^2. In such an extended space the self-dual Yang–Mills or Einstein equations admit an interpretation as Cauchy–Riemann conditions. In a close analogy, the constraints of $N = 2$ matter, Yang–Mills and supergravity theories in harmonic superspace are reformulated as Cauchy–Riemann conditions, both concerning the dependence on the Grassmann and harmonic variables. Let us very briefly discuss this remarkable similarity.

In the harmonic superspace approach the sphere S^2 is introduced as the coset space $SU(2)_A/U(1)$, where the numerator is the automorphism group of the $N = 2$ supersymmetry algebra. In the twistor theory in Minkowski space the sphere S^2 is related to the Lorentz group $SO(3,1) \simeq SL(2,\mathbb{C})$:

$$S^2 = SL(2,\mathbb{C})/B\,, \tag{1.53}$$

where the denominator is the Borel subgroup of the Lorentz group. In the Euclidean version the 'Lorentz' group becomes $SO(4) \simeq SU(2)_L \times SU(2)_R$. This suggests to use a spinor formalism with undotted and dotted indices belonging to $SU(2)_L$ and to $SU(2)_R$, respectively. In this case S^2 corresponds to the coset $SU(2)_L/U(1)$ (or $SU(2)_R/U(1)$) and one deals with the harmonic space

$$\mathbb{H}^{4+2} = \mathbb{R}^4 \times S^2 = (x^{\alpha\dot\alpha}, u^{\pm}_{\dot\alpha})\,, \tag{1.54}$$

where $x^{\alpha\dot\alpha} = x^a\sigma_a^{\alpha\dot\alpha}$ are the $\mathbb{R}^4$ coordinates in the spinor notations and $u^{\pm}_{\dot\alpha}$ are harmonic coordinates of the sphere S^2.

To get a general feeling of how harmonic analyticity works in the $N = 0$ case, here we briefly discuss its application to solving the self-duality Yang–Mills equations in $\mathbb{R}^4$. The commutator of two covariant derivatives $D_{\alpha\dot\alpha} = \partial_{\alpha\dot\alpha} + iA_{\alpha\dot\alpha}(x)$,

$$[D_{\alpha\dot\alpha}, D_{\beta\dot\beta}] = \epsilon_{\alpha\beta}F_{\dot\alpha\dot\beta}(x) + \epsilon_{\dot\alpha\dot\beta}F_{\alpha\beta}(x)\,, \tag{1.55}$$

introduces the Yang–Mills field strength $F_{\alpha\beta}$, $F_{\dot\alpha\dot\beta}$. In the spinor formalism the self-duality equation reads

$$F_{\dot\alpha\dot\beta}(x) = 0\,, \tag{1.56}$$

or, equivalently,

$$[D_{\alpha(\dot\alpha}, D_{\beta\dot\beta)}] = 0\,, \tag{1.57}$$

where $(\dot\alpha\dot\beta)$ means symmetrization. Now, multiplying eq.(1.57) by the harmonics $u^{+\dot\alpha}$, $u^{+\dot\beta}$ we obtain

$$[D^+_\alpha, D^+_\beta] = 0\,, \tag{1.58}$$

where

$$D^+_\alpha = u^{+\dot\alpha}D_{\alpha\dot\alpha}\,. \tag{1.59}$$

From this definition it follows that

$$[D^{++}, D^+_\alpha] = 0 \tag{1.60}$$

(recall (1.28)).

The pair of equations (1.58) and (1.60) is equivalent to the self-duality eq. (1.56). The first of them, eq. (1.58), admits a general solution in the form of a 'pure gauge':

$$D^+_\alpha = h\partial^+_\alpha h^{-1} = \partial^+_\alpha + h(\partial^+_\alpha h^{-1})\,, \tag{1.61}$$

where the arbitrary harmonic function $h = h(x, u)$ is an element of the gauge group. Like in $N = 2$ Yang–Mills, this allows one to pass to an 'analytic frame' where the covariant derivative D^+_α becomes short:

$$D^+_a = u^{+\dot\alpha}\partial_{\alpha\dot\alpha} = \frac{\partial}{\partial x^{-\alpha}} \equiv \partial_{\alpha-}\,, \tag{1.62}$$

where the space-time coordinates have been harmonic-projected, $x^{\pm\alpha} = x^{\alpha\dot\alpha}u^\pm_{\dot\alpha}$. At the same time, the harmonic derivative D^{++} acquires a connection:

$$D^{++} \quad \Rightarrow \quad \mathcal{D}^{++} = D^{++} + iV^{++}\,, \tag{1.63}$$

where

$$V^{++} = -ih^{-1}D^{++}h\,. \tag{1.64}$$

So, this frame rotation solves the first equation of the pair, (1.58), and the second, (1.60), becomes the *Cauchy–Riemann condition for* $N = 0$ *analyticity*

$$\frac{\partial}{\partial x^{-\alpha}}V^{++} = 0\,. \tag{1.65}$$

It means that the harmonic connection has to be analytic, $V^{++} = V^{++}(x^{+\alpha}, u^\pm)$.

The conclusion is that the solutions of the self-dual Yang–Mills equations are encoded in the analytic potentials V^{++}. Instantons and monopoles are special solutions of the self-duality equation which possess finite action and finite energy. In [K3, O1] they have been described in the harmonic space language. As an example, here is the potential for the one-instanton solution of the $SU(2)$ Yang–Mills theory first obtained in [B5]:

$$(V^{++})^i_j = -\frac{i}{\rho^2}x^{+i}x^+_j\,, \tag{1.66}$$

where ρ is the instanton size.

The same basic ideas can be readily extended to other systems of non-linear differential constraints, like self-dual gravity, etc. Chapter 11 contains more examples and a detailed treatment of hyper-Kähler manifolds, along the above lines.

We close this somewhat eclectic introductory overview with a more formal list of chapters.

1.15 Chapters of the book and their abstracts

Chapter 1. Introductory overview

Origins, motivations, basic ideas and results of the harmonic superspace (and space) approach are overviewed. Its deep affinity to the twistor approach is emphasized. The major aim is to give the reader a preliminary impression of the subject before immersion into the main body of the book.

Chapter 2. Elements of supersymmetry

In this chapter we give some useful information about space-time supersymmetry algebras, their representations and realizations in terms of fields. We need all that as a necessary background for what follows. Of course, these rather sketchy remarks are by no means a systematic introduction to supersymmetry. The reader who is not familiar with the foundations of supersymmetry is referred to the available reviews [F2, F3, F4, F7, N2, O4, S13, V1] and books [B17, G26, W7, W12].

Chapter 3. Superspace

A manifestly supersymmetric description requires an adequate space, superspace, just as the natural realization of Poincaré symmetry is achieved in Minkowski space. A systematic way of finding such spaces and symmetry realizations is the coset space method due to E. Cartan. Chapter 3 begins with a reminder of the main points of Cartan's construction, which is then applied to deriving the superspaces of N-extended Poincaré supersymmetry and finding relevant coordinate realizations of the latter. $N = 2$ harmonic superspace and its analytic subspace (most important for what follows) are treated in great detail. In particular, the appropriate covariant derivatives (spinor and harmonic) are deduced using the powerful technique of Cartan forms.

Chapter 4. Harmonic analysis

In this chapter we translate the basic notions of harmonic analysis on $SU(2)$ (harmonic series, invariant integrals, etc.) into the parametrization-independent language of harmonic coordinates. A new object not present in the standard textbook presentations is the harmonic derivative. We define harmonic distributions (delta functions and Green's functions) and discuss the problem of solving differential equations on S^2. The explicit formulas and rules of this chapter are used intensively in the following chapters, especially in Chapter 8 devoted to quantization in harmonic superspace.

Chapter 5. $N = 2$ matter with infinite sets of auxiliary fields

A consistent description of matter supermultiplets (those containing only spins 0 and 1/2 on shell) is a necessary ingredient of any approach to supersymmetry.

In the harmonic superspace approach there are two unconstrained forms of the off-shell $N = 2$ matter supermultiplets, the q^+ and ω hypermultiplets. They basically differ only in the $SU(2)$ assignments of the physical fields, being duality-equivalent to each other. Their description by unconstrained analytic superfields (containing infinite sets of auxiliary fields) allows one to write down the most general self-couplings in the form of the arbitrary potential $L^{+4}(q^+, \tilde{q}^+, u)$. The corresponding sigma models have complex hyper-Kähler manifolds as their target manifolds. So, L^{+4} is the hyper-Kähler prepotential. We consider some simple explicit examples of such potentials which, upon elimination of the auxiliary fields give rise of the well-known Taub–NUT and Eguchi–Hanson metrics.

Chapter 6. $N = 2$ matter multiplets with a finite number of auxiliary fields. $N = 2$ duality transformations

In this chapter we demonstrate that all off-shell $N = 2$ matter multiplets with finite sets of auxiliary fields (including all such multiplets known before the invention of harmonic superspace) are described by Grassmann analytic superfields with properly chosen harmonic constraints. We define the $N = 2$ superfield duality transformation and use it to show explicitly that the general self-couplings of these constrained $N = 2$ matter superfields are reduced to particular classes of the q^+ hypermultiplet self-coupling. Such $N = 2$ multiplets with finite sets of auxiliary fields are used as compensators in passing from conformal $N = 2$ supergravity to its Einstein descendants. So, the relevant constrained harmonic superfields are necessary ingredients of the harmonic superspace formulation of $N = 2$ supergravities (Chapter 10).

Chapter 7. Supersymmetric Yang–Mills theories

In this chapter we describe the formulation of $N = 2$ SYM theory in harmonic superspace. In the first part of the chapter we show how one can arrive at the idea of an analytic unconstrained prepotential V^{++} for $N = 2$ SYM starting from very simple considerations of minimal gauge coupling to $N = 2$ supersymmetric matter. As an introduction to this method we first briefly discuss the $N = 0$ and $N = 1$ theories. In the second part we develop a suitable geometric framework. There we begin with a set of constraints on the superspace differential geometry and derive the prepotential as the solution to these constraints. All the curvature tensors can then be expressed in terms of the prepotential and an off-shell action can be constructed. Once again, the $N = 1$ case serves as an example of the general approach.

Chapter 8. Harmonic supergraphs

In this chapter we develop a manifestly supersymmetric quantization scheme for $N = 2$ matter and SYM theories. Using harmonic distributions in the propagators might in principle cause specific harmonic divergences, but we show that this is actually not the case. A number of explicit examples of one-loop calculations as well as an example of a finite two-loop four-point correlation function are worked out in detail. An important application of the supergraph techniques is a direct power-counting proof of the ultraviolet finiteness of two-dimensional $N = 4$ sigma models.

Chapter 9. Conformal invariance in $N = 2$ harmonic superspace

This chapter is preparatory for the next. Using Cartan's techniques we derive the realizations of the rigid $N = 2$ superconformal group $SU(2, 2|2)$ in ordinary $N = 2$ superspace, its harmonic extension and the analytic subspace of the latter. To do this in a systematic way, we identify these superspaces with the appropriate coset spaces of the extended supergroup $SU(2, 2|2) \times SU(2)_A$ where $SU(2)_A$ is the group of outer automorphisms of $SU(2, 2|2)$. The superconformal properties of the basic $N = 2$ multiplets (including those used as supergravity compensators) are examined.

Chapter 10. Supergravity

We begin this chapter by recalling the basic facts about the geometric formulation of $N = 1$ conformal and Einstein supergravities in superspace and explain on this simple example the method of conformal compensators. Next we define the gauge group and prepotentials of conformal $N = 2$ supergravity in harmonic superspace. The gauge group is a local version of the analytic superspace realization of the rigid superconformal group $SU(2, 2|2)$ discussed in the preceding chapter. The prepotentials arise as the Grassmann analytic vielbeins covariantizing the harmonic derivative D^{++}. The next important step is the construction of the harmonic superspace action of Einstein $N = 2$ supergravity. Its first basic ingredient is the action for a Maxwell compensating superfield in the background of conformal supergravity. To find it one needs a superspace density that we build out of simply transforming building blocks. As a second compensator we first consider a q^+ hypermultiplet. This gives rise to an essentially new version of Einstein $N = 2$ supergravity. It contains an infinite number of auxiliary fields (coming from the q^+ compensator) and is the only one allowing for the most general matter couplings. All the other versions correspond to using multiplets with finite sets of auxiliary fields as compensators and lead to rather restricted matter couplings. We present the harmonic superspace formulation of these versions as well. In most cases these are related to the first, 'principal' version by an $N = 2$ duality transformation.

Chapter 11. Hyper-Kähler geometry in harmonic space

The aim of this chapter is to show that the concept of harmonic analyticity has deep implications not only in $N = 2$ (and $N > 2$) supersymmetry, but also in purely bosonic ($N = 0$) gauge theories. Namely, this concept allows one to obtain unconstrained geometric formulations of self-dual Yang–Mills theory and hyper-Kähler geometry. These formulations closely parallel those of $N = 2$ Yang–Mills theory and $N = 2$ supergravity in harmonic superspace. The basic objects are unconstrained prepotentials defined on an analytic subspace of the harmonic extension of the original space (in the general case it is $\mathbb{R}^{4n} \times S^2$). They encode all the information about the quantities present in conventional formulations, self-dual Yang–Mills connections in the first case and hyper-Kähler metrics in the second.

The inherent relevance of $SU(2)$ harmonics both to hyper-Kähler geometry and $N = 2$ supersymmetry makes very clear why $N = 2$ supersymmetric sigma models necessarily have hyper-Kähler manifolds as their target spaces. We show that the harmonic-analytic prepotential of the most general hyper-Kähler manifold serves as the Lagrangian of the most general $N = 2$ sigma model (after identifying the coordinates of the analytic subspace of this manifold with the analytic superfields q^+ describing $N = 2$ matter). This establishes a direct, one-to-one correspondence between $N = 2$ sigma models and hyper-Kähler manifolds.

Chapter 12. $N = 3$ supersymmetric Yang–Mills theory

The method of harmonic superspace provides the solution to one of the outstanding problems in supersymmetry. The off-shell formulation of $N = 3$ SYM theory is only possible within the framework of harmonic superspace. The reason is that one cannot define this theory off shell with a finite number of auxiliary fields. Infinite towers of auxiliary fields naturally arise in a harmonic expansion, just as in the case of the $N = 2$ complex hypermultiplet. After that, constructing an off-shell action for $N = 3$ SYM theory as an integral over the analytic subspace of $N = 3$ harmonic superspace becomes possible. Besides, $N = 3$ SYM theory has a number of very unusual features which are studied in detail.

2

Elements of supersymmetry

In this chapter we give some information about space-time supersymmetry algebras, their representations and realization in terms of fields. Here, as well as in the rest of the book we mainly deal with four-dimensional theories. The purpose of this chapter is to provide readers with the necessary background and to let them get accustomed to our notations and conventions. Of course, these rather sketchy remarks are by no means a systematic introduction to supersymmetry. For this we refer the reader to a number of available books [B17, G26, W7, W12] and reviews [F2, F3, F4, F7, N2, O4, S13, V1].

We start by recalling some elementary facts about Poincaré and conformal symmetry ('$N = 0$ supersymmetry') in four dimensions. Then we discuss the group-theoretical background of simple ($N = 1$) and extended ($N \geq 2$) supersymmetry.

2.1 Poincaré and conformal symmetries

2.1.1 Poincaré group

The Poincaré group $ISO(1,3)$ can be realized in Minkowski space $M^4 = (x^a)$, $a = 0, 1, 2, 3$,* by linear transformations:

$$x'^a = \Lambda^a_b x^b + c^a , \tag{2.1}$$

which preserve the space-time interval

$$ds^2 = \eta_{ab}\, dx^a\, dx^b , \qquad \eta_{ab} = \mathrm{diag}(1, -1, -1, -1) . \tag{2.2}$$

The subgroup of homogeneous transformations (i.e., those with parameters Λ^a_b) is the Lorentz group $SO(1,3)$. The invariance of ds^2 implies

$$\eta_{ab}\Lambda^a_c\Lambda^b_d = \eta_{cd} \quad \Rightarrow \quad \det \Lambda = \pm 1 \tag{2.3}$$

* In different chapters of this book we shall denote the four-vector indices sometimes by $a, b, \ldots$, and sometimes by $m, n, \ldots$, in order to avoid confusion with other types of indices.

(as usual, we take the proper subgroup with $\det \Lambda = 1$). The infinitesimal Lorentz transformations can be defined as

$$\Lambda_{ab} \equiv \eta_{ac}\Lambda^c_b = \eta_{ab} - \lambda_{ab}\,, \qquad \lambda_{ab} = -\lambda_{ba}\,. \tag{2.4}$$

The infinitesimal form of the transformations (2.1) adapted to this convention is given by

$$\delta x^a = i\left[c^b P_b - \frac{1}{2}\lambda^{ab}\hat{L}_{ab},\, x^a\right], \tag{2.5}$$

where the differential operators

$$P_b = -i\partial_b\,, \quad \hat{L}_{ab} = i(x_a\partial_b - x_b\partial_a) \tag{2.6}$$

are generators of the Poincaré group in the particular realization on the coordinates x^a.

The Poincaré algebra is defined by the commutators of the generators of infinitesimal translations P_a and Lorentz rotations L_{ab}:

$$\begin{aligned} [P_a, P_b] &= 0\,, \\ [L_{ab}, P_c] &= i(\eta_{bc}P_a - \eta_{ac}P_b)\,, \\ [L_{ab}, L_{cd}] &= i(\eta_{bc}L_{ad} - \eta_{ac}L_{bd} + \eta_{ad}L_{bc} - \eta_{bd}L_{ac})\,. \end{aligned} \tag{2.7}$$

It is easy to check the validity of these relations for the particular realization (2.6).

2.1.2 Conformal group

In what follows we shall also use the conformal group which is an extension of the Poincaré group. It is the group of transformations (not only linear) leaving invariant the interval (2.2) up to an x-dependent factor. Besides the Poincaré generators, its algebra contains the generators of dilatation D and conformal boosts K_a. The part of the algebra involving the new generators is

$$\begin{aligned} [D, P_a] &= iP_a\,, \qquad [D, K_a] = -iK_a\,, \qquad [D, L_{ab}] = 0\,, \\ [L_{ab}, K_c] &= i(\eta_{bc}K_a - \eta_{ac}K_b)\,, \\ [P_a, K_b] &= -2i(\eta_{ab}D + L_{ab})\,, \qquad [K_a, K_b] = 0\,. \end{aligned} \tag{2.8}$$

Note that the algebra of the four-dimensional conformal group is isomorphic to the algebra of the six-dimensional pseudo-orthogonal group $SO(2,4)$ as well as to the algebra of the pseudo-unitary group $SU(2,2)$.*

* In the n-dimensional case with the Lorentz group $SO(1, n-1)$ the conformal group is $SO(2,n)$. All the relations (2.1)–(2.8) above remain valid with $a = 0, 1, \ldots, n-1$.

2.1.3 Two-component spinor notation

In four dimensions it is often very convenient to use two-component spinor notation, not only for spinor representations, but also for vectors, tensors, etc. The Lorentz group $SO(1,3)$ is locally isomorphic to the group $SL(2,\mathbb{C})$ of 2×2 complex unimodular matrices. The latter has two inequivalent fundamental representations. One of them is described by a pair of complex numbers,

$$\psi_\alpha = \begin{pmatrix} \psi_1 \\ \psi_2 \end{pmatrix}, \tag{2.9}$$

with the transformation law:

$$\psi'_\alpha = \Lambda_\alpha^\beta \psi_\beta , \qquad \Lambda \in SL(2,\mathbb{C}). \tag{2.10}$$

It is referred to as the representation (1/2,0), or as a left-handed spinor. The other fundamental representation (0,1/2) is obtained by complex conjugation:

$$\bar\psi'_{\dot\alpha} = \bar\psi_{\dot\beta} \bar\Lambda_{\dot\alpha}^{\dot\beta} , \qquad \bar\Lambda_{\dot\alpha}^{\dot\beta} = \overline{(\Lambda_\alpha^\beta)} . \tag{2.11}$$

In $SL(2,\mathbb{C})$ notation a four-component Dirac spinor is represented by a pair of left- and right-handed spinors:

$$\Psi_{\hat\alpha} = \begin{pmatrix} \psi_\alpha \\ \bar\chi^{\dot\alpha} \end{pmatrix} \tag{2.12}$$

(for a Majorana spinor $\bar\chi_{\dot\alpha} = \overline{(\psi_\alpha)}$). A four-vector x^a belongs to the representation (1/2,1/2) and can be replaced by a Hermitian matrix:

$$x_{\alpha\dot\beta} = (\sigma_a)_{\alpha\dot\beta} x^a , \qquad x^{\dot\alpha\beta} = (\tilde\sigma_a)^{\dot\alpha\beta} x^a . \tag{2.13}$$

The infinitesimal form of the $SL(2,\mathbb{C})$ matrices in (2.10), (2.11) is as follows:

$$\Lambda_\alpha^\beta = \delta_\alpha^\beta + \frac{i}{4}\lambda^{ab}(\sigma_{ab})_\alpha{}^\beta , \qquad \bar\Lambda_{\dot\alpha}^{\dot\beta} = \delta_{\dot\alpha}^{\dot\beta} - \frac{i}{4}\lambda^{ab}(\tilde\sigma_{ab})_{\dot\alpha}{}^{\dot\beta} . \tag{2.14}$$

For details of our notational conventions and, in particular, the definition of the 2×2 matrices $(\sigma_a)_{\alpha\dot\beta}$, $(\sigma_{ab})_\alpha{}^\beta$ and $(\tilde\sigma_{ab})_{\dot\alpha}{}^{\dot\beta}$ see Appendix A.1.

2.2 Poincaré and conformal superalgebras

2.2.1 $N=1$ Poincaré superalgebra

The Poincaré superalgebra is an extension of the Poincaré algebra with spinor generators. The generators of supersymmetry (supertranslations) Q_α, $\bar Q_{\dot\alpha}$ form a Majorana spinor of the Lorentz group:

$$\begin{aligned} \left[L_{ab}, Q_\alpha\right] &= -\frac{1}{2}(\sigma_{ab})_\alpha^{\ \beta} Q_\beta\,, \\ \left[L_{ab}, \bar{Q}_{\dot\alpha}\right] &= \frac{1}{2}(\tilde\sigma_{ab})_{\dot\alpha}^{\ \dot\beta} \bar{Q}_{\dot\beta}\,, \\ \bar{Q}_{\dot\alpha} &= (Q_\alpha)^\dagger\,. \end{aligned} \tag{2.15}$$

They commute with the translations:

$$[Q_\alpha, P_a] = [\bar{Q}_{\dot\alpha}, P_a] = 0\,. \tag{2.16}$$

Among themselves the supersymmetry generators satisfy *anticommutation* rather than commutation relations:

$$\begin{aligned} \left\{Q_\alpha, Q_\beta\right\} &= \left\{\bar{Q}_{\dot\alpha}, \bar{Q}_{\dot\beta}\right\} = 0\,, \\ \left\{Q_\alpha, \bar{Q}_{\dot\alpha}\right\} &= 2(\sigma^a)_{\alpha\dot\alpha} P_a\,. \end{aligned} \tag{2.17}$$

The correct relation between spin and statistics is maintained owing to the use of anticommutators (instead of commutators) between the fermionic generators. In other words, the supersymmetry algebra is a Z_2 graded algebra: The bosonic generators P_a, L_{ab} have grading $p = 0$, and the fermionic ones Q_α, $\bar{Q}_{\dot\alpha}$ have grading $p = 1$.

One can argue [H1] that the superalgebra (2.7), (2.15), (2.16), (2.17) is the unique extension of the Poincaré algebra involving one Majorana spinor generator. This superalgebra is usually referred to as the simple (or non-extended) supersymmetry algebra.

2.2.2 Extended supersymmetry

In N-extended supersymmetry there are N Majorana spinor generators Q^i_α, $\bar{Q}_{\dot\alpha i} = (Q^i_\alpha)^\dagger$, $i = 1, 2, \ldots, N$.* Its superalgebra contains obvious modifications of (2.15) and (2.16) whereas (2.17) become

$$\begin{aligned} \left\{Q^i_\alpha, Q^j_\beta\right\} &= \epsilon_{\alpha\beta} Z^{ij}\,, \qquad \left\{\bar{Q}_{\dot\alpha i}, \bar{Q}_{\dot\beta j}\right\} = \epsilon_{\dot\alpha\dot\beta} \bar{Z}_{ij}\,, \\ \left\{Q^i_\alpha, \bar{Q}_{\dot\alpha j}\right\} &= 2\delta^i_j (\sigma^a)_{\alpha\dot\alpha} P_a\,. \end{aligned} \tag{2.18}$$

The generators $Z^{ij} = -Z^{ji}$, $\bar{Z}_{ij} = (Z^{ij})^\dagger$ are called central charges, since they commute with all the generators of the superalgebra:

$$[Z, Z] = [Z, P] = [Z, Q] = [Z, L] = 0\,. \tag{2.19}$$

* In this sense the Poincaré algebra is $N = 0$ supersymmetry, and the superalgebra (2.16), (2.17) corresponds to $N = 1$.

This superalgebra has an *outer* automorphism group $U(N)$ (also known as R symmetry). The spinor generators Q^i_α and $\bar{Q}_{\dot{\alpha}i}$ transform according to the fundamental N and $\bar{N}$ representations of $U(N)$, respectively. We use the following convention for the algebra of $SU(N)$:

$$[T^i_j, T^k_l] = i(\delta^i_l T^k_j - \delta^k_j T^i_l), \qquad (T^i_j)^\dagger = -T^j_i, \qquad T^i_i = 0 \quad (i, j = 1, \ldots, N). \tag{2.20}$$

In [H1] it has been shown that the superalgebra (2.18), (2.19) is the only non-trivial spinor extension of the Poincaré algebra within the context of local field theory with non-vanishing masses. In the massless case the symmetry can be enlarged to conformal symmetry.

2.2.3 Conformal supersymmetry

The conformal superalgebra $SU(2,2|N)$ is an extension of the conformal algebra $SO(2,4) \sim SU(2,2)$ (2.7), (2.8), and at the same time of the N-extended Poincaré superalgebra (2.18) (with vanishing central charges, $Z^{ij} = 0$). In addition to the above generators it also contains the generators I^j_i, R of the *inner* automorphism group $U(N)$ and the 'special' (or conformal) supersymmetry generators $S_{\alpha i}$, $\bar{S}^j_{\dot{\alpha}}$. The latter are counterparts of the usual supersymmetry generators Q^i_α, $\bar{Q}_{\alpha j}$ (like the conformal boosts K_a are counterparts of the translation generators P_a). The (anti)commutation relations are given by (2.7), (2.8), (2.15), (2.16), (2.18) (with $Z = 0$) and

$$\begin{aligned}
\{S_{\alpha k}, \bar{S}^i_{\dot{\beta}}\} &= 2\delta^i_k\, (\sigma^a)_{\alpha\dot{\beta}} K_a\,, \\
\{Q^i_\alpha, S^\beta_j\} &= -\delta^i_j (\sigma^{ab})_\alpha{}^\beta L_{ab} - 4i\delta^\beta_\alpha I^i_j - 2i\delta^\beta_\alpha \delta^i_j D + \frac{2(4-N)}{N}\delta^\beta_\alpha \delta^i_j R\,, \\
[Q^i_\alpha, K_a] &= -(\sigma_a)_{\alpha\dot{\alpha}} \bar{S}^{\dot{\alpha}i}, \qquad [\bar{Q}_{\dot{\alpha}i}, K_a] = (\sigma_a)_{\alpha\dot{\alpha}} S^\alpha_i\,, \\
[S_{\alpha i}, P_a] &= -(\sigma_a)_{\alpha\dot{\alpha}} \bar{Q}^{\dot{\alpha}}_i\,, \qquad [\bar{S}^i_{\dot{\alpha}}, P_a] = (\sigma_a)_{\alpha\dot{\alpha}} Q^{\alpha i}\,,
\end{aligned} \tag{2.21}$$

$$[D, Q] = \frac{i}{2}Q\,, \qquad [D, \bar{Q}] = \frac{i}{2}\bar{Q}\,, \qquad [D, S] = -\frac{i}{2}S\,, \qquad [D, \bar{S}] = -\frac{i}{2}\bar{S}\,,$$

$$[R, Q] = -\frac{1}{2}Q\,, \qquad [R, \bar{Q}] = \frac{1}{2}\bar{Q}, \qquad [R, S] = \frac{1}{2}S\,, \qquad [R, \bar{S}] = -\frac{1}{2}\bar{S}\,,$$

together with the standard $SU(N)$ relations

$$[I^i_j, Q^k] = \delta^k_j Q^i - \frac{1}{N}\delta^i_j Q^k\,, \qquad [I^i_j, \bar{Q}_k] = -\delta^i_k \bar{Q}_j + \frac{1}{N}\delta^i_j \bar{Q}_k\,, \tag{2.22}$$

and similarly for S.

2.2.4 Central charges from higher dimensions

It is often useful to give the central charges appearing in (2.18) the interpretation of extra components of the momentum in a higher-dimensional space. We explain this on the simplest example of $N = 2$ supersymmetry which is the main subject of this book. In this case the algebra (2.18) becomes

$$\begin{aligned} \{Q^i_\alpha, Q^j_\beta\} &= 2\epsilon_{\alpha\beta}\epsilon^{ij} Z\,, \qquad \{\bar{Q}_{\dot\alpha i}, \bar{Q}_{\dot\beta j}\} = -2\epsilon_{\dot\alpha\dot\beta}\epsilon_{ij}\bar{Z}\,, \\ \{Q^i_\alpha, \bar{Q}_{\dot\alpha j}\} &= 2\delta^i_j (\sigma^a)_{\alpha\dot\alpha} P_a\,, \end{aligned} \tag{2.23}$$

where Z is a complex central charge.

The algebra (2.23) can be rewritten in the form of $N = 1$ Poincaré supersymmetry algebra in six dimensions [H13, K11, Z2]. The spinor representations of the six-dimensional Lorentz group are classified by $SU^*(4)$ (for details see, e.g., [H13]). There are two types of spinor indices $\hat\alpha = 1, 2, 3, 4$, 'left-handed' $\psi^{\hat\alpha}$ and 'right-handed' $\psi_{\hat\alpha}$. Correspondingly, there are two types of gamma matrices, $\Gamma^{\hat m}_{\hat\alpha\hat\beta}$ and $\Gamma^{\hat m\hat\alpha\hat\beta}$ ($\hat m = 0, 1, 2, 3, 5, 6$), which are related by raising and lowering the pair of antisymmetric spinor indices $\hat\alpha\hat\beta$:

$$\Gamma^{\hat m\hat\alpha\hat\beta} = \frac{1}{2}\epsilon^{\hat\alpha\hat\beta\hat\gamma\hat\delta}\Gamma^{\hat m}_{\hat\gamma\hat\delta}\,, \qquad \Gamma^{\hat m}_{\hat\alpha\hat\beta} = \frac{1}{2}\epsilon_{\hat\alpha\hat\beta\hat\gamma\hat\delta}\Gamma^{\hat m\hat\gamma\hat\delta}\,. \tag{2.24}$$

The epsilon symbol is defined by $\epsilon^{1234} = \epsilon_{1234} = 1$. The generators $Q_{i\hat\alpha}$ of six-dimensional $N = 1$ supersymmetry (sometimes it is called (0, 1) supersymmetry because of the chiral nature of the spinors) has a right-handed spinor index of $SU^*(4)$ and an $SU(2)$ index i. Besides the Weyl condition (which is taken care of by the above notation), these spinors satisfy a pseudo-Majorana reality condition (see [H13]). Thus they have eight real components, just as many as the generators of $N = 2$ supersymmetry in four dimensions. Their algebra is

$$\{Q_{\hat\alpha i}, Q_{\hat\beta j}\} = 2\epsilon_{ij}\Gamma^{\hat m}_{\hat\alpha\hat\beta}P_{\hat m}\,, \tag{2.25}$$

where $P_{\hat m} = -i(\partial/\partial x^{\hat m})$ and $x^{\hat m}$ are the coordinates of six-dimensional space.

The algebra (2.25) is manifestly covariant under the six-dimensional Lorentz group $SO(1, 5) \sim SU^*(4)$. It can be rewritten in a way in which only the four-dimensional Lorentz symmetry $SO(1, 3)$ is manifest. To this end one replaces the four-component spinor index $\hat\alpha$ of $SU^*(4)$ by a pair of an undotted and a dotted spinor indices of $SL(2, C)$, $\hat\alpha = (\alpha, \dot\alpha)$. The gamma matrices $\Gamma^{\hat m}$ then split into

$$\begin{aligned} &\Gamma^a_{\alpha\dot\beta} = \sigma^a_{\alpha\dot\beta}\,, && \Gamma^a_{\alpha\beta} = \Gamma^a_{\dot\alpha\dot\beta} = 0\ (a = 0, 1, 2, 3)\,; \\ &\Gamma^5_{\alpha\dot\beta} = 0\,, && \Gamma^5_{\alpha\beta} = -\epsilon_{\alpha\beta}\,, \qquad \Gamma^5_{\dot\alpha\dot\beta} = -\epsilon_{\dot\alpha\dot\beta}\,; \\ &\Gamma^6_{\alpha\dot\beta} = 0\,, && \Gamma^6_{\alpha\beta} = i\epsilon_{\alpha\beta}\,, \qquad \Gamma^6_{\dot\alpha\dot\beta} = -i\epsilon_{\dot\alpha\dot\beta}\,. \end{aligned} \tag{2.26}$$

The four-index epsilon symbol splits into a product of two two-index ones:

$$\epsilon_{\hat{\alpha}\hat{\beta}\hat{\gamma}\hat{\delta}} \quad \to \quad \epsilon_{\alpha\beta}\epsilon_{\dot{\gamma}\dot{\delta}} \tag{2.27}$$

(the other components vanish or are obtained by permutations).

Accordingly, the algebra (2.25) becomes

$$\begin{aligned} \{Q^i_\alpha, \bar{Q}_{\dot{\alpha}j}\} &= 2\delta^i_j \sigma^a_{\alpha\dot{\alpha}} P_a\,, \\ \{Q^i_\alpha, Q^j_\beta\} &= -2i\epsilon^{ij}\epsilon_{\alpha\beta}(\partial_5 - i\partial_6)\,, \\ \{\bar{Q}_{\dot{\alpha}i}, \bar{Q}_{\dot{\beta}j}\} &= 2i\epsilon_{ij}\epsilon_{\dot{\alpha}\dot{\beta}}(\partial_5 + i\partial_6)\,. \end{aligned} \tag{2.28}$$

Comparing (2.28) with (2.23), one can identify the central charge generators as follows:

$$Z = \frac{1}{i}(\partial_5 - i\partial_6) \equiv \frac{1}{i}\partial_z, \qquad \bar{Z} = \frac{1}{i}(\partial_5 + i\partial_6) \equiv \frac{1}{i}\partial_{\bar{z}}\,, \tag{2.29}$$

where

$$z = \frac{1}{2}(x^5 + ix^6)\,.$$

2.3 Representations of Poincaré supersymmetry

A comprehensive discussion of the massive and massless representations of N extended Poincaré superalgebra with and without central charges can be found, e.g., in [R1, S13, S14]. We summarize it in this section.

The irreducible representations of Poincaré supersymmetry are constructed from those of the Poincaré group, so we start by recalling a few points about the latter.

2.3.1 Representations of the Poincaré group

The representations of the Poincaré group are classified by the eigenvalues of its Casimir operators. The first one is the mass operator $P^2 = P_a P^a$. For $P^2 = m^2 > 0$ the second Casimir operator is the square $W^2 = W_a W^a$ of the Pauli–Lubanski vector

$$W_a = \frac{1}{2}\epsilon_{abcd}L^{bc}P^d\,. \tag{2.30}$$

In the rest frame

$$P_a = (m, 0, 0, 0) \tag{2.31}$$

W^2 is reduced to

$$W^2 = -m^2\vec{L}^2\,, \tag{2.32}$$

where $\vec{L}$ is the angular momentum operator. The latter takes the usual eigenvalues:

$$\vec{L}^2 = j(j+1)\,, \tag{2.33}$$

and so does $-1/m^2W^2$, which can be called the spin operator. So, the massive irreducible representations are labeled by their mass and spin.

In the massless case $P^2 = 0$ one can choose a Lorentz frame in which

$$P_a = (p, 0, 0, p)\,. \tag{2.34}$$

There the Pauli–Lubanski vector is proportional to the momentum operator:

$$W_a = \lambda P_a\,, \tag{2.35}$$

where λ is the helicity of the massless state, i.e., the projection of the spin along the direction of momentum. It can take integer or half-integer eigenvalues.

For $P^2 > 0$ the fermionic states have half-integer spins while the bosonic ones have integer spins (or helicities if $P^2 = 0$). Each irreducible representation of supersymmetry (supermultiplet) contains both bosonic states and fermionic states. The spinor generator Q^i_α maps fermionic (bosonic) states onto bosonic (fermionic) ones. The anticommutator $\{Q^i_\alpha, \bar{Q}_{\dot\beta j}\}$ is proportional to the momentum P_a (2.18) which maps fermionic (bosonic) states onto fermionic (bosonic) ones. So we arrive at the fundamental spin rule: In any supermultiplet of non-vanishing four-momentum there are equal numbers of fermionic and bosonic degrees of freedom. See [F9] for further spin rules that follow from the extended supersymmetry algebra.

2.3.2 *Poincaré superalgebra representations. Massive case*

The massive representations of the Poincaré superalgebra are obtained using the rest frame (2.31). There, in the absence of central charges the anticommutation relations (2.18) become the Clifford algebra of fermionic creation and annihilation operators:

$$\begin{aligned}\{Q^i_\alpha, Q^j_\beta\} &= \{\bar{Q}_{\dot\alpha i}, \bar{Q}_{\dot\beta j}\} = 0\,,\\ \{Q^i_\alpha, \bar{Q}_{\dot\alpha j}\} &= 2m\,\delta^i_j(\sigma^0)_{\alpha\dot\alpha} = 2m\,\delta^i_j\delta_{\alpha\dot\alpha}\,.\end{aligned} \tag{2.36}$$

Then one defines a Clifford vacuum $|s\rangle$ as an irreducible representation of the Poincaré algebra of mass m and spin s, such that

$$Q^i_\alpha|s\rangle = 0\,. \tag{2.37}$$

An irreducible representation of the supersymmetry algebra (supermultiplet) is obtained by the successive action of the creation operators $\bar{Q}_{\dot\alpha i}$ on the vacuum. At each step the spin changes by $\pm 1/2$. For the case $N = 1$ the results of this operation are listed below:

State	Spin	# of components
$\vert s\rangle$	s	$2s+1$
$\bar{Q}_{\dot\alpha}\vert s\rangle$	$s \pm 1/2$	$4s+2$
$(\bar{Q})^2\vert s\rangle$	s	$2s+1$

(2.38)

Here $(\bar{Q})^2 \equiv \bar{Q}_{\dot{\alpha}}\bar{Q}^{\dot{\alpha}}$. The total number of states is $2^2(2s+1)$, half of them being fermionic and the other half being bosonic. Note that the dimension of the Clifford vacuum is $d_{|s\rangle} = 2s+1$.

We see that the massive $N = 1$ supermultiplets are characterized by the spin s of their vacuum states. This quantum number is referred to as the superspin Y of the supermultiplet. For example, the $N = 1$ matter multiplet (chiral multiplet) has superspin $Y = 0$ and contains states with spins $1/2$, $(0)^2$; the off-shell Yang–Mills supermultiplet has superspin $Y = 1/2$ and describes states 1, $(1/2)^2$, 0. Note also that the superspin is related to the eigenvalue of the Casimir operator obtained as a superextension of the Pauli–Lubanski vector.

One can construct similar representations for any $N > 1$. We concentrate on the case $N = 2$ only, because it is the main subject of this book. Now the Clifford vacuum $|Y, I\rangle$ is labeled by its spin $s = Y$ and isospin I (the isospin characterizes the $SU(2)_A$ properties of the state). The full list of states in the supermultiplet is:*

State	Spin	Isospin
$\vert Y, I\rangle$	Y	I
$\bar{Q}_{\dot{\alpha}i}\vert Y, I\rangle$	$Y \pm 1/2$	$I \pm 1/2$
$\bar{Q}^i_{(\dot{\alpha}}\bar{Q}_{\dot{\beta})i}\vert Y, I\rangle$	$Y \pm 1, Y$	I
$\bar{Q}^{(i}_{\dot{\alpha}}\bar{Q}^{\dot{\alpha}j)}\vert Y, I\rangle$	Y	$I \pm 1, I$
$(\bar{Q})^3_{\dot{\alpha}i}\vert Y, I\rangle$	$Y \pm 1/2$	$I \pm 1/2$
$(\bar{Q})^4\vert Y, I\rangle$	Y	I

(2.39)

So, a massive supermultiplet with quantum numbers Y and I contains $8(2Y+1)(2I+1)$ fermionic and $8(2Y+1)(2I+1)$ bosonic states. Note that the Clifford vacuum has dimension $d_{|Y,I\rangle} = (2Y+1)(2I+1)$. In general the ratio of the dimension of a massive supermultiplet to that of its Clifford vacuum is 2^N.

The quantum numbers Y and I are called the superspin and superisospin of the supermultiplet. Here are some examples, which we shall often encounter in this book. The supermultiplet $Y = 0$, $I = 0$ contains states

$$|0,0\rangle^2\,, \qquad |1,0\rangle\,, \qquad |0,1\rangle\,, \qquad |1/2,1/2\rangle^2\,, \tag{2.40}$$

i.e., eight bosonic and eight fermionic components. In Chapter 7 we shall see that it corresponds to the $N = 2$ off-shell Yang–Mills multiplet.

The $N = 2$ massive matter multiplet (see Chapter 5) has $Y = 0$ and $I = 1/2$. This corresponds to the following 16 bosonic + 16 fermionic states:

$$|0,1/2\rangle^3\,, \qquad |0,3/2\rangle\,, \qquad |1,1/2\rangle\,, \qquad |1/2,0\rangle^2\,, \qquad |1/2,1\rangle^2\,. \tag{2.41}$$

* Note that for small values of Y and I there can be obvious deviations from this table, according to the standard spin multiplication rules.

Our final example is the multiplet with $Y = 1$ and $I = 0$. It describes 24 + 24 states arranged as follows:

$$|0,0\rangle\,,\quad |1,0\rangle^3\,,\quad |1,1\rangle\,,\quad |2,0\rangle\,,\quad |1/2,1/2\rangle^2\,,\quad |3/2,1/2\rangle^2\,. \tag{2.42}$$

In Chapter 10 we shall see that it corresponds to $N = 2$ off-shell conformal supergravity.

2.3.3 *Poincaré superalgebra representations. Massless case*

The massless representations of the Poincaré superalgebra are shorter than the massive ones. In the frame (2.34) the basic anticommutator (2.18) takes the form

$$\{Q^i_\alpha, \bar{Q}_{\dot{\alpha}j}\} = 2\delta^i_j(1+\sigma_3)_{\alpha\dot{\alpha}}p\,. \tag{2.43}$$

In fact, (2.43) is equivalent to the Clifford algebra

$$\{Q^i_1, \bar{Q}_{\dot{1}j}\} = 4\delta^i_j p\,, \tag{2.44}$$

with all the other anticommutators vanishing. Now one can define a vacuum state with helicity λ (the internal symmetry quantum numbers are not indicated):

$$Q^i_1|\lambda\rangle = Q^i_2|\lambda\rangle = \bar{Q}_{\dot{2}i}|\lambda\rangle = 0\,. \tag{2.45}$$

Then a massless multiplet is obtained by applying the creation operators $\bar{Q}_{\dot{1}i}$ (helicity $-1/2$) to the vacuum:

$$\begin{array}{ccc}
\text{State} & \text{Helicity} & \#\text{ of components} \\
|\lambda\rangle & \lambda & 1 \\
\bar{Q}_{\dot{1}i}|\lambda\rangle & \lambda-1/2 & N \\
\bar{Q}_{\dot{1}i}\bar{Q}_{\dot{1}j}|\lambda\rangle & \lambda-1 & N(N-1)/2 \\
\vdots & \vdots & \vdots \\
(\bar{Q})^N|\lambda\rangle & \lambda-N/2 & 1
\end{array} \tag{2.46}$$

In particular, for $N = 1$ the massless supermultiplets are pairs of states $|\lambda\rangle$, $|\lambda-1/2\rangle$, for $N = 2$ they become $|\lambda\rangle$, $|\lambda-1/2\rangle^2$, $|\lambda-1\rangle$, etc. Helicities of opposite sign are obtained by PCT conjugation. Of special interest are the self-conjugate multiplets, i.e., those containing the full range of helicities from λ to $-\lambda$. These are the

$N = 2$ matter multiplet:

$$1/2,\ (0)^2,\ -1/2\,; \tag{2.47}$$

$N = 4$ gauge multiplet:

$$1,\ (1/2)^4,\ (0)^6,\ (-1/2)^4,\ -1\,; \tag{2.48}$$

$N = 8$ supergravity multiplet:

$$2,\ (3/2)^8,\ (1)^{28},\ (1/2)^{56},\ (0)^{70},\ (-1/2)^{56},\ (-1)^{28},\ (-3/2)^8,\ -2\,. \tag{2.49}$$

Note that for $N > 8$ any supermultiplet would contain helicities greater than 2, therefore such supersymmetries are considered unphysical.

In what follows we shall mainly deal with $N = 2$ theories. The massless matter on-shell multiplet is shown in (2.47), the on-shell gauge multiplet is $1, (1/2)^2, (0)^2, (-1/2)^2, -1$ and the on-shell supergravity multiplet is $2, (3/2)^2, 1, -1, (-3/2)^2, -2$. We would like to stress that off shell the number of states in the supermultiplets is determined by the massive representations discussed above.

2.3.4 Representations with central charge

As we saw above, the massless representations of the Poincaré superalgebra are half the size of the massive ones. Extended supersymmetry ($N > 1$) provides another way to realize 'half-size' representations. They are massive, but in addition have a *central charge equal to the mass*. Let us explain this mechanism in the simplest case of $N = 2$ supersymmetry with central charge (2.23). In order to construct the Fock space of a *massive* representation, we need to rearrange the spinor generators as follows:

$$A^{\pm}_{\alpha} = \frac{1}{2}(Q^1_{\alpha} \pm \bar{Q}^{\dot{\alpha}}_2), \qquad \bar{A}^{\pm}_{\dot{\alpha}} = \frac{1}{2}(\bar{Q}_{\dot{\alpha}1} \pm Q^{\alpha 2}). \tag{2.50}$$

Note that in the rest frame (2.31) where the Lorentz group is reduced to the little group $SO(3) \sim SU(2)$, dotted and undotted spinor indices can freely be mixed, as in (2.50). It is easy to check that the new generators satisfy the following algebra:

$$\{A^{\pm}_{\alpha}, A^{\pm}_{\dot{\alpha}}\} = \delta_{\alpha\dot{\alpha}}\left[m \pm \frac{1}{2}(Z + \bar{Z})\right], \tag{2.51}$$

where Z is the eigenvalue of the complex central charge. It is then clear that in the special case

$$m = \frac{1}{2}(Z + \bar{Z}) \tag{2.52}$$

the only generators forming a Clifford algebra are A^+:

$$\{A^{+}_{\alpha}, A^{+}_{\dot{\alpha}}\} = 2m\delta_{\alpha\dot{\alpha}}. \tag{2.53}$$

Then we can define a vacuum state of spin Y:

$$A^{-}_{\alpha}|Y\rangle = A^{-}_{\dot{\alpha}}|Y\rangle = A^{+}_{\alpha}|Y\rangle = 0. \tag{2.54}$$

The multiplet is obtained by applying the creation operator $A^{+}_{\dot{\alpha}}$ (spin 1/2) to the vacuum:

$$|Y\rangle, \qquad A^{+}_{\dot{\alpha}}|Y\rangle, \qquad A^{+}_{\dot{\alpha}}A^{+}_{\dot{\beta}}|Y\rangle. \tag{2.55}$$

This gives rise to massive states of spins $Y - 1/2, Y, Y, Y + 1/2$, i.e., only half the number of states in a massive multiplet without central charge (see (2.39)). In particular, by choosing $Y = 0$, we obtain a very short multiplet containing only spins 0 and 1/2. It is the massive matter multiplet of $N = 2$ supersymmetry with central charge. It should be stressed that such a short massive multiplet *does not exist without central charge*.

In conclusion, we make a few comments. Firstly, the condition (2.52) fixes only the real part of the central charge. The imaginary part remains as a free, mass-like parameter. Since it has no direct physical meaning,* it is customary to fix it, for instance, by demanding

$$Z = \bar{Z} = m \,. \tag{2.56}$$

Secondly, an alternative to (2.50) combination of Q and $\bar{Q}$ would lead to a condition on the imaginary, rather than the real part of z. This is a consequence of the presence of a $U(1)$ factor in the automorphism group of the algebra (2.23), which rotates the central charges (this $U(1)_{\rm R}$ is broken when the central charges are fixed). Finally, the structure of the representations (2.55) resembles that of the massless multiplets (2.46) because both Fock spaces are constructed from two isomorphic Clifford algebras.

2.4 Realizations of supersymmetry on fields. Auxiliary fields

In the preceding sections we have seen how supersymmetry is realized on *on-shell* physical states. This section is devoted to the *off-shell* realizations of supersymmetry on fields. As an illustration of the main issues we first consider the simplest example of $N = 1$ supersymmetric matter.

2.4.1 $N = 1$ matter multiplet

The $N = 1$ chiral supermultiplet has superspin 0, so on shell it contains two spin 0 particles (in fact, one scalar and one pseudoscalar) and one spin 1/2 particle. They can be described by the real and imaginary parts of a complex scalar field $\phi(x)$ and by a left-handed spinor field $\psi_\alpha(x)$. Off shell one can add to that set of fields another complex scalar $F(x)$, which is called an auxiliary field (its rôle is explained in the next subsection). The $N = 1$ supersymmetry transformations with constant Grassmann parameter $\epsilon_\alpha, \bar{\epsilon}_{\dot\alpha} = \overline{\epsilon_\alpha}$ are realized on the above fields as follows:

$$\delta\phi(x) \;\; = \;\; -\epsilon^\alpha \psi_\alpha(x) \,,$$

* Both $N = 2$ central charges can become active in a theory with interaction; for instance, in the spontaneously broken $N = 2$ Yang–Mills theory they are identified with some unbroken ('electric') $U(1)$ charge and topological ('magnetic') charge, respectively [W17].

$$\begin{aligned}\delta\psi_\alpha(x) &= -2i(\sigma^a_{\alpha\dot\alpha})\bar\epsilon^{\dot\alpha}\partial_a\phi(x) - 2\epsilon_\alpha F(x)\,, \\ \delta F(x) &= -i\bar\epsilon^{\dot\alpha}(\sigma^a_{\alpha\dot\alpha})\partial_a\psi^\alpha(x)\,.\end{aligned} \tag{2.57}$$

It is easy to check that the commutator of two such supertransformations with parameters ϵ_1 and ϵ_2 produces a space-time translation with composite parameter $i(\epsilon_1\sigma^a\bar\epsilon_2 - \epsilon_2\sigma^a\bar\epsilon_1)$ on each of the fields of the multiplet, in accordance with the supersymmetry algebra (2.17).

The field-theoretical model describing the interaction of the above fields invariant under supersymmetry is known by the name of the Wess–Zumino model [W9]. It is historically the first example of a four-dimensional supersymmetric field theory:*

$$\begin{aligned}S = \int d^4x \Big\{ &-\frac{i}{2}\psi\sigma^a\partial_a\bar\psi - \bar\phi\Box\phi + \bar F F \\ &+ \Big[m\Big(\phi F - \frac{1}{4}\psi\psi\Big) + g\Big(\phi^2 F - \frac{1}{2}\phi\psi\psi\Big) + \text{c.c.}\Big]\Big\}\,.\end{aligned} \tag{2.58}$$

Using the transformation laws (2.57), it is a straightforward exercise to check that the Lagrangian in (2.58) transforms by a total space-time derivative, so the action is invariant.

Actually, both the invariant action (2.58) and transformation rules (2.57) can be derived by a systematic procedure involving superfields. This will be explained in detail in Chapter 3.

The auxiliary field $F(x)$ has dimension $[F] = 2$ and it enters the action (2.58) without derivatives. Therefore, its equation of motion is purely algebraic:

$$\frac{\delta L}{\delta\bar F} = F + m\bar\phi + g\bar\phi^2 = 0\,. \tag{2.59}$$

This allows one to eliminate $F(x)$ both from the action (2.58):

$$\begin{aligned}S = \int d^4x \Big[&-\frac{i}{2}\psi\sigma^a\partial_a\bar\psi - \bar\phi\Box\phi \\ &-\frac{1}{2}\Big(\frac{1}{2}m\psi\psi + g\psi\psi\phi + \text{c.c.}\Big) - (m\bar\phi + g\bar\phi^2)(m\phi + g\phi^2)\Big]\end{aligned} \tag{2.60}$$

and from the supersymmetry transformations (2.57):

$$\begin{aligned}\delta\phi(x) &= -\epsilon^\alpha\psi_\alpha(x)\,, \\ \delta\psi_\alpha(x) &= -2i(\sigma^a_{\alpha\dot\alpha})\bar\epsilon^{\dot\alpha}\partial_a\phi(x) + 2\epsilon_\alpha(m\bar\phi + g\bar\phi^2)\,.\end{aligned} \tag{2.61}$$

We see on this example that upon elimination of the auxiliary fields (the scalar $F(x)$ here) the number of fermionic fields is *not equal* to that of the bosonic

* With the *linearly* realized supersymmetry.

ones. In addition, unlike the original supersymmetry transformations (2.57), the new ones (2.61) are *non-linear* and their form is influenced by the form of the action. In other words, they become model-dependent because one has used the equations of motion to eliminate the auxiliary fields. Finally, the algebra of (2.61) *does not close off shell*:

$$\begin{aligned}(\delta_1\delta_2 - \delta_2\delta_1)\psi_\alpha &= -2i(\epsilon_1\sigma^a\bar{\epsilon}_2)\partial_a\psi_\alpha \\ &\quad + 2\epsilon_{1\alpha}\bar{\epsilon}_{2\dot{\alpha}}\left[-i\partial\!\!\!/^{\dot{\alpha}}_{\alpha}\psi^\alpha + m\bar{\psi}^{\dot{\alpha}} + 2g\bar{\phi}\bar{\psi}^{\dot{\alpha}}\right] - (1 \leftrightarrow 2)\,.\end{aligned} \tag{2.62}$$

The second term in the right-hand side of (2.62) seems to violate the supersymmetry algebra (2.17). However, it is proportional to the field equation for ψ_α, so it *vanishes on shell*.

One can summarize the above discussion of the Wess–Zumino model by the following general statements concerning supersymmetric theories in which the auxiliary fields have been eliminated:

(i) In such theories the supersymmetry transformations are *non-linear* in terms of the fields. As a consequence, checking the supersymmetry invariance is in general not easy. *The numbers of boson and fermion fields do not match.*

(ii) The form of the supersymmetry transformations is *model dependent*. Only the action as a whole is supersymmetric, *the kinetic, mass or interaction terms are not separately invariant.*

(iii) The algebra of the transformations *closes only on shell*, i.e., with the help of the *equations of motion.*

All this makes the search for supersymmetric interacting theories in the absence of auxiliary fields a non-trivial task. The rather involved Noether procedure (see, e.g., [B13, V1]) has to be used in order to find both the form of the invariant action and the on-shell transformation rules.

On the other hand, the introduction of auxiliary fields makes the supersymmetry transformation laws linear, model-independent and closed off shell, without changing the physical content of the theory. This considerably simplifies the construction of supersymmetric models, especially in such complicated theories as supergravity. In addition, the auxiliary fields are crucial for the spontaneous breaking of supersymmetry [C12, F5], they facilitate quantization, etc. In what follows we shall adopt the point of view that every 'respectable' supersymmetric theory should have its auxiliary fields, although they are not always very easy to find.*

Before discussing a systematic approach to off-shell supersymmetric theories, we give a few more well-known examples.

* The concept of auxiliary degrees of freedom is indeed not very new. In electrodynamics the Coulomb field is actually an auxiliary one. It can be (non-locally) expressed in terms of the charged fields, but this results in the loss of manifest Lorentz invariance.

2.4.2 $N = 1$ gauge multiplet

The $N = 1$ gauge supermultiplet [F12, S2, W10] is described by a real vector (gauge) field $A_a(x)$, a Majorana spinor $\psi_\alpha(x)$, $\bar{\psi}_{\dot\alpha}(x)$ and an auxiliary real scalar $D(x)$. All these fields are Lie-algebra valued, e.g., $A_a(x) = A^i_a(x)t_i$ where t_i are the generators of the adjoint representation of the gauge group. Their supersymmetry transformation laws are

$$\begin{aligned}
\delta F_{ab} &= -i(\epsilon\sigma_b \mathcal{D}_a\bar{\psi} + \bar{\epsilon}\tilde{\sigma}_b\mathcal{D}_a\psi) - (a \leftrightarrow b)\,, \\
\delta\psi_\alpha &= \frac{i}{2}(\sigma^{ab}\epsilon)_\alpha F_{ab} + i\epsilon_\alpha D\,, \\
\delta D &= -\epsilon\sigma^a\mathcal{D}_a\bar{\psi} + \bar{\epsilon}\tilde{\sigma}^a\mathcal{D}_a\psi\,.
\end{aligned} \tag{2.63}$$

Here $F_{ab} = \partial_a A_b - \partial_b A_a + i[A_a, A_b]$ is the field strength and $\mathcal{D}_a\psi(x) = \partial_a\psi(x) + i[A_a, \psi(x)]$ is the gauge covariant derivative. Under gauge transformations one has

$$\delta A_a = \mathcal{D}_a\rho(x)\,, \qquad \delta\psi_\alpha = i[\psi_\alpha, \rho(x)]\,, \qquad \delta D = i[D, \rho(x)]\,, \tag{2.64}$$

where $\rho(x) = \rho^i(x)t_i$ are the Yang–Mills group parameters.

The supersymmetric action for this theory is

$$S = \int d^4x \mathrm{Tr}\left[-\frac{1}{4}F^{ab}F_{ab} - i\psi\sigma^a\mathcal{D}_a\bar{\psi} + \frac{1}{2}D^2\right]. \tag{2.65}$$

The rôle of the auxiliary field $D(x)$ here is very similar to the one in the Wess–Zumino model.

The auxiliary fields for $N = 1$ supergravity [D11, F16] are also well known [F11, S22].* So, all the basic $N = 1$ supersymmetric theories admit (sometimes more than one) sets of auxiliary fields.

2.4.3 Auxiliary fields and extended supersymmetry

The auxiliary field problem for extended supersymmetric theories is much more difficult. Before 1984 the auxiliary fields for some of the $N = 2$ theories were known, but other theories were believed not to possess them at all. In this subsection we recall the arguments (the so-called 'no-go' theorem [H18, S21]) against the possibility to formulate the Fayet–Sohnius massless hypermultiplet (one of the versions of the $N = 2$ supersymmetric matter theory) off shell. More precisely, one can argue that *a finite set of auxiliary fields* does not exist for this theory. Similar arguments apply to a number of higher N supersymmetric theories. The solution to this problem, at least for all the $N = 2$ theories and for $N = 3$ super-Yang–Mills theory, was found in 1984 in the framework of

* In fact, there exist several versions of that theory [G26], which differ just by their sets of auxiliary fields.

harmonic superspace. The key point there is the use of *infinite sets of auxiliary fields*.

The Fayet–Sohnius $N = 2$ massless matter hypermultiplet [F1, S12] (to be referred to as the q^+ hypermultiplet in what follows) contains two (necessarily complex) scalar fields forming an isodoublet of the automorphism group $SU(2)_A$, $f^i(x)$, $\bar{f}_i(x) \equiv \overline{f^i(x)}$ $i = 1, 2$ and an isosinglet Dirac spinor field $\psi_\alpha(x)$, $\bar{\kappa}^{\dot{\alpha}}(x)$. The free action for this multiplet,

$$S = \int d^4x \left(\partial^a f^i \partial_a \bar{f}_i - \frac{i}{2} \bar{\psi} \partial\!\!\!/ \psi - \frac{i}{2} \bar{\kappa} \partial\!\!\!/ \kappa \right), \tag{2.66}$$

is invariant under the following $N = 2$ supersymmetry transformations:

$$\begin{aligned} \delta f^i &= -(\epsilon^i \psi + \bar{\epsilon}^i \bar{\kappa}), \\ \delta \psi_\alpha &= -2i(\sigma^a \bar{\epsilon}^i)_\alpha \partial_a f_i, \\ \delta \kappa_\alpha &= -2i(\sigma^a \bar{\epsilon}^i)_\alpha \partial_a \bar{f}_i. \end{aligned} \tag{2.67}$$

These transformations form an open algebra. For instance,

$$\begin{aligned} (\delta_2\delta_1 - \delta_1\delta_2)\psi_\alpha &= 2i(\epsilon_1^i \sigma^a \bar{\epsilon}_{2i})\partial_a \psi_\alpha - 2i\epsilon_{2\alpha i}(\bar{\epsilon}_1^i \bar{\partial\!\!\!/} \psi) - 4i(\bar{\epsilon}_1 \bar{\epsilon}_2)(\bar{\partial\!\!\!/} \bar{\kappa})_\alpha \\ &\quad -(1 \leftrightarrow 2). \end{aligned} \tag{2.68}$$

As in the case of the $N = 1$ matter multiplet (see (2.62)), the extra terms in the right-hand side of (2.68) are proportional to the field equations, so they vanish on shell. However, unlike the $N = 1$ case, this time adding any finite number of auxiliary fields to the action does not help to close the algebra off shell. This is the content of one of the so-called 'no-go theorems' in extended supersymmetry [H18, R2, S8, S21].

The argument goes as follows. Suppose that such a finite set of auxiliary fields exists. First of all, the auxiliary fermions (if any) must have half-integer dimensions (this follows from the dimension $[\epsilon] = -1/2$ of the supersymmetry parameters). So, they should enter the action in pairs with dimensions, e.g., 3/2 and 5/2. Thus each member of a pair will be a Lagrange multiplier for the other one, preventing it from propagation. Next, from the $SU(2)_A$ assignments in (2.67) it is clear that the auxiliary fermion fields should carry integer isospins. This implies that they should form Dirac spinors (like the physical ones) in order to preserve off shell the invariance of the on-shell action (2.66) under parity.

Let us now count the number of fermions. A Dirac spinor has eight real degrees of freedom off shell, and a representation of isospin I has dimension $2I+1$. So, the number of fermionic components in an auxiliary pair will be $2 \times 8 \times (2I+1)$. In addition, one has eight physical off-shell fermionic components. Altogether this gives

$$16q + 8\,, \qquad q \text{ integer}\,, \tag{2.69}$$

fermionic degrees of freedom.

On the other hand, a complex irreducible off-shell $N = 2$ supermultiplet contains at least $2^4 \times d_{|0,J\rangle} = 2^4 \times 2(2J + 1)$ bosonic and fermionic states (see (2.39)). Indeed, in our case the Clifford vacuum should have spin 0 (because the supermultiplet under consideration has superspin 0) and should belong to a complex representation of $SU(2)_\mathrm{A}$ with isospin J (because the physical bosons do so). Thus, in a multiplet of this kind one finds $\frac{1}{2} \times 2^4 \times 2(2J + 1)$ fermions. Taking a number of such multiplets (one physical + several auxiliary), one obtains the total number of fermions

$$16p\,, \qquad p \text{ integer}\,. \tag{2.70}$$

Clearly, the two ways of counting, (2.69) and (2.70), are incompatible.

We would like to emphasize the following two crucial points:

(i) This proof is based on a counting argument, i.e., one of the basic assumptions is that one deals with *finite sets of fields*. This is precisely the loophole in all the 'no-go' theorems in extended supersymmetry. It can be successfully circumvented in harmonic superspace where one naturally deals with *infinite sets of fields* (see Section 3.4 below);

(ii) If the physical fields belong to a real representation of $SU(2)_\mathrm{A}$ then the theorem ceases to be valid. In this case in a formula of the type (2.70) one would have $8p$ instead of $16p$. Examples of such real supermultiplets having finite sets of auxiliary fields are the tensor and relaxed hypermultiplets discussed in Chapter 6.

3
Superspace

In Chapter 2 we considered representations of the supersymmetry algebra in terms of ordinary fields (bosons and fermions). A manifestly supersymmetric description would require an adequate space, superspace, as in the case of Poincaré symmetry, which is realized in Minkowski space. A systematic way of finding such spaces and symmetry realizations is the coset space method proposed by Cartan [C4]. Some years later this method was (independently) rediscovered by physicists [C11, O3, V3, W5] and was applied to spontaneous symmetry breaking, current algebra, etc. We begin by recalling the main points of this construction.

3.1 Coset space generalities

Consider a group $G \equiv \{X_i, Y_a\}$ (generators X_i, $i = 1, \ldots, n$ and Y_a, $a = 1, \ldots, m$) having a subgroup $H \equiv \{X_i\}$ (generators X_i). The Lie algebra of G has the following structure:

$$[X, X] \sim X\,, \qquad [X, Y] \sim Y\,, \qquad [Y, Y] \sim X + Y\,. \tag{3.1}$$

The coset space G/H is parametrized by coordinates ξ^a corresponding to the generators Y_a. In the exponential parametrization an element of this coset has the form

$$G/H: \qquad \Omega = \exp\{i\xi^a Y_a\} \tag{3.2}$$

(other parametrizations are equally admissible). Any element $g \in G$ can be uniquely split into G/H and H factors (in the vicinity of the identity):

$$g = \exp\{ic^a Y_a\}\exp\{i\lambda^k X_k\}\,, \qquad g \in G\,. \tag{3.3}$$

Then the left action of G on the coset G/H is defined as follows:

$$g\ \exp\{i\xi^a Y_a\} = \exp\{i\xi'^a(\xi, g)Y_a\}h(\xi, g) \quad \Rightarrow \quad \Omega' = g\,\Omega\,h^{-1}\,, \tag{3.4}$$

where $h(\xi, g) = \exp\{if^i(\xi, g)X_i\}$ is some induced element of the subgroup H with composite parameters $f^i(\xi, g)$. The new parameters ξ'^a and f^i in (3.4) can be derived using the Baker–Campbell–Hausdorf formula

$$e^A e^B = \exp\left\{A + B + \frac{1}{2}[A, B] + \frac{1}{12}([A, [A, B]] + [[A, B], B]) + \cdots\right\} \tag{3.5}$$

and the algebra (3.1).

Equation (3.4) not only gives the action of the group G on the coset space coordinates ξ^a, but allows one to define covariant fields on G/H. They are classified according to the irreps of the subgroup H. In other words, the field $\psi_k(\xi)$ transforms under the action of G as a representation of H (k is the matrix index of that representation) with the composite parameters $f^i(\xi, g)$ from (3.4):

$$G: \qquad \psi'_k(\xi') = \left(e^{if^i(\xi,g)X_i}\right)_k^l \psi_l(\xi). \tag{3.6}$$

In particular, if one takes a $g \in H$, one finds that $\psi_k(\xi)$ transforms linearly (i.e., f^i does not depend on ξ). The rest of the group G is realized non-linearly.

Note that the non-linear realizations formalism always yields the so-called co-adjoint action of the group G on the coset coordinates ξ^a. Namely, a finite G transformation of ξ^a can be represented in the form:

$$G: \qquad (\xi^a)' = \tilde{g}^{-1}\, \xi^a\, \tilde{g}\,, \qquad \tilde{g} = \exp\{ic^a\tilde{Y}_a\}\exp\{i\lambda^k\tilde{X}_k\}, \tag{3.7}$$

where the generators $\tilde{Y}_a$, $\tilde{X}_k$ are appropriate differential operators ('vector fields') on the coset manifold. On the other hand, eq. (3.6) implies the standard action of G on the fields $\psi_l(\xi)$:

$$G: \qquad \psi'_k(\xi) = g_\psi\ \psi_k(\xi)\ g_\psi^{-1} \tag{3.8}$$

(the relevant generators are basically the differential operators $\tilde{Y}_a$, $\tilde{X}_k$ with the appropriate matrix parts added; the latter act only on the external H-irrep index of the field and can be read off from (3.6)). One should keep in mind these peculiarities when determining the precise form of the G generators in the coset realizations.

The ordinary (partial) derivatives $\partial/\partial\xi^a$ of $\psi(\xi)$ do not transform covariantly. One can define covariant derivatives in the following way. Consider the Cartan form

$$\Omega^{-1}d\Omega = i(\omega^a Y_a + \omega^i X_i). \tag{3.9}$$

From (3.4) one can derive its transformation law:

$$\left(\Omega^{-1}d\Omega\right)' = h\left(\Omega^{-1}d\Omega\right)h^{-1} + h\,dh^{-1}. \tag{3.10}$$

Comparing (3.10) with (3.9) one sees that the 1-forms ω_a transform homogeneously:

$$\omega^{a\prime} Y_a = h\omega^a Y_a h^{-1}\,, \tag{3.11}$$

so they can be interpreted as covariant differentials of the coset parameters ξ^a. Further, the inhomogeneously transforming 1-forms ω^i define connections for the subgroup H in the coset space G/H; they are needed because the transformations (3.6) depend on the coordinates. Indeed, the covariant differential

$$(D\psi)_k = \left[(d + i\omega^i X_i)\psi\right]_k \tag{3.12}$$

transforms homogeneously under G, just as the field ψ_k itself. Correspondingly, the covariant derivative D_a of ψ_k can be defined as follows:

$$(D\psi)_k \equiv \omega^a (D_a\psi)_k\,. \tag{3.13}$$

3.2 Coset spaces for the Poincaré and super Poincaré groups

It is instructive to begin with Minkowski space as the simplest and best known example of a coset space. There $G = \{L_{ab}, P_a\}$ is the Poincaré group and $H = \{L_{ab}\}$ is the Lorentz group. Then the coset parameters are the familiar coordinates x^a, $a = 0, 1, 2, 3$:

$$M^4 = \frac{\{L_{ab}, P_a\}}{\{L_{ab}\}} = (x^a)\,. \tag{3.14}$$

From the transformation law (3.4) with $g = \exp\{ic^a P_a\}$ one easily obtains the translation $x'^a = x^a + c^a$, $h = \mathbb{I}$. A Lorentz rotation $g = \exp\{\frac{i}{2}\lambda^{ab} L_{ab}\}$ with parameters $\lambda^{ab} = -\lambda^{ba}$ produces, via the relations (2.7),

$$x'^a = \left[\exp\left\{-\frac{i}{2}\lambda^{cd}\hat{L}_{cd}\right\}\right]^a_{\ b} x^b \equiv \Lambda^a_b x^b$$

(see (2.5), (2.6)), and $h = \exp\{\frac{i}{2}\lambda^{ab} L_{ab}\}$. In this case $\omega^a = dx^a$, $\omega^{ab} = 0$, so $D_a = \partial/\partial x^a$.

The main application of Cartan's construction for our purposes is to find a superspace, i.e., a space where the Poincaré supersymmetry algebra (2.7), (2.15)–(2.19) can be realized in the same way as the Poincaré algebra (2.7) is realized in Minkowski space. The most straightforward way to do this is to copy (3.14), i.e., to choose H to be the group of automorphisms $SL(2, \mathbb{C}) \times SU(N)$. Thus one obtains a 'real' superspace:

$$\mathbb{R}^{4|4N} = \frac{\{L_{ab}, su(N), P_a, Q^i_\alpha, \bar{Q}_{\dot{\alpha}i}\}}{\{L_{ab}, su(N)\}} = \left(x^a, \theta^\alpha_i, \bar{\theta}^{\dot{\alpha}i}\right) \equiv (X^A)\,. \tag{3.15}$$

Here and hereafter we use the notation $su(N)$ for the algebra of $SU(N)$, x^a are even (commuting) and θ^α_i, $\bar{\theta}^{\dot{\alpha}i}$ are odd (anticommuting) coordinates associated with translations and supertranslations. They have the following properties under complex conjugation:

$$\overline{x^a} = x^a\,, \qquad \overline{\theta_i^\alpha} = \bar{\theta}^{\dot{\alpha}i}\,, \tag{3.16}$$

which explains the name 'real superspace' for $\mathbb{R}^{4|4N}$.

The application of Cartan's rules is once again very easy. We choose

$$\Omega = \exp i\{-x^a P_a + \theta_i^\alpha Q_\alpha^i + \bar{\theta}_{\dot{\alpha}}^i \bar{Q}_i^{\dot{\alpha}}\}\,. \tag{3.17}$$

Clearly, the new coordinates θ transform as Lorentz spinors (just by the laws (2.10), (2.11)) and according to the $SU(N)$ fundamental representations. The supersymmetry transformations are generated by

$$g = \exp i\{\epsilon_i^\alpha Q_\alpha^i + \bar{\epsilon}_{\dot{\alpha}}^i \bar{Q}_i^{\dot{\alpha}}\} \tag{3.18}$$

and have the following form (with infinitesimal parameters ϵ):

$$\begin{aligned} \delta x^a &= i(\epsilon^i \sigma^a \bar{\theta}_i - \theta^i \sigma^a \bar{\epsilon}_i)\,, \\ \delta\theta_{\alpha i} &= \epsilon_{\alpha i}\,, \qquad \delta\bar{\theta}_{\dot{\alpha}}^i = \bar{\epsilon}_{\dot{\alpha}}^i\,. \end{aligned} \tag{3.19}$$

Using (3.19), it is easy to express the corresponding generators as differential operators in superspace (recall eq. (3.7)):

$$\begin{aligned} Q_\alpha^i &= i\frac{\partial}{\partial\theta_i^\alpha} + \bar{\theta}^{\dot{\alpha}i}(\sigma^a)_{\alpha\dot{\alpha}}\frac{\partial}{\partial x^a}\,, \\ \bar{Q}_{i\dot{\alpha}} &= -i\frac{\partial}{\partial\bar{\theta}^{\dot{\alpha}i}} - \theta_i^\alpha(\sigma^a)_{\alpha\dot{\alpha}}\frac{\partial}{\partial x^a}\,, \\ P_a &= \frac{1}{i}\frac{\partial}{\partial x^a}\,. \end{aligned} \tag{3.20}$$

The coset part of the Cartan forms defined according to (3.9) is given by

$$\begin{aligned} \omega^a &= dx^a - i(\sigma^a)_{\alpha\dot{\alpha}}(d\theta_i^\alpha\bar{\theta}^{\dot{\alpha}i} + d\bar{\theta}^{\dot{\alpha}i}\theta_i^\alpha)\,, \\ \omega_i^\alpha &= d\theta_i^\alpha\,, \qquad \omega_{\dot{\alpha}}^i = d\bar{\theta}_{\dot{\alpha}}^i\,, \end{aligned} \tag{3.21}$$

while the H-part is vanishing in the present case. Then from (3.13) one finds the covariant derivatives:

$$\begin{aligned} D_a &= \frac{\partial}{\partial x^a}\,, \\ D_\alpha^i &= \frac{\partial}{\partial\theta_i^\alpha} + i\bar{\theta}^{\dot{\alpha}i}(\sigma^a)_{\alpha\dot{\alpha}}\frac{\partial}{\partial x^a}\,, \\ \bar{D}_{\dot{\alpha}i} &= -\frac{\partial}{\partial\bar{\theta}^{\dot{\alpha}i}} - i\theta_i^\alpha(\sigma^a)_{\alpha\dot{\alpha}}\frac{\partial}{\partial x^a}\,. \end{aligned} \tag{3.22}$$

They obey the following algebra:

$$\{D^i_\alpha, D^j_\beta\} = \{\bar{D}_{\dot{\alpha}i}, \bar{D}_{\dot{\beta}j}\} = 0\,, \qquad \{D^i_\alpha, \bar{D}_{\dot{\beta}j}\} = -2i\delta^i_j(\sigma^a)_{\alpha\dot{\beta}}D_a\,. \tag{3.23}$$

Superfields are defined as superfunctions in $\mathbb{R}^{4|4N}$, $\Phi^{ij\ldots}_{\alpha\dot{\beta}\ldots}(x,\theta,\bar{\theta})$, carrying $SL(2,\mathbb{C})$ and $SU(N)$ indices. They transform as scalars under translations and supertranslations:

$$\Phi'(X') = \Phi(X)\,, \tag{3.24}$$

and according to their index structure under the automorphism groups. Since the Grassmann coordinates $\theta, \bar{\theta}$ are nilpotent, the θ-expansion of a superfield is always finite, e.g., a scalar superfield

$$\Phi(x,\theta,\bar{\theta}) = \phi(x) + \theta^\alpha_i \psi^i_\alpha(x) + \bar{\theta}^i_{\dot{\alpha}}\bar{\chi}^{\dot{\alpha}}_i(x) + \cdots + \theta^{2N}\bar{\theta}^{2N}D(x) \tag{3.25}$$

describes a set of 2^{4N} component fields. If a superfield belongs to a real representation of $SL(2,\mathbb{C}) \times SU(N)$, a reality condition can be imposed:

$$\overline{\Phi_{\ldots}(X)} = \Phi_{\ldots}(X)\,. \tag{3.26}$$

For instance, if the scalar superfield (3.25) is real, its components satisfy the conditions:

$$\overline{\phi(x)} = \phi(x)\,, \qquad \overline{\psi^i_\alpha(x)} = \bar{\chi}_{\dot{\alpha}i}(x)\,, \qquad \ldots, \qquad \overline{D(x)} = D(x)\,. \tag{3.27}$$

Comparing the expansion (3.25) with the transformation laws (3.24), (3.19) one can find out how the components transform, e.g.,

$$\delta\phi(x) = -\epsilon^\alpha_i\psi^i_\alpha(x) - \bar{\epsilon}^i_{\dot{\alpha}}\bar{\chi}^{\dot{\alpha}}_i\,, \qquad \text{etc.} \tag{3.28}$$

Note that the closure of the supersymmetry algebra has to be explicitly verified in the component form (3.28) whereas it is automatic in the superfield form (3.24). The superfield (3.25) has 2^{4N} components in its θ-expansion. In general, this is too many for describing irreducible supermultiplets of matter or gauge fields like those in Section 2.4. For example, in the $N = 1$ case the full expansion

$$\begin{aligned}\Phi &= \phi + \theta^\alpha\psi_\alpha + \bar{\theta}_{\dot{\alpha}}\bar{\chi}^{\dot{\alpha}} + \theta\theta\, M + \bar{\theta}\bar{\theta}\, N + \theta\sigma^a\bar{\theta}\, A_a \\ &\quad + \bar{\theta}\bar{\theta}\,\theta^\alpha\rho_\alpha + \theta\theta\,\bar{\theta}_{\dot{\alpha}}\bar{\lambda}^{\dot{\alpha}} + \theta\theta\,\bar{\theta}\bar{\theta}\, D\end{aligned} \tag{3.29}$$

contains more fields than either the $N = 1$ matter or $N = 1$ Maxwell multiplets (2.57), (2.63). In other words, superfields are in general highly reducible representations of supersymmetry. To make them suit the purpose of describing irreducible supermultiplets, one has to impose constraints or introduce gauge invariances. A simple example of an irreducibility condition is the so-called chirality condition:

$$\bar{D}_{\dot{\alpha}}\Phi(x,\theta,\bar{\theta}) = 0\,. \tag{3.30}$$

It can be easily solved (see (3.22)):

$$\Phi_L = \Phi_L(x_L, \theta)\,, \qquad x_L^a = x^a + i\theta\sigma^a\bar{\theta}\,. \tag{3.31}$$

The chiral superfield (3.31) has a shorter decomposition:

$$\Phi_L = \phi(x_L) + \theta^\alpha\psi_\alpha(x_L) + \theta\theta F(x_L)\,, \tag{3.32}$$

which exactly coincides with the content of the $N = 1$ matter multiplet (2.57).

In fact, the existence of such 'short' superfields can be explained naturally in Cartan's scheme. The point is that the choice of the subgroup H made in (3.15) is not unique. Another possibility is to add half of the supertranslation generators, e.g., $\bar{Q}_{\dot{\alpha}i}$, to the automorphism generators of $SL(2,\mathbb{C})$ and $SU(N)$. Thus one obtains a new, *chiral superspace*:

$$\mathbb{C}^{4|2N} = \frac{\{L_{ab}, su(N), P_a, Q_\alpha^i, \bar{Q}_{\dot{\alpha}i}\}}{\{L_{ab}, su(N), \bar{Q}_{\dot{\alpha}i}\}} = \left(x_L^a, \theta_{L\,i}^\alpha\right) \equiv (\zeta_L^A)\,. \tag{3.33}$$

One can arrive at the idea of a chiral superspace in the following way. Take the representative (3.17) of the coset $\mathbb{R}^{4|4N}$ and rewrite it as follows:

$$\exp i(-x^a P_a + \theta_i^\alpha Q_\alpha^i + \bar{\theta}_{\dot{\alpha}}^i \bar{Q}_i^{\dot{\alpha}}) = \exp i(-x_L^a P_a + \theta_{L\,i}^\alpha Q_\alpha^i) \exp i(\bar{\theta}_{L\,\dot{\alpha}}^i \bar{Q}_i^{\dot{\alpha}})\,. \tag{3.34}$$

This factorization corresponds to sending $\bar{Q}$ to the denominator in (3.33). From (3.34) and the supersymmetry algebra one can easily derive:

$$x_L^a = x^a + i\theta_{L\,i}\sigma^a\bar{\theta}_L^i\,, \qquad \theta_{L\,i}^\alpha = \theta_i^\alpha\,, \qquad \bar{\theta}_{L\,\dot{\alpha}}^i = \bar{\theta}_{\dot{\alpha}}^i\,. \tag{3.35}$$

One can call $(x_L, \theta_L, \bar{\theta}_L)$ (3.35) a (left-handed) chiral basis in $\mathbb{R}^{4|4N}$. The crucial property of chiral superspace, following from the coset construction above, is that it is closed under the full supersymmetry algebra:

$$\delta x_L^a = -2i\theta_L^i\sigma^a\bar{\epsilon}_i\,, \qquad \delta\theta_{L\,i}^\alpha = \epsilon_i^\alpha\,. \tag{3.36}$$

This allows one to consider 'short' or chiral superfields $\Phi(x_L, \theta_L)$.

Strictly speaking, to correctly define $\mathbb{C}^{4|2N}$ one needs to identify it with the coset of the complex extension of the supersymmetry algebra, i.e., to regard the generators P_a in (3.33) as complex $(P_{L\,a})$ and Q, $\bar{Q}$ as not related by conjugation $(Q_L, \bar{Q}_L)$:

$$\mathbb{C}^{4|2N} = \frac{\{L_{ab}, su(N), P_{L\,a}, Q_{L\,\alpha}^i, \bar{Q}_{L\,\dot{\alpha}i}\}}{\{L_{ab}, su(N), \bar{Q}_{L\,\dot{\alpha}i}\}} = \left(x_L^a, \theta_{L\,i}^\alpha\right). \tag{3.37}$$

Then, repeating the factorization procedure (3.34), we have to start with *complex* even coordinates ξ^a and *non-conjugated* odd ones. Then the shift

$$x_L^a = \xi^a + i\theta_{L\,i}\sigma^a\bar{\theta}_L^i \tag{3.38}$$

is just a change of variables in the complex superspace. In this context the chiral superspace $\mathbb{C}^{4|2N}$ is a true subspace of the complexification of $\mathbb{R}^{4|4N}$. Inversely, the *real* superspace $\mathbb{R}^{4|4N}$ can be interpreted as a real hypersurface (i.e., a real subspace) in the chiral one $\mathbb{C}^{4|2N}$. This hypersurface is defined by the embedding conditions:

$$\mathrm{Re}\, x_L^a = x^a\,, \qquad \mathrm{Im}\, x_L^a = \theta_i \sigma^a \bar{\theta}^i\,, \qquad \theta^\alpha_{L\,i} = \theta^\alpha_i\,, \qquad \bar{\theta}_L^{\dot{\alpha}i} = \bar{\theta}^{\dot{\alpha}i} = \overline{(\theta^\alpha_i)}\,. \tag{3.39}$$

These conditions are covariant only with respect to the real supersymmetry algebra, but our aim was just to find a suitable realization of this. It is in this sense that one should understand the term 'chiral basis' of $\mathbb{R}^{4|4N}$.

In the chiral basis (3.35), using the standard Cartan technique, one can find the expressions for the covariant derivatives (3.22):

$$\begin{aligned} D_a &= \frac{\partial}{\partial x_L^a}\,, \\ D^i_\alpha &= \frac{\partial}{\partial \theta^\alpha_i} + 2i\bar{\theta}^{\dot{\alpha}i}(\sigma^a)_{\alpha\dot{\alpha}}\frac{\partial}{\partial x_L^a}\,, \\ \bar{D}_{\dot{\alpha}i} &= -\frac{\partial}{\partial \bar{\theta}^{\dot{\alpha}i}}\,. \end{aligned} \tag{3.40}$$

It is remarkable that in this chiral basis the derivative $\bar{D}_{\dot{\alpha}i}$ becomes just a partial derivative. Consequently, the chirality condition (3.30) takes the form of a condition of independence of the variable $\bar{\theta}^{\dot{\alpha}i}$.

Now it becomes clear why the chirality condition (3.30) could be solved as in (3.31). In the chiral basis (3.35) it simply means that the superfield Φ_L is holomorphic in θ (i.e., it does not depend on $\bar{\theta}$). Therefore, it is quite natural to refer to eq. (3.30) as the Cauchy–Riemann condition for an analyticity of a new kind – *Grassmann analyticity* [G8]. It just means that the 'analytic' superfield Φ_L depends on θ and is independent of its conjugate $\bar{\theta}$. The above example is the simplest one. In the rest of this book the generalization of the concept of Grassmann analyticity will play a fundamental rôle in the description of $N = 1, 2$ matter and of $N = 1, 2, 3$ gauge theories.

3.3 $N = 2$ harmonic superspace

Our aim is to find manifestly supersymmetric unconstrained off-shell formulations of all the $N = 2$ supersymmetric theories. The standard superspaces described above are fully adequate for all the $N = 1$ theories, but not in the $N = 2$ case. The reason is as follows. Generic superfields involve a number of unphysical fields. To eliminate them one has to impose appropriate constraints. However, some of these constraints turn out to be compatible only with the free equations of motion. So, one encounters problems trying to describe interactions. The reader already knows from Section 2.4 that the $N = 2$ matter hypermultiplet in a complex representation cannot be described off shell in terms of a finite set of fields, i.e., in terms of standard superfields. So, the standard superspaces discussed in the preceding section are not well adapted to the description of $N = 2$ supersymmetric theories.

The guideline announced in the Introduction was: *'Explore all possible superspaces'*. In the case of extended supersymmetry one has new possibilities, indeed. One should search for an extension of Poincaré superspace such that infinite sets of fields can be accommodated in it. In this subsection we introduce the new harmonic $N = 2$ superspace that provides the desired extension adequate for all the $N = 2$ theories.

A natural generalization of the real superspace $\mathbb{R}^{4|8}$ is obtained by keeping only the $U(1)$ part of the automorphism group $SU(2)$ in the subgroup H:

$$\mathbb{HR}^{4+2|8} = \frac{\{L_{ab}, P_a, Q^i_\alpha, \bar{Q}_{\dot{\alpha}i}, su(2)\}}{\{L_{ab}, u(1)\}}\,. \tag{3.41}$$

We call it 'harmonic superspace' for reasons which will become clear later on. Besides the familiar coordinates $x^a, \theta^\alpha_i, \bar{\theta}^{i\dot{\alpha}}$, this superspace contains two new even coordinates corresponding to the coset $SU(2)/U(1)$ which is well known to be the two-dimensional sphere S^2. So, one can conclude that $\mathbb{HR}^{4+2|8}$ is obtained by tensoring the real standard superspace $\mathbb{R}^{4|8}$ with the sphere:

$$\mathbb{HR}^{4+2|8} = \mathbb{R}^{4|8} \times S^2\,. \tag{3.42}$$

Indeed, one can parametrize the coset (3.41) in the following way:

Central basis:

$$\Omega = \exp i\left\{-x^a P_a + \theta^\alpha_i Q^i_\alpha + \bar{\theta}^i_{\dot{\alpha}} \bar{Q}^{\dot{\alpha}}_i\right\} \exp i\left\{\xi T^{++} + \bar{\xi} T^{--}\right\} \tag{3.43}$$

(the name 'central basis' will become clear soon). Here $T^{\pm\pm}$ are the coset generators for $SU(2)/U(1)$. Together with the $U(1)$ generator T^0 they form the algebra of $SU(2)$:*

$$\left[T^{++}, T^{--}\right] = T^0\,, \quad \left[T^0, T^{\pm\pm}\right] = \pm 2T^{\pm\pm}, \quad T^{\pm\pm} = T^1 \pm i T^2, \quad T^0 = 2T^3. \tag{3.44}$$

The choice (3.43) is motivated by the fact that the new coordinates ξ do not transform under translations and supertranslations, so one can covariantly drop the ξ dependence and go back to the original superspace $\mathbb{R}^{4|8}$ (in other words, one can factor out the generators $T^{\pm\pm}$ and obtain the coset (3.17)).

It is not hard to derive the transformation rules for the coordinates in (3.43). Under supersymmetry $x^a, \theta_i, \bar{\theta}^i$ transform as before (recall (3.19)) whereas ξ stays inert. Under $SU(2)$ $\theta_i, \bar{\theta}^i$ behave as isospinors, and ξ transforms non-linearly.

---

* They are related to the generators introduced in (2.20) as follows: $iT^k_l = (\tau^a)^k_l T^a$ $(a = 1, 2, 3)$, where τ^a are the Pauli matrices.

The main advantage of the harmonic superspace (3.41) is the existence of a new invariant subspace containing only half of the original Grassmann variables (a different half as compared to the chiral case). This is only possible in the framework extended by harmonic variables and is not possible in the standard $N = 2$ superspace $\mathbb{R}^{4|8}$ (3.15) (at least, not if one wants to preserve the $SU(2)$ automorphism). The crucial observation is that there exists an alternative choice of the subgroup H in (3.41). Indeed, one can add half of the supersymmetry generators to those of $SL(2,\mathbb{C})$ and $U(1)$ appearing in the denominator. This time, however, it will not be the right-handed chiral half (as in (3.33)), but rather the $+$ projection of these generators with respect to $U(1)$. We mean the following. The $U(1)$ charge generator is represented on Q^i by the matrix τ^3:

$$\left[T^0, Q^i\right] = (\tau^3)^i_j Q^j\,, \qquad \tau^3 = \begin{pmatrix} 1 & 0 \\ 0 & -1 \end{pmatrix}, \tag{3.45}$$

while the generators T^{++} and T^{--} are represented by the matrices

$$\tau^{++} = \frac{1}{2}(\tau^1 + i\tau^2) = \begin{pmatrix} 0 & 1 \\ 0 & 0 \end{pmatrix} \quad \text{and} \quad \tau^{--} = \frac{1}{2}(\tau^1 - i\tau^2) = \begin{pmatrix} 0 & 0 \\ 1 & 0 \end{pmatrix}, \tag{3.46}$$

respectively:

$$\left[T^{\pm\pm}, Q^i\right] = (\tau^{\pm\pm})^i_j Q^j\,. \tag{3.47}$$

One can introduce the notation

$$Q^1_\alpha \equiv Q^+_\alpha\,, \qquad Q^2_\alpha \equiv Q^-_\alpha\,, \qquad \bar{Q}_{1\dot\alpha} \equiv \bar{Q}^-_{\dot\alpha}\,, \qquad \bar{Q}_{2\dot\alpha} \equiv -\bar{Q}^+_{\dot\alpha}\,. \tag{3.48}$$

Then

$$\begin{aligned} \left[T^0, Q^+\right] &= Q^+\,, & \left[T^0, Q^-\right] &= -Q^-\,, \\ \left[T^{++}, Q^+\right] &= 0\,, & \left[T^{++}, Q^-\right] &= Q^+\,, \\ \left[T^{--}, Q^+\right] &= Q^-\,, & \left[T^{--}, Q^-\right] &= 0\,, \end{aligned} \tag{3.49}$$

and the same relations for $\bar{Q}^\pm$. The supersymmetry algebra becomes

$$\begin{aligned} \{Q^+_\alpha, \bar{Q}^+_{\dot\alpha}\} &= \{Q^-_\alpha, \bar{Q}^-_{\dot\alpha}\} = 0\,, \\ \{Q^+_\alpha, \bar{Q}^-_{\dot\alpha}\} &= -\{Q^-_\alpha, \bar{Q}^+_{\dot\alpha}\} = 2(\sigma^a_{\alpha\dot\alpha})P_a\,. \end{aligned} \tag{3.50}$$

Now it is clear that L_{ab}, T^0 and Q^+_α, $\bar{Q}^+_{\dot\alpha}$ form a closed subalgebra. Therefore, one can define the following new superspace:

$$\mathbb{HA}^{4+2|4} = \frac{\{L_{ab}, P_a, Q^i_\alpha, \bar{Q}_{\dot\alpha i}, T^{\pm\pm}, T^0\}}{\{L_{ab}, T^0, Q^+_\alpha, \bar{Q}^+_{\dot\alpha}\}}\,. \tag{3.51}$$

We make heavy use of it under the name *analytic superspace*.

In particular, one can regard the superspace (3.51) as a subspace of the full harmonic superspace $\mathbb{HR}^{4+2|8}$ (3.42). To see this one should choose a new parametrization of $\mathbb{HR}^{4+2|8}$ that will be referred to in what follows as the *analytic basis* (as opposed to the *central basis* (3.43)):

$$\begin{aligned} &\text{Analytic basis:} \\ &\Omega = \exp i\left\{\xi T^{++} + \bar{\xi} T^{--}\right\} \exp i\left\{-x_A^a P_a - \theta_A^{+\alpha} Q_\alpha^- - \bar{\theta}_A^{+\dot{\alpha}} \bar{Q}_{\dot{\alpha}}^-\right\} \\ &\quad \times \exp i\left\{\theta_A^{-\alpha} Q_\alpha^+ + \bar{\theta}_A^{-\dot{\alpha}} \bar{Q}_{\dot{\alpha}}^+\right\}. \end{aligned} \tag{3.52}$$

The third exponent in (3.52) corresponds to the possibility of factoring out Q^+, $\bar{Q}^+$, according to (3.51). The particular position of the $SU(2)/U(1)$ factor can be motivated in the following way. Under the left action of $SU(2)$ the representative $\exp i\left\{\xi T^{++} + \bar{\xi} T^{--}\right\}$ of the coset $SU(2)/U(1)$ transforms with a $U(1)$ factor on its right. Consequently, all the coordinates to the right of that factor will behave as $U(1)$ representations under $SU(2)$.

The main feature of the new basis (3.52) is that it enables us to define 'short' superfields depending only on $\theta^+, \bar{\theta}^+$, but not on $\theta^-, \bar{\theta}^-$. The first two exponents in (3.52) on their own constitute a coset realization of $N = 2$ supersymmetry and so represent the coset space (3.51). Thus, the coordinate set $(\xi, \bar{\xi}, x_A, \theta_A^+, \bar{\theta}_A^+)$ is closed under the left action of the $N = 2$ Poincaré supergroup and its $SU(2)$ automorphism group.* As a result, the functions of these coordinates transform into themselves, i.e., they are $N = 2$ superfields.

The existence of such analytic $N = 2$ superfields can also be understood from another point of view. Following the coset routine it is not hard to find the expressions for the spinor covariant derivatives in the analytic basis:

$$\begin{aligned} D_\alpha^+ &= \frac{\partial}{\partial \theta^{-\alpha}}, & \bar{D}_{\dot{\alpha}}^+ &= \frac{\partial}{\partial \bar{\theta}^{-\dot{\alpha}}}, \\ D_\alpha^- &= -\frac{\partial}{\partial \theta^{+\alpha}} + 2i\bar{\theta}^{-\dot{\alpha}} \partial_{\alpha\dot{\alpha}}, & \bar{D}_{\dot{\alpha}}^- &= -\frac{\partial}{\partial \bar{\theta}^{+\dot{\alpha}}} - 2i\theta^{-\alpha} \partial_{\alpha\dot{\alpha}}. \end{aligned} \tag{3.53}$$

These derivatives satisfy the following algebra:

$$\begin{aligned} \{D_\alpha^+, D_\beta^+\} &= \{D_\alpha^+, \bar{D}_{\dot{\alpha}}^+\} = \{\bar{D}_{\dot{\alpha}}^+, \bar{D}_{\dot{\beta}}^+\} = 0, \\ \{D_\alpha^-, D_\beta^-\} &= \{D_\alpha^-, \bar{D}_{\dot{\alpha}}^-\} = \{\bar{D}_{\dot{\alpha}}^-, \bar{D}_{\dot{\beta}}^-\} = 0, \\ \{D_\alpha^+, \bar{D}_{\dot{\alpha}}^-\} &= -\{D_\alpha^-, \bar{D}_{\dot{\alpha}}^+\} = -2i\partial_{\alpha\dot{\alpha}}. \end{aligned} \tag{3.54}$$

Of crucial importance is the fact that the spinor derivatives D^+ and $\bar{D}^+$ form an anticommuting subset in the algebra (3.54). It implies the consistency of the following covariant conditions:

$$D_\alpha^+ \Phi(x_A, \theta_A^\pm, \bar{\theta}_A^\pm, \xi, \bar{\xi}) = \bar{D}_{\dot{\alpha}}^+ \Phi(x_A, \theta_A^\pm, \bar{\theta}_A^\pm, \xi, \bar{\xi}) = 0. \tag{3.55}$$

* It is also closed under yet another chiral $U(1)$ (R symmetry) automorphism group which acts as conjugate phase transformations of Q, $\bar{Q}$, and, respectively, of $\theta, \bar{\theta}$; this symmetry can be broken in the presence of central charges in the $N = 2$ superalgebra (see Section 2.3.4).

The analytic basis form (3.53) of the spinor derivatives allows one to interpret (3.55) as a Grassmann Cauchy–Riemann condition with the obvious solution

$$\Phi = \Phi(x_A, \theta^+_A, \bar\theta^+_A, \xi, \bar\xi)\,. \tag{3.56}$$

So, we arrive at a new kind of Grassmann analyticity (cf. the chiral case in Section 3.2), *harmonic Grassmann analyticity*. It will be extremely useful in what follows.

At this point we should recall that the chiral superspace $\mathbb{C}^{4|2N}$ (3.33) is a complex superspace. Similarly, the coset $\mathbb{HA}^{4+2|4}$ is complex since only the generators Q^+, $\bar Q^+$ but not their complex conjugates Q^-, $\bar Q^-$ are present in the denominator. As a consequence, the bosonic coordinates x_A become, strictly speaking, complex (see (3.57) in the next section). However, unlike the chiral case, here we have the possibility to define a real structure within the analytic superspace itself. This important point will be explained in detail in Section 3.7.

3.4 Harmonic variables

In this section we pass from the newly introduced coordinates ξ, $\bar\xi$ to an equivalent description of S^2 in terms of harmonic variables. We start by studying the relation between the two parametrizations, (3.43) and (3.52), of $\mathbb{HR}^{4+2|8}$. Using the supersymmetry and $SU(2)$ algebras, one finds

$$\begin{aligned} x^a_A &= x^a - 2i\theta^{(i}\sigma^a\bar\theta^{j)}u^+_i u^-_j\,, \\ \theta^\pm_{A\,\alpha} &= u^\pm_i\theta^i_\alpha\,, \qquad \bar\theta^\pm_{A\,\dot\alpha} = u^\pm_i\bar\theta^i_{\dot\alpha}\,, \end{aligned} \tag{3.57}$$

where

$$\begin{aligned} u^+_i &= \left[\exp i(\xi\tau^{++} + \bar\xi\tau^{--})\right]^1_i\,, \\ u^-_i &= \left[\exp i(\xi\tau^{++} + \bar\xi\tau^{--})\right]^2_i\,, \\ u^{+i}u^-_i &= 1\,. \end{aligned} \tag{3.58}$$

The spinor covariant derivatives in the two bases, i.e., the expressions (3.22) and (3.53), are related by

$$D^\pm_\alpha = u^\pm_i D^i_\alpha\,, \qquad \bar D^\pm_{\dot\alpha} = u^\pm_i \bar D^i_{\dot\alpha}\,. \tag{3.59}$$

From (3.19) and (3.57) one can derive the transformation rules for the analytic basis coordinates under supersymmetry transformations:

$$\begin{aligned} \delta x^a_A &= -2i(\epsilon^i\sigma^a\bar\theta^+ + \theta^+\sigma^a\bar\epsilon^i)u^-_i\,, \\ \delta\theta^\pm_{A\,\alpha,\dot\alpha} &= u^\pm_i\epsilon^i_{\alpha,\dot\alpha}\,, \\ \delta u^\pm_i &= 0\,. \end{aligned} \tag{3.60}$$

Of course, these laws can be equivalently deduced by considering the left action of the $N = 2$ supergroup on the representatives of the coset (3.51), i.e., on the first two exponents in (3.52).

As before, the coset coordinates ξ transform non-linearly under $SU(2)$. However, the newly defined quantities $u^\pm_i$ (3.58) have a much simpler transformation law (see (3.4)):

$$SU(2): \qquad (u^\pm_i)' = \Lambda^j_i u^\pm_j e^{\pm i\psi(\Lambda,\xi,\bar{\xi})}\,, \tag{3.61}$$

where Λ^j_i is an element of $SU(2)$, and $\psi(\Lambda, \xi, \bar{\xi})$ is the local phase of an induced $U(1)$ transformation. Consequently, the new coordinates x_A, $\theta^\pm_A$ (3.57) transform only under this induced $U(1)$ group according to their charge (0 for x_A and ± 1 for $\theta^\pm_A$). One could say that $u^\pm_i$ play the rôle of 'zweibeins' converting $SU(2)$ indices into $U(1)$ indices. Actually, if $u^\pm_i$ were true zweibeins, they would transform under a $U(1)$ group independent of $SU(2)$ (like the vierbeins e^a_μ in general relativity, which transform under the diffeomorphism group and an independent Lorentz group). As we shall see later, one can indeed upgrade $u^\pm_i$ (3.58) to full $SU(2) \to U(1)$ zweibeins by adding an extra phase degree of freedom.

A better understanding of the meaning of $u^\pm_i$ is achieved by realizing that they are the lowest (i.e., fundamental) spherical harmonics and, consequently that their symmetrized products

$$u^+_{(i_1} \ldots u^+_{i_m} u^-_{j_1} \ldots u^-_{j_n)} \tag{3.62}$$

form a complete set of spherical harmonics on the sphere $SU(2)/U(1) \sim S^2$, in terms of which any square-integrable function on S^2 can be expanded (see Chapter 4). Thus, any such function of charge $q \geq 0$ can be expanded as follows:

$$f^{(q)}(u) = \sum_{n=0}^{\infty} f^{(i_1\ldots i_{n+q} j_1 \ldots j_n)} u^+_{i_1} \ldots u^+_{i_{n+q}} u^-_{j_1} \ldots u^-_{j_n}\,, \tag{3.63}$$

and similarly for $q < 0$. Clearly, the expansion (3.63) has the advantage of manifest $SU(2)$ covariance, i.e., the coefficients $f^{(i_1 \ldots j_n)}$ are irreducible representations of $SU(2)$ with isospin $n + \frac{q}{2}$. Another important property of (3.63) is its $U(1)$ covariance. Indeed, the induced $U(1)$ transformations (3.61) of $u^\pm$ combine into a $U(1)$ transformation of $f^{(q)}$ according to its charge q. Moreover, one sees that if one multiplies $u^\pm$ by arbitrary phase factors:

$$u^+_i \quad \to \quad u^+_i e^{i\psi}\,, \qquad u^-_i \quad \to \quad u^-_i e^{-i\psi}\,, \tag{3.64}$$

$f^{(q)}$ acquires the corresponding factor:

$$f^{(q)} \quad \to \quad f^{(q)} e^{iq\psi}\,. \tag{3.65}$$

The change (3.64) can be rewritten in a matrix form

$$\| u \| \equiv \begin{pmatrix} u_1^+ & u_1^- \\ u_2^+ & u_2^- \end{pmatrix} \to \| u \| \, e^{i\psi\tau^3} , \tag{3.66}$$

and so it amounts to replacing the matrix $\| u \|$ defined on the coset $SU(2)/U(1)$ by an arbitrary $SU(2)$ matrix. Further, the function $f^{(q)}(\xi, \bar{\xi}, \psi)$ is now defined on the entire group $SU(2)$, and is homogeneous of degree q in $e^{i\psi}$. In fact, it is well known that one can describe functions on a coset space G/H by considering those functions on G which are homogeneous in the coordinates of H. Thus we arrive at the conclusion that all the square-integrable functions on S^2 are given by an expansion of the form (3.63) where $u_i^\pm$ are arbitrary $SU(2)$ matrices:

$$\| u \| \in SU(2) : \quad u^{+i} u_i^- = 1 , \quad \overline{u^{+i}} = u_i^- . \tag{3.67}$$

In what follows we consider (3.63) as the definition of the functions on S^2. This approach, besides its manifest $SU(2)$ covariance, also has the advantage of avoiding the use of an explicit parametrization of S^2 or $SU(2)$. It is very important that in doing so we deal with quantities that are globally defined on S^2. As we have already said, the expansion (3.63) is a harmonic expansion on S^2, and $u_i^\pm$ are the (isospinor) spherical harmonics. Therefore, we call $u_i^\pm$ *harmonic variables*, $f^{(q)}(u)$ *harmonic functions* of charge q, and eq. (3.63) a *harmonic expansion*.*

The reader can appreciate the advantage of the harmonic coordinates when looking at the expansion of, e.g., an analytic superfield (3.56) of $U(1)$ charge q:

$$\begin{aligned} \Phi^{(q)}&(x_A, \theta^+, \bar{\theta}^+, u) \\ &= \phi^{(q)}(x_A, u) + \theta^+ \psi^{(q-1)}(x_A, u) + \bar{\theta}^+ \bar{\chi}^{(q-1)}(x_A, u) \\ &\quad + (\theta^+)^2 M^{(q-2)}(x_A, u) + (\bar{\theta}^+)^2 N^{(q-2)}(x_A, u) \\ &\quad + \theta^+ \sigma^a \bar{\theta}^+ A_a^{(q-2)}(x_A, u) + (\bar{\theta}^+)^2 \theta^+ \lambda^{(q-3)}(x_A, u) \\ &\quad + (\theta^+)^2 \bar{\theta}^+ \kappa^{(q-3)}(x_A, u) + (\theta^+)^2 (\bar{\theta}^+)^2 D^{(q-4)}(x_A, u) . \end{aligned} \tag{3.68}$$

The components in the θ expansion above have $U(1)$ charges varying from q to $q-4$, so that the superfield as a whole has charge q. Each of the harmonic fields in (3.68) has to be expanded in the harmonic variables, e.g.,

$$\phi^{(q)}(x_A, u) = \phi^{i_1 \dots i_q}(x_A) u_{i_1}^+ \dots u_{i_q}^+ + \phi^{i_1 \dots i_{q+2}}(x_A) u_{(i_1}^+ \dots u_{i_{q+1}}^+ u_{i_{q+2})}^- + \cdots , \tag{3.69}$$

thus giving rise to an *infinite set* of ordinary, u independent fields. The latter are irreps of $SU(2)$. One sees that the harmonic variables allow one to work

* The harmonic analysis on S^2 is treated in many textbooks (see, e.g., [V2]). However, in most of them an explicit coordinate parametrization of S^2 is used. Respectively, the harmonic expansions are formulated in terms of the spherical harmonic functions $Y^{lm}(\theta, \phi)$, the Jacobi polynomials, etc. See Chapter 4 for a discussion of the relationship to this customary approach.

with superfields which transform as $U(1)$ representations, but nevertheless have $SU(2)$ covariant components.

The presence of infinite sets of fields is a crucial new feature of the harmonic superspace formalism. As we shall see later on, the number of *physical* fields always remains finite, the infinite sets are either auxiliary or pure gauge degrees of freedom.

Here it is appropriate to examine the content of the analytic superfield (3.68) with respect to superspin Y and superisospin I (these quantum numbers characterizing the irreps of $N = 2$ supersymmetry were defined in Section 2.3.2). Comparing the spin content of the superfield (3.68) with the table (2.39), we see that it carries a single superspin $Y = 0$. However, due to the harmonic expansion the superfield contains an infinite tower of superisospins. To find their values recall that the superisospin of a given irreducible $N = 2$ multiplet coincides with the isospin of the highest spin state in it (see eq. (2.39)). In our case the highest spin 1 is carried by the vector field $A_a^{(q-2)}(x_A, u)$. According to (3.63), it contains isospins $|q/2 - 1|$,$|q/2 - 1| + 1$, etc. Therefore, the superisospin content of $\Phi^{(q)}$ is given by the formula

$$\Phi^{(q)}(x_A, \theta^+, \bar{\theta}^+, u): \qquad I = \left|\frac{q}{2} - 1\right| + n\,, \qquad n = 0, 1, 2, \ldots. \tag{3.70}$$

Clearly, the same analysis applies to analytic superfields with Lorentz indices. The only difference is that they may contain several irreducible representations with respect to the superspin Y, each having the same superisospin content. Later on we shall see that in practice only the lowest superisospin survives, the rest are either auxiliary or gauge degrees of freedom.

To illustrate the above discussion we present the superspin-superisospin content of some analytic superfields to be used in what follows:

$$\begin{array}{rlll} \Phi: & Y = 0; & I = 1,2,3,\ldots \\ \Phi^+: & Y = 0; & I = 1/2,3/2,5/2,\ldots \\ \Phi^{++}: & Y = 0; & I = 0,1,2,\ldots \\ \Phi^{++a}: & Y = 1,0; & I = 0,1,2,\ldots \\ \Phi^{+\alpha,\dot{\alpha}}: & Y = 1/2; & I = 1/2,3/2,5/2,\ldots \\ \Phi^{+++\alpha,\dot{\alpha}}: & Y = 1/2; & I = 1/2,3/2,5/2,\ldots \end{array} \tag{3.71}$$

Before going on, we summarize the various harmonic superspaces introduced in this section. The analytic harmonic superspace (3.51) is parametrized by the following set of coordinates:

$$\mathbb{HA}^{4+2|4} = (x^a_A, \theta^+_\alpha, \bar{\theta}^+_{\dot{\alpha}}, u^\pm_i) \equiv (\zeta, u)\,. \tag{3.72}$$

The real harmonic superspace $\mathbb{HR}^{4+2|8}$ (3.41) has two parametrizations which we call *central* and *analytic* bases:

$$\text{CB:} \quad (x^a, \theta_{\alpha i}, \bar{\theta}^i_{\dot{\alpha}}, u^\pm_i) \equiv (X, u)\,, \tag{3.73}$$

$$\text{AB:} \quad (x^a_A, \theta^\pm_\alpha, \bar{\theta}^\pm_{\dot{\alpha}}, u^\pm_i) \equiv (X_A, u) = (\zeta, \theta^-, \bar{\theta}^-, u)\,. \tag{3.74}$$

The relation between CB and AB was given in (3.57).

3.5 Harmonic covariant derivatives

In the preceding section we defined the harmonic variables $u^{\pm}_i$. Here we find out how one can covariantly differentiate with respect to them. The covariant harmonic derivatives are an essential element in the construction of invariant actions for harmonic-analytic superfields and for many other purposes. Their form is again given by the coset method, used earlier to derive the covariant spinor derivatives. We start by computing the Cartan forms on S^2 in terms of the harmonics (3.58). The general formula (3.9) specialized to this particular case becomes

$$e^{-i(\xi\tau^{++}+\bar{\xi}\tau^{--})}de^{i(\xi\tau^{++}+\bar{\xi}\tau^{--})} = \frac{i}{2}\omega^i\tau^i \equiv \frac{i}{2}\omega^3\tau^3 + \frac{i}{2}\left(\omega^{--}\tau^{++} + \omega^{++}\tau^{--}\right),$$
$$\omega^{\pm\pm} = \omega^1 \pm i\omega^2 . \tag{3.75}$$

Using the identification (3.58) one easily obtains

$$\omega^{\pm\pm} = \mp 2iu^{\pm j}du^{\pm}_j , \qquad \omega^3 = 2iu^{+j}du^-_j = 2iu^{-j}du^+_j \tag{3.76}$$

(the two equivalent expressions for ω^3 follow from the condition $u^{+i}u^-_i = 1$).

For the covariant differential of a function $f^{(q)}(u)$ on S^2 one has, in accordance with (3.12),

$$Df^{(q)}(u) = \left(d - \frac{iq}{2}\omega^3\right) f^{(q)}(u) \tag{3.77}$$

(the minus sign as compared to the general formula (3.12) is dictated by our definition of the harmonic $U(1)$ charge, eq. (3.61), and the fact that under the non-linear realization of $SU(2)$ the fields and the harmonics are multiplied by the induced $U(1)$ factor and its inverse, respectively). Using the evident identity

$$df^{(q)}(u) = \left(du^{+i}\frac{\partial}{\partial u^{+i}} + du^{-i}\frac{\partial}{\partial u^{-i}}\right) f^{(q)}(u) \tag{3.78}$$

and the completeness relation

$$u^{+i}u^-_j - u^{-i}u^+_j = \delta^i_j \tag{3.79}$$

(equivalent to the condition $u^{+i}u^-_i = 1$ (3.67)), it is easy to rewrite (3.77) in terms of the 1-forms (3.76):

$$Df^{(q)}(u) = \frac{i}{2}\left[\omega^3(D^0 - q) + \omega^{++}\partial^{--} + \omega^{--}\partial^{++}\right] f^{(q)}(u) . \tag{3.80}$$

Here

$$D^0 = u^{+i}\frac{\partial}{\partial u^{+i}} - u^{-i}\frac{\partial}{\partial u^{-i}}\,, \qquad \partial^{\pm\pm} = u^{\pm i}\frac{\partial}{\partial u^{\mp i}}\,. \tag{3.81}$$

It is important to always use the harmonic derivatives (3.81) and not the partial ones $\partial/\partial u^{\pm}$, because only the former are compatible with the defining relation $u^{+i}u_i^- = 1$:

$$D^0(u^{+i}u_i^-) = 0\,, \qquad \partial^{++}(u^{+i}u_i^-) = 0\,, \qquad \partial^{--}(u^{+i}u_i^-) = 0\,.$$

From the general decomposition (3.63) it follows that $f^{(q)}(u)$ satisfies the condition

$$D^0 f^{(q)}(u) = q f^{(q)}(u) \tag{3.82}$$

(this is actually the definition of a function on the coset $SU(2)/U(1)$). Thus, from the general definition (3.13) one obtains the following expressions for the covariant derivatives on S^2:

$$\begin{aligned} Df^{(q)}(u) &= \frac{i}{2}\left[\omega^{++}D^{--} + \omega^{--}D^{++}\right] f^{(q)}(u)\,, \\ &\Rightarrow \quad D^{\pm\pm} = \partial^{\pm\pm} = u^{\pm i}\frac{\partial}{\partial u^{\mp i}}\,. \end{aligned} \tag{3.83}$$

Note that for general functions on $SU(2) \sim S^3$ the relation (3.82) does not hold, so there D^0 appears as a covariant derivative along the third direction on S^3, on an equal footing with $D^{\pm\pm}$.

The expressions (3.81), (3.82) are applicable only in the central basis of harmonic superspace where the group exponents are arranged as in (3.43). In the analytic basis (3.52) the harmonic derivatives acquire additional terms involving differentiation with respect to x_A and θ. These are due to extra terms in the Cartan forms ω^{μ}, $\omega^{\pm\alpha}$, $\omega^{\pm\dot\alpha}$ coming from the S^2 exponent in (3.52) (the S^2 1-forms do not change). We leave it to the reader to check that in the analytic basis the harmonic covariant derivatives have the following expressions:

$$\begin{aligned} D^0_A &= \partial^0 + \theta^{+\alpha}\frac{\partial}{\partial\theta^{+\alpha}} - \theta^{-\alpha}\frac{\partial}{\partial\theta^{-\alpha}} + \bar\theta^{+\dot\alpha}\frac{\partial}{\partial\bar\theta^{+\dot\alpha}} - \bar\theta^{-\dot\alpha}\frac{\partial}{\partial\bar\theta^{-\dot\alpha}}\,, \\ D^{++}_A &= \partial^{++} - 2i\theta^+\sigma^a\bar\theta^+\frac{\partial}{\partial x^a_A} + \theta^{+\alpha}\frac{\partial}{\partial\theta^{-\alpha}} + \bar\theta^{+\dot\alpha}\frac{\partial}{\partial\bar\theta^{-\dot\alpha}}\,, \\ D^{--}_A &= \partial^{--} - 2i\theta^-\sigma^a\bar\theta^-\frac{\partial}{\partial x^a_A} + \theta^{-\alpha}\frac{\partial}{\partial\theta^{+\alpha}} + \bar\theta^{-\dot\alpha}\frac{\partial}{\partial\bar\theta^{+\dot\alpha}}\,. \end{aligned} \tag{3.84}$$

Of course, the same expressions can be obtained by making the straightforward change of variables (3.57) in the central basis derivatives (3.81).

It should be mentioned that the derivatives D^{++}, D^{--}, D^0 form an $SU(2)$ algebra

$$\left[D^{++}, D^{--}\right] = D^0\,, \qquad \left[D^0, D^{\pm\pm}\right] = \pm 2D^{\pm\pm}\,. \tag{3.85}$$

They can be regarded as the generators of right $SU(2)$ rotations acting on the indices $\pm$ of the harmonics $u_i^{\pm}$:

$$D^0 u_i^{\pm} = \pm u_i^{\pm}\,, \qquad D^{\pm\pm} u_i^{\mp} = u_i^{\pm}\,, \qquad D^{\pm\pm} u_i^{\pm} = 0\,. \tag{3.86}$$

A very important feature of the harmonic derivative D^{++} is that it forms a commutative algebra together with the Grassmann-analyticity defining derivatives $D^+_\alpha, \bar{D}^+_{\dot\alpha}$:

$$\begin{aligned} &\{D^+_\alpha, D^+_\beta\} = \{D^+_\alpha, \bar{D}^+_{\dot\alpha}\} = \{\bar{D}^+_{\dot\alpha}, \bar{D}^+_{\dot\beta}\} = 0\,, \\ &[D^{++}, D^+_\alpha] = [D^{++}, \bar{D}^+_{\dot\alpha}] = 0\,. \end{aligned} \tag{3.87}$$

This implies, in particular, that D^{++} of an analytic superfield is again an analytic superfield.

The remaining commutators involving harmonic derivatives are

$$\begin{aligned} [D^{++}, D^-_{\alpha,\dot\alpha}] &= D^+_{\alpha,\dot\alpha}\,, \\ [D^{--}, D^+_{\alpha,\dot\alpha}] &= D^-_{\alpha,\dot\alpha}\,, \\ [D^{--}, D^-_{\alpha,\dot\alpha}] &= 0\,, \\ [D^0, D^\pm_{\alpha,\dot\alpha}] &= \pm D^\pm_{\alpha,\dot\alpha}\,. \end{aligned} \tag{3.88}$$

3.6 $N = 2$ superspace with central charge coordinates

In this section we briefly explain how the above superspace constructions can be extended to the case of $N = 2$ supersymmetry with central charges (2.23). The real superspace (3.15) is now replaced by the following coset:

$$\mathbb{R}^{6|4N} = \frac{\{L_{ab}, su(N), P_a, Z, \bar{Z}, Q^i_\alpha, \bar{Q}_{\dot\alpha i}\}}{\{L_{ab}, su(N)\}} = \left(x^a, z, \bar{z}, \theta^\alpha_i, \bar\theta^{\dot\alpha i}\right) \equiv (X^A, z, \bar{z})\,. \tag{3.89}$$

Correspondingly, the coset representative (3.17) becomes

$$\Omega\ \exp i\{-zZ - \bar{z}\bar{Z}\}\,. \tag{3.90}$$

The left action of $N = 2$ supersymmetry on the central charge coordinates is given by

$$\delta z = -i\epsilon^i\theta_i\,, \qquad \delta\bar{z} = i\bar\epsilon^i\bar\theta_i\,. \tag{3.91}$$

There also appear two new Cartan forms associated with the generators $Z, \bar{Z}$:

$$\omega_z = dz + i\, d\theta_i\, \theta^i\,, \qquad \omega_{\bar z} = \overline{(\omega_z)}\,. \tag{3.92}$$

The spinor covariant derivatives (3.22) acquire additional central charge terms:

$$\begin{aligned} D^i_\alpha &= \frac{\partial}{\partial\theta^\alpha_i} + i\bar\theta^{\dot\alpha i}(\sigma^a)_{\alpha\dot\alpha}\frac{\partial}{\partial x^a} - i\theta^i_\alpha\frac{\partial}{\partial z}\,, \\ \bar{D}_{\dot\alpha i} &= -\frac{\partial}{\partial\bar\theta^{\dot\alpha i}} - i\theta^\alpha_i(\sigma^a)_{\alpha\dot\alpha}\frac{\partial}{\partial x^a} - i\bar\theta_{\dot\alpha i}\frac{\partial}{\partial\bar z}\,. \end{aligned} \tag{3.93}$$

Correspondingly, the first two relations in the algebra (3.23) are modified as follows:

$$\{D^i_\alpha, D^j_\beta\} = -2i\epsilon_{\alpha\beta}\epsilon^{ij}\partial_z\,, \qquad \{\bar{D}_{\dot\alpha i}, \bar{D}_{\dot\beta j}\} = 2i\epsilon_{\dot\alpha\dot\beta}\epsilon_{ij}\partial_{\bar z}\,. \tag{3.94}$$

The harmonic extension of this central charge superspace is straightforward. The analytic subspace (3.72) acquires two new coordinates:

$$\mathbb{HA}^{6+2|4} = (x^a_A, z_A, \bar z_A, \theta^+_\alpha, \bar\theta^+_{\dot\alpha}, u^\pm_i) \equiv (\zeta, u)\,, \tag{3.95}$$

where

$$z_A = z + i\theta^+\theta^-\,, \qquad \bar z_A = \bar z - i\bar\theta^+\bar\theta^-\,. \tag{3.96}$$

In the analytic basis the spinor derivatives $D^+_{\alpha,\dot\alpha}$ are again short, as in (3.53), which allows one to define analytic superfields with central charge:

$$\Phi^{(q)}(\zeta, z_A, \bar z_A, u)\,. \tag{3.97}$$

Finally, the harmonic derivative D^{++} in the analytic basis is

$$\begin{aligned} D^{++}_A \;=\; & \partial^{++} - 2i\theta^+\sigma^a\bar\theta^+\frac{\partial}{\partial x^a_A} + i(\theta^+)^2\frac{\partial}{\partial z_A} - i(\bar\theta^+)^2\frac{\partial}{\partial \bar z_A} \\ & + \theta^{+\alpha}\frac{\partial}{\partial\theta^{-\alpha}} + \bar\theta^{+\dot\alpha}\frac{\partial}{\partial\bar\theta^{-\dot\alpha}}\,.\,. \end{aligned} \tag{3.98}$$

In conclusion, we should mention that in many applications one only uses a single real central charge. This can be achieved by demanding that the superfields do not depend, e.g., on $x^6 = -i(z - \bar z)$:

$$\partial_6\Phi = 0 \quad\Rightarrow\quad \Phi = \Phi(X, x^5, u)\,, \qquad x^5 = z + \bar z\,. \tag{3.99}$$

As a consequence, on such superfields one has the identification

$$\frac{\partial}{\partial z} = \frac{\partial}{\partial \bar z} = \frac{\partial}{\partial x_5}\,. \tag{3.100}$$

The condition (3.99) can be imposed in the central as well as the analytic bases. The corresponding modifications of the above relations are evident.

3.7 Reality properties

This section deals with the reality properties of the harmonic functions $f^{(q)}(u)$. If $q \neq 0$, they are necessarily complex, their complex conjugates have charge $-q$. At the same time, if q is even, i.e., if the isospins in the harmonic expansion take integer values, one can arrange for the tensor coefficients in the expansion to be real. To this end one has to accompany the complex conjugation by a new involution acting on the harmonics only and changing their charge. Indeed, such an involution exists:

$$(u^{+i})^\star = u^{-i}, \qquad (u^{-i})^\star = -u^{+i}. \tag{3.101}$$

It has the simple geometric meaning of an antipodal map, sending each point of the sphere to the diametrically opposite one. As follows from (3.101), it squares to $-\mathbb{I}$ on the harmonics, as well as on any odd product of harmonics, and to $\mathbb{I}$ on even products.

Then it is natural to define a new, generalized conjugation $\widetilde{\ }$ as the product of ordinary complex conjugation and the antipodal map $\widetilde{\ } = \bar{\ }^{\star}$. It acts on the coefficients in the harmonic expansions $f^{i_1\ldots i_n}$, on the superspace coordinates in the central basis and on the harmonics in the following way:

$$\begin{aligned} \widetilde{f^{i_1\ldots i_n}} &= \overline{f^{i_1\ldots i_n}} &\equiv \bar{f}_{i_1\ldots i_n}\,, \\ \widetilde{\theta_{\alpha i}} &= \overline{\theta_{\alpha i}} &= \bar{\theta}^i_{\dot\alpha}\,, \\ \widetilde{u^\pm_i} &= u^{\pm i} &\equiv \epsilon^{ij} u^\pm_j\,. \end{aligned} \tag{3.102}$$

Repeating it twice we obtain

$$\widetilde{\widetilde{f^{i_1\ldots i_n}}} = f^{i_1\ldots i_n}\,, \qquad \widetilde{\widetilde{u^\pm_i}} = -u^\pm_i\,. \tag{3.103}$$

For the harmonic functions this implies

$$\widetilde{\widetilde{f^{(q)}(u)}} = (-1)^q f^{(q)}(u)\,. \tag{3.104}$$

Therefore, for even values of q one can impose the reality condition:

$$\widetilde{f^{(2k)}(u)} = f^{(2k)}(u) \quad\Rightarrow\quad \overline{f^{i_1\ldots i_{2k}}} = \epsilon_{i_1 j_1}\ldots\epsilon_{i_{2k} j_{2k}} f^{j_1\ldots j_{2k}}\,. \tag{3.105}$$

The existence of the conjugation (3.102) is of crucial importance. Owing to it the analytic superspace (3.51)

$$\mathbb{HA}^{4+2|4}: \qquad (x^a_A, \theta^+_\alpha, \bar\theta^+_{\dot\alpha}, u^\pm_i) \tag{3.106}$$

turns out to be real. Indeed, using (3.57) and (3.102) one finds

$$\widetilde{x^a_A} = x^a_A\,, \qquad \widetilde{\theta^+_\alpha} = \bar\theta^+_{\dot\alpha}\,, \qquad \widetilde{\bar\theta^+_{\dot\alpha}} = -\theta^+_\alpha\,. \tag{3.107}$$

In this context one observes an interesting difference from the case of chiral superspace. The latter is essentially complex, so the real superspace can only be interpreted as a real hypersurface (see Section 3.2). With respect to ordinary complex conjugation the relationship between $\mathbb{HR}^{4+2|8}$ and $\mathbb{HA}^{4+2|4}$ is similar: The former is a real hypersurface in the latter. However, with respect to the generalized conjugation (3.107) the situation is just the opposite: $\mathbb{HA}^{4+2|4}$ is a *real* subspace of $\mathbb{HR}^{4+2|8}$.

The analytic superfields of even or odd charge have the same properties as the harmonic functions in (3.104). So, one can impose a reality condition on analytic superfields of even charge:

$$\widetilde{\Phi^{(q)}}(x_A, \theta^+, \bar\theta^+, u) = \Phi^{(q)}(x_A, \theta^+, \bar\theta^+, u)\,, \qquad q = 2n\,. \tag{3.108}$$

For the component fields (see (3.68)) this implies

$$\widetilde{\phi^{(q)}}(x_A, u) = \phi^{(q)}(x_A, u), \quad \ldots, \quad \widetilde{A_a^{(q-2)}}(x_A, u) = -A_a^{(q-2)}(x_A, u), \quad \ldots, \quad \text{etc.} \tag{3.109}$$

Note that the antipodal map (3.101) is just the so-called Weyl reflection on the sphere S^2. It can be represented as a particular right $SU(2)$ rotation of the harmonics:

$$\begin{aligned} u^{+i\prime} &= \exp\left\{-\frac{\pi}{2}(D^{++} - D^{--})\right\} u^{+i} \exp\left\{\frac{\pi}{2}(D^{++} - D^{--})\right\} = u^{-i}\,, \\ u^{-i\prime} &= \exp\left\{-\frac{\pi}{2}(D^{++} - D^{--})\right\} u^{-i} \exp\left\{\frac{\pi}{2}(D^{++} - D^{--})\right\} = -u^{+i}.. \end{aligned} \tag{3.110}$$

With the help of (3.110) one can recast the operation $\widetilde{\ }$ in a 'differential' form

$$\widetilde{f^{(q)}}(\xi, u) = \exp\left\{-\frac{\pi}{2}(D^{++} - D^{--})\right\} \overline{f^{(q)}(\xi, u)} \exp\left\{\frac{\pi}{2}(D^{++} - D^{--})\right\}. \tag{3.111}$$

A summary of the harmonic superspace conjugation rules is given in Appendix A.4.

3.8 Harmonics as square roots of quaternions

The intrinsic relationships between supersymmetry and the hypercomplex numbers have been widely discussed in the literature from various points of view (see, e.g., [G8, G43, K12]). The simplest example is $N = 1$ supersymmetry associated with complex numbers (note the fundamental rôle played by chirality, the simplest form of Grassmann analyticity, in $N = 1$ theories). Following this analogy, $N = 2$ supersymmetry has been believed to be related to quaternions. Then one might expect that $N = 2$ Grassmann analyticity, being the natural generalization of $N = 1$ chirality, is somehow related to quaternions. Here we show the existence of such a connection. More precisely, following ref. [E10], we can argue that in a general reference frame quaternions can be considered as bilinear combinations of the $SU(2)$ harmonics (3.58).

To see this, we unify the $U(1)$ charges of the harmonics into one index $\alpha = (+, -)$, $u_i^\pm \to u_i^\alpha$, and introduce the inverse harmonic matrix

$$u_\alpha^i = -\epsilon^{ik}\epsilon_{\alpha\beta}u_k^\beta\,. \tag{3.112}$$

Then the defining constraint (3.58) and the completeness relation (3.79) acquire a symmetric form

$$u^{\beta}_{i} u^{i}_{\alpha} = \delta^{\beta}_{\alpha}\,, \qquad u^{\alpha}_{i} u^{k}_{\alpha} = \delta^{i}_{k}\,. \tag{3.113}$$

Note that the harmonic derivatives $D^{\pm\pm}$ and the $U(1)$ charge-counting operator D^0 are just the generators of the 'right' $SU(2)$ acting on the indices α, β.

Now the whole two-parameter family of quaternionic imaginary units that are arbitrarily oriented in the three-dimensional isotopic space in which this extra $SU(2)$ acts as the rotation group is given by

$$(e_n)^k_i = -i\, u^{\alpha}_{i} (\tau_n)^{\beta}_{\alpha} u^{j}_{\beta}\,. \tag{3.114}$$

Indeed, using (3.113), it is easy to check that

$$e_n\, e_m = -\delta_{nm} + \epsilon_{nmp}\, e_p\,. \tag{3.115}$$

The $N = 2$ supersymmetry automorphism group $SU(2)_{\rm A}$ acting on the indices i, k is recognized as the diagonal in the tensor product of two $SU(2)$ groups realized as left and right multiplications of the quaternion $(e_0)^k_i = \delta^k_i,\ (e_n)^k_i$ by the unit norm quaternions.

Alternatively, one can construct a two-parameter family of quaternionic units with α, β as the external matrix indices:

$$(\hat{e}_n)^{\beta}_{\alpha} = -i\, u^{i}_{\alpha} (\tau_n)^{k}_{i}\, u^{\beta}_{k}\,. \tag{3.116}$$

Such units are transformed as 3-vectors under $SU(2)_{\rm A}$, while the extra $SU(2)$ is realized as the diagonal subgroup in the product of left and right quaternionic multiplications.

So, *harmonics are square roots of quaternions.* This explains why they are relevant to the problems where manifolds possessing quaternionic structures appear: $N = 2$ supersymmetric theories, self-dual $N = 0$ gauge theories, hyper-Kähler and quaternionic-Kähler geometries, etc., as we shall demonstrate in the rest of this book.

To reveal one more link with quaternions, let us compare the relation between the analytic and central basis bosonic coordinates, eq. (3.57), with the relation between the bosonic coordinates of the real and chiral $N = 1$ superspaces, eq. (3.35). The central basis coordinate appears as the real part of the complex coordinate of chiral superspace, and the nilpotent bilinear in the odd coordinates as its imaginary part. In (3.57) we see a similar structure, but the nilpotent part now involves $\theta^{(i}\sigma^a\bar{\theta}^{j)}$ which is an $SU(2)_{\rm A}$ triplet. This suggests a quaternionic interpretation of relation (3.57). Defining

$$x^a_m = \theta^{(i}\sigma^a\bar{\theta}_{j)}(\tau_m)^j_i\,, \qquad x^a_0 = x^a\,, \tag{3.117}$$

we can construct the real quaternionic coordinate

$$(\hat{x}^a)^{\alpha}_{\beta} = x^a_0 (\hat{e}_0)^{\alpha}_{\beta} + x^a_m (\hat{e}_m)^{\alpha}_{\beta}\,, \tag{3.118}$$

where

$$(\hat{e}_0)^{\alpha}_{\beta} = \delta^{\alpha}_{\beta} \tag{3.119}$$

and $\hat{e}_m$ is defined by (3.116). It is straightforward to check that the diagonal components of $\hat{x}^a$ are just x^a_A given by (3.57) and its conjugate (with respect to the standard complex conjugation).

Thus, just like the bosonic coordinate of $N = 1$ chiral superspace is a complex quantity, the bosonic coordinate of $N = 2$ analytic superspace can be viewed as a quaternion. The real part of this quaternion is the central basis bosonic coordinate, while the imaginary part is an $SU(2)_{\mathrm{A}}$ triplet made out of Grassmann coordinates. The relevance of such a nilpotent triplet to the off-shell description of the Fayet–Sohnius hypermultiplet [F1, S12] was pointed out for the first time in [G9] at the level of the $N = 1$ superfield formalism.* As we shall see in the next chapter, the Fayet–Sohnius multiplet is described off shell by the analytic superfield $q^+(\zeta, u)$, which explains this early observation.

Finally, we would like to point out that $N = 2$ Grassmann analyticity, though revealing some relations to quaternions as described above, is an essentially different generalization of the concept of complex analyticity compared, e.g., to the previously discussed quaternionic, so-called 'Fueter analyticity' (see, e.g., [G44] and references therein). Instead of considering functions valued in the non-commutative algebra of quaternions, this generalization deals with functions defined on the sphere $SU(2)/U(1)$ parametrized by harmonics which are treated as new bosonic coordinates u on an equal footing with the Minkowski space coordinates x. The interplay between the harmonic and customary quaternionic analyticities is discussed in detail in [E10].

* Historically, this work, along with [G8, I3, O6], was one of the milestones on the road toward harmonic superspace.

4
Harmonic analysis

The harmonic analysis on group and coset manifolds (in particular, on $S^2 = SU(2)/U(1)$) is the subject of many mathematical books (see, e.g., [G31, H5, H6, V2]). Our aim here is to formulate the basic notions of this analysis on the harmonic sphere $S^2 = SU(2)_A/U(1)$ defined in the previous chapter, using the parametrization-independent language of the harmonic variables $u_i^{\pm}$, $u^{+i}u_i^- = 1$.

4.1 Harmonic expansion on the two-sphere

In Chapter 3 we saw the necessity to extend the ordinary $N = 2$ superspace by the sphere $S^2 = SU(2)_A/U(1)$ where $SU(2)_A$ is the automorphism group of $N = 2$ supersymmetry. In order to keep this very important $SU(2)$ symmetry manifest, we chose to describe functions on S^2 as functions on $SU(2)$ with a fixed $U(1)$ charge. Moreover, we found it very convenient to use a manifestly $SU(2)$ covariant parametrization of $SU(2)$ in terms of the harmonic variables $u_i^{\pm}$. In this language any function on $SU(2)$ $f^{(q)}(u)$ with a given $U(1)$ charge q is defined by its harmonic expansion:

$$f^{(q)}(u) = \sum_{n=0}^{\infty} f^{(i_1 \dots i_{n+q} j_1 \dots j_n)} u_{i_1}^+ \dots u_{i_{n+q}}^+ u_{j_1}^- \dots u_{j_n}^- \tag{4.1}$$

for $q \geq 0$, and similarly for $q < 0$. Of course, given any specific parametrization of $SU(2)$, $f^{(q)}$ becomes an ordinary function of these parameters. For instance, one can choose the well-known Euler-angle parametrization:

$$\| u \| = \begin{pmatrix} u_1^+ & u_1^- \\ u_2^+ & u_2^- \end{pmatrix} = \begin{pmatrix} \cos\frac{\theta}{2} e^{\frac{i(\phi+\rho)}{2}} & i \sin\frac{\theta}{2} e^{\frac{i(\phi-\rho)}{2}} \\ i \sin\frac{\theta}{2} e^{-\frac{i(\phi-\rho)}{2}} & \cos\frac{\theta}{2} e^{-\frac{i(\phi+\rho)}{2}} \end{pmatrix},$$

$$0 \leq \theta \leq \pi\,, \qquad 0 \leq \phi \leq 2\pi\,, \qquad -2\pi \leq \rho \leq 2\pi\,. \tag{4.2}$$

For the purpose of defining the harmonic integral, deriving Green's functions, etc. we find it convenient to use stereographic coordinates:

$$\| u \| = \begin{pmatrix} u_1^+ & u_1^- \\ u_2^+ & u_2^- \end{pmatrix} = \frac{1}{\sqrt{1+t\bar{t}}} \begin{pmatrix} e^{i\psi} & -\bar{t}e^{-i\psi} \\ te^{i\psi} & e^{-i\psi} \end{pmatrix}, \qquad 0 \leq \psi < 2\pi \, . \tag{4.3}$$

The phase ψ is the coordinate on the subgroup $U(1)$. Our functions (4.1) are by definition homogeneous in $e^{i\psi}$:

$$f^{(q)}(t, \bar{t}, \psi) = e^{iq\psi} f^{(q)}(t, \bar{t}) \, , \tag{4.4}$$

thus they are equivalent to functions on the coset $SU(2)/U(1) \sim S^2$ (we use the notation $f(t, \bar{t})$ to distinguish an arbitrary function on the complex plane from a holomorphic $f(t)$ or antiholomorphic $f(\bar{t})$ one). In this parametrization the * involution defined in (3.101) becomes the antipodal map

$$t^\star = -\bar{t}^{-1} \, , \qquad (e^{i\psi})^\star = -\sqrt{\frac{\bar{t}}{t}} \, e^{-i\psi} \tag{4.5}$$

and the $\widetilde{\ }$ conjugation (3.102) is given by

$$\tilde{t} = -t^{-1} \, , \qquad e^{i\tilde{\psi}} = -\sqrt{\frac{t}{\bar{t}}} \, e^{i\psi} \, . \tag{4.6}$$

Note that the set of coordinates (4.3) only covers part of the sphere (excluding the north pole); a complementary set of coordinates is obtained from (4.3) by the antipodal map (4.5).

The traditional way of expanding such functions on the sphere, e.g., in the parametrization (4.3), is rather cumbersome:

$$\begin{aligned} f^{(q)}(t, \bar{t}, \psi) &= e^{iq\psi} \sum_{n=0}^{\infty} \sum_{p=0}^{2n+q} f^{2n+q}_{p-n-q/2} (-1)^{n+q} \sqrt{\frac{n!(n+q)!}{(2n+q-p)!p!}} \\ &\quad \times \left(\frac{t}{\bar{t}}\right)^{\frac{n-p}{2}} P^{n+q/2}_{p-n-q/2,\, q/2}\left(\frac{t\bar{t}-1}{t\bar{t}+1}\right), \end{aligned} \tag{4.7}$$

where $P^{n+q/2}_{p-n-q/2,\, q/2}(x)$ are the Jacobi polynomials [V2] and

$$f^{2n+q}_{p-n-q/2} = f^{(\overbrace{1...1}^{2n+q-p} \; \overbrace{2...2}^{p})} \, .$$

Note that the $SU(2)$ properties are not manifest in this expansion.

4.2 Harmonic integrals

In the parametrization (4.3) the invariant integral on $SU(2)$ has the form

$$\int du \; f^{(q)}(u) \equiv \frac{i}{4\pi^2} \int_0^{2\pi} d\psi \int \frac{dt \wedge d\bar{t}}{(1+t\bar{t})^2} f^{(q)}(t, \bar{t}, \psi) \, . \tag{4.8}$$

The invariant measure in (4.8) can be derived by taking the wedge product of the forms $\omega^{\pm\pm}$ (3.76). Note that a sufficient (but not necessary) condition for a function of the type (4.4) to have an expansion like (4.1) (i.e., to be a 'harmonic function') is to be square-integrable on the sphere S^2 [V2]:

$$\int du\, |f^{(q)}(u)|^2 = \frac{i}{2\pi}\int \frac{dt \wedge d\bar{t}}{(1+t\bar{t})^2}\, |f^{(q)}(t,\bar{t})|^2 < \infty\,. \tag{4.9}$$

Such functions are defined on the whole of S^2, i.e., they are *globally defined.*

Obviously, if $q \neq 0$ the ψ integral in (4.8) vanishes, so we derive our first integration rule:

$$(i) \qquad \int du\, f^{(q)}(u) = 0 \qquad \text{if } q \neq 0\,. \tag{4.10}$$

The second rule is

$$(ii) \qquad \int du\, 1 = 1, \tag{4.11}$$

which is just the normalization condition for the integral in (4.8). Finally, our third rule is

$$(iii) \qquad \int du\, u^+_{(i_1} \ldots u^+_{i_n} u^-_{j_1} \ldots u^-_{j_n)} = 0 \qquad \text{for } n \geq 1\,. \tag{4.12}$$

It follows from the fact that the harmonic variables $u^\pm_i$ transform as isodoublets under $SU(2)$, while the measure is $SU(2)$ invariant; thus the left-hand side in (4.12) must be an invariant constant totally symmetric tensor of $SU(2)$, which does not exist. Summarizing, we can say that the three rules (4.10), (4.11), (4.12) are equivalent to the original definition of the invariant integral (4.8). In other words, given a harmonic function $f^{(q)}(u)$ in the form of a harmonic expansion (4.1), one can apply the formal integration rules (i), (ii), (iii) without referring to any parametrizations like (4.2) or (4.3).

The symmetrized products of $u^\pm$:

$$(u^+)^{(m}(u^-)^{n)} \equiv u^{+(i_1} \ldots u^{+i_m} u^{-j_1} \ldots u^{-j_n)} \tag{4.13}$$

form an orthonormal basis on S^2:

$$\int du\, (u^+)^{(m}(u^-)^{n)}(u^+)_{(k}(u^-)_{l)} = \frac{(-1)^n m!n!}{(m+n+1)!}\delta^{(i_1}_{(j_1} \ldots \delta^{i_{m+n})}_{j_{m+n})}\delta_{ml}\delta_{nk}\,. \tag{4.14}$$

This can be verified with the help of the reduction identities

$$\begin{aligned} u^+_i u^+_{(j_1} \ldots u^+_{j_n} u^-_{k_1} \ldots u^-_{k_m)} &= u^+_{(i} u^+_{j_1} \ldots u^-_{k_m)} \\ &\quad + \frac{m}{m+n+1}\epsilon_{i(k_1} u^+_{j_1} \ldots u^+_{j_n} u^-_{k_2} \ldots u^-_{k_m)}\,, \\ u^-_i u^+_{(j_1} \ldots u^+_{j_n} u^-_{k_1} \ldots u^-_{k_m)} &= u^-_{(i} u^+_{j_1} \ldots u^-_{k_m)} - \frac{n}{m+n+1}\epsilon_{i(j_1} u^+_{j_2} \ldots u^-_{k_m)}\,. \end{aligned} \tag{4.15}$$

The property (4.14) allows one to find the coefficients of the harmonic expansion (4.1) of a function $f^{(q)}(u)$ on S^2 (just as one does in ordinary Fourier analysis):

$$f^{(i_1 \ldots i_{n+q} j_1 \ldots j_n)} = \frac{(-1)^{n+q}(2n+q+1)!}{(n+q)!n!} \int du \, (u^+)^{(n}(u^-)^{n+q)} f^{(q)}(u) \, . \quad (4.16)$$

An important property of the harmonic integral (4.8) is the vanishing of the integral of a total derivative:

$$\int du \, D^{++} f^{(-2)}(u) = \int du \, D^{--} f^{(+2)}(u) = 0 \, . \quad (4.17)$$

It follows from the facts that the integral projects out the singlet part of the integrand whereas the charged functions $f^{(-2)}$ or $f^{(+2)}$ in (4.17) do not contain $SU(2)$ singlets. A more direct proof can be given using the parametrization (4.3) of the harmonic variables, the invariant measure (4.8) and the expression for D^{++} in this parametrization:

$$D^{++} = e^{2i\psi} \left[-(1+t\bar{t})\frac{\partial}{\partial \bar{t}} + \frac{it}{2}\frac{\partial}{\partial \psi} \right] \quad (4.18)$$

which can be obtained following the general Cartan's scheme described in Chapter 3. Substituting (4.8), (4.18) and (4.4) into (4.17), one finds

$$\begin{aligned} \int du \, D^{++} f^{(-2)}(u) &= \frac{i}{4\pi^2} \int_0^{2\pi} d\psi \int \frac{dt \wedge d\bar{t}}{(1+t\bar{t})^2} \left[-(1+t\bar{t})\frac{\partial f}{\partial \bar{t}} + tf \right] \\ &= \frac{1}{2\pi i} \int dt \wedge d\bar{t} \, \frac{\partial}{\partial \bar{t}} \left(\frac{f}{1+t\bar{t}} \right) . \end{aligned} \quad (4.19)$$

The latter integral vanishes since the function $f(t, \bar{t})$ satisfies the integrability condition (4.9).

4.3 Differential equations on S^2

A common problem in harmonic analysis is to solve differential equations on S^2. The simplest example is the first-order linear equation:

$$D^{++} f^{(q)}(u) = 0 \, . \quad (4.20)$$

Inspecting the harmonic expansion (4.1) one sees that this equation has the following solutions:

$$f^{(q)} = 0 \qquad \text{if } q < 0 \, ; \quad (4.21)$$

$$f^{(q)} = u^+_{i_1} \ldots u^+_{i_q} c^{(i_1 \ldots i_q)} \qquad \text{if } q \geq 0 \, , \quad (4.22)$$

where $c^{(i_1 \dots i_q)}$ is a constant $SU(2)$ tensor. This conclusion can be confirmed by a direct argument. Using (4.4) and (4.18) one can rewrite (4.20) as a differential equation on the complex plane:

$$D^{++} f^{(q)} = -e^{(q+2)i\psi} \left[(1+t\bar{t}) \frac{\partial f^{(q)}}{\partial \bar{t}} + \frac{qt}{2} f^{(q)} \right] = 0 . \qquad (4.23)$$

It has the general solution:

$$f^{(q)}(t, \bar{t}) = (1+t\bar{t})^{-\frac{q}{2}} f_0(t) , \qquad (4.24)$$

where $f_0(t)$ is an arbitrary holomorphic function. Now recall that we are looking for solutions satisfying the square-integrability condition (4.9):

$$\frac{i}{2\pi} \int \frac{dt \wedge d\bar{t}}{(1+t\bar{t})^{q+2}} | f_0(t)|^2 < \infty . \qquad (4.25)$$

This implies that $f_0(t)$ must have the large $|t|$ asymptotic behavior $| f_0(t)| \leq |t|^\alpha$ where $\alpha < q+1$. Then, applying Liouville's theorem, we see that for $q < 0$ the only solution is $f_0 = 0$ (cf. (4.21)) and for $q \geq 0$ $f_0(t)$ must be a polynomial of degree q, whose $SU(2)$ covariant form is just (4.22).

The generalization to higher-order differential equations of the form

$$(D^{++})^n f^{(q)}(u) = 0 \qquad (4.26)$$

is straightforward. Using the harmonic expansion (4.1) one quickly obtains the general solution of (4.26). Similar results hold for the complex conjugate derivative D^{--}. These examples clearly show that the harmonic description of functions on S^2 is much easier to use in practice than any explicit parametrization like (4.2) or (4.3).

4.4 Harmonic distributions

Another typical problem of harmonic analysis is to find the solutions of inhomogeneous differential equations, e.g.,

$$D^{++} f^{(q)}(u) = J^{(q+2)}(u) , \qquad (4.27)$$

where $J^{(q+2)}$ is a given source. As usual, the general solution to (4.27) can be written down in terms of a Green's function for the operator D^{++}:

$$f^{(q)}(u) = f_0^{(q)}(u) + \int dv \; G^{(q,-q-2)}(u, v) \; J^{(q+2)}(v) . \qquad (4.28)$$

Here $f_0^{(q)}$ is the solution of the homogeneous eq. (4.20), and the Green's function should obey the equation:

$$D_1^{++} G^{(q,-q-2)}(u_1, u_2) = \delta^{(q+2,-q-2)}(u_1, u_2) . \qquad (4.29)$$

The harmonic delta function in (4.29) is defined by the property:

$$\int dv\ \delta^{(q,-q)}(u,v)\ f^{(q)}(v) = f^{(q)}(u)\,, \tag{4.30}$$

where $f^{(q)}(u)$ is an arbitrary test function. In the parametrization (4.3) the harmonic delta function has the form

$$\delta^{(q,-q)}(u_1,u_2) = \pi e^{iq(\psi_1-\psi_2)}(1+t_1\bar{t}_1)^2\delta(t_1-t_2)\,. \tag{4.31}$$

Indeed, with the help of (4.31), (4.4) and (4.8) the definition (4.30) is reduced to the definition of the delta function in the complex plane.

One can prove a number of useful properties of the harmonic delta functions. First of all,

$$\delta^{(q,-q)}(u_1,u_2) = (u_1^+u_2^-)^q\delta^{(0,0)}(u_1,u_2)\,. \tag{4.32}$$

Here (see (4.3))

$$u_1^+u_2^- \equiv u_1^{+i}u_{2i}^- = e^{i(\psi_1-\psi_2)}\frac{1+t_1\bar{t}_2}{\sqrt{(1+t_1\bar{t}_1)(1+t_2\bar{t}_2)}}\,, \tag{4.33}$$

so (4.32) is the same as (4.31). In a similar way one can show that

$$\delta^{(q,-q)}(u_1,u_2) = (u_1^+u_2^-)\delta^{(q-1,1-q)}(u_1,u_2)\,, \qquad \text{etc.} \tag{4.34}$$

Other useful identities are

$$\delta^{(q,-q)}(u_1,u_2) = \delta^{(-q,q)}(u_2,u_1)\,, \tag{4.35}$$

$$f^{(p)}(u_2)\delta^{(q,-q)}(u_1,u_2) = f^{(p)}(u_1)\delta^{(q-p,p-q)}(u_1,u_2)\,, \tag{4.36}$$

$$(u_1^+u_2^+)\delta^{(q,-q)}(u_1,u_2) = (u_1^-u_2^-)\delta^{(q,-q)}(u_1,u_2) = 0\,, \tag{4.37}$$

$$\widetilde{\delta^{(q,-q)}}(u_1,u_2) = \delta^{(q,-q)}(u_1,u_2)\,. \tag{4.38}$$

The harmonic delta functions can be differentiated in a natural way, e.g.,

$$\begin{aligned}&\int du_2\ D_2^{++}\delta^{(q,-q)}(u_1,u_2)\ f^{(q-2)}(u_2)\\ &\quad = -\int du_2\ \delta^{(q,-q)}(u_1,u_2)\ D_2^{++}f^{(q-2)}(u_2) = -D_1^{++}f^{(q-2)}(u_1)\\ &\quad \Rightarrow\quad D_2^{++}\delta^{(q,-q)}(u_1,u_2) = -D_1^{++}\delta^{(q-2,2-q)}(u_1,u_2)\,.\end{aligned} \tag{4.39}$$

Let us now go back to the Green's function eq. (4.29). We first consider the case $q=-1$. It is not hard to see that the Green's function $G^{(-1,-1)}(u_1,u_2)$ is given by the expression:

$$G^{(-1,-1)}(u_1,u_2) = \frac{1}{u_1^+u_2^+}\,, \tag{4.40}$$

where

$$u_1^+ u_2^+ \equiv u_1^{+i} u_{2i}^+ = e^{i(\psi_1+\psi_2)} \frac{t_2 - t_1}{\sqrt{(1+t_1\bar{t}_1)(1+t_2\bar{t}_2)}} \,. \tag{4.41}$$

Indeed, with the help of (4.18) one gets

$$\begin{aligned} D_1^{++} \frac{1}{u_1^+ u_2^+} &= e^{i(\psi_1-\psi_2)} (1+t_1\bar{t}_1)^{3/2} (1+t_2\bar{t}_2)^{1/2} \frac{\partial}{\partial \bar{t}_1} \frac{1}{t_1 - t_2} \\ &= \pi e^{i(\psi_1-\psi_2)} (1+t_1\bar{t}_1)^2 \delta(t_1 - t_2) \\ &= \delta^{(1,-1)}(u_1, u_2) \,. \end{aligned} \tag{4.42}$$

Here we assume that the singular distribution t^{-1} is defined so that the relation

$$\frac{\partial}{\partial \bar{t}} t^{-1} = \pi \delta(t) \tag{4.43}$$

holds.

The above result can be immediately generalized to the case $q \geq -1$:

$$G^{(q,-q-2)}(u_1, u_2) = \frac{(u_1^+ u_2^-)^{q+1}}{u_1^+ u_2^+} \,, \qquad q \geq -1 \,. \tag{4.44}$$

To check this one uses (4.42) and (4.32). Note that for $q \leq -2$ the expression is

$$G^{(q,-q-2)}(u_1, u_2) = \frac{(u_1^- u_2^+)^{|q|-1}}{u_1^+ u_2^+} (-1)^{|q|-1} \,, \qquad q \leq -2 \,, \tag{4.45}$$

but in this case the source $J^{(q+2)}$ in eq. (4.27) has to be restricted by the condition that the first term $(u)^{(|q|-2)}$ in its harmonic expansion is absent.

The harmonic distribution $(u_1^+ u_2^+)^{-1}$ used above can be generalized to $(u_1^+ u_2^+)^{-n}$. The latter has a number of useful properties. Firstly, acting on it with D_1^{--} one finds

$$D_1^{--} \frac{1}{(u_1^+ u_2^+)^n} = -n \frac{u_1^- u_2^+}{(u_1^+ u_2^+)^{n+1}} \,, \tag{4.46}$$

which follows from (4.41) and the standard rule $(\partial/\partial t) t^{-n} = -n t^{-n-1}$. Secondly, one can show that

$$D_1^{++} \frac{1}{(u_1^+ u_2^+)^n} = \frac{1}{(n-1)!} (D_1^{--})^{n-1} \delta^{(n,-n)}(u_1, u_2) \,. \tag{4.47}$$

Take, for instance, $n = 2$. Multiplying eq. (4.42) by $u_1^+ u_2^-$ and using (4.32) and the fact that $D_1^{++}(u_1^+ u_2^-) = 0$, one obtains

$$D_1^{++} \frac{u_1^+ u_2^-}{u_1^+ u_2^+} = \delta^{(2,-2)}(u_1, u_2) \,. \tag{4.48}$$

Further, using the commutator $[D^{++}, D^{--}] = D^0$ (3.85), the fact that the distribution $(u_1^+ u_2^-)(u_1^+ u_2^+)^{-1}$ has charge 0 with respect to u_1, and (4.46), one finds

$$D_1^{--} D_1^{++} \frac{u_1^+ u_2^-}{u_1^+ u_2^+} = D_1^{++} D_1^{--} \frac{u_1^+ u_2^-}{u_1^+ u_2^+}$$

$$= D_1^{++} \left(\frac{(u_1^- u_2^-)}{(u_1^+ u_2^+)} - \frac{(u_1^+ u_2^-)(u_1^- u_2^+)}{(u_1^+ u_2^+)^2} \right) = D_1^{++} \frac{1}{(u_1^+ u_2^+)^2} , \tag{4.49}$$

which is in agreement with (4.47). Here we have applied the useful identity (see (3.79)):

$$(u_1^+ u_2^+)(u_1^- u_2^-) - (u_1^+ u_2^-)(u_1^- u_2^+) = 1 . \tag{4.50}$$

Other properties of the distributions $(u_1^+ u_2^+)^{-n}$ are their (anti)symmetry (see (4.41)):

$$\frac{1}{(u_1^+ u_2^+)^n} = (-1)^n \frac{1}{(u_2^+ u_1^+)^n} \tag{4.51}$$

and reality:

$$\widetilde{\left(\frac{1}{(u_1^+ u_2^+)^n} \right)} = \frac{1}{(u_1^+ u_2^+)^n} . \tag{4.52}$$

5

$N = 2$ matter with infinite sets of auxiliary fields

A consistent description of matter supermultiplets (those containing only spins 0 and 1/2 on shell) is a necessary ingredient of any approach to supersymmetry. In the harmonic superspace one can introduce two *unconstrained* forms of the *off shell* $N = 2$ matter supermultiplets, q^+ and ω hypermultiplets. They are related by an off-shell duality transformation. On shell they only differ in the $SU(2)$ assignment of the physical fields. Their description in terms of unconstrained *analytic* superfields (containing infinitely many auxiliary fields) allows one to write down the most general self-couplings. The associated sigma models have complex hyper-Kähler manifolds as their target manifolds. The corresponding analytic superfield Lagrangian plays the rôle of a hyper-Kähler potential, the basic unconstrained object of hyper-Kähler geometry. A simple example of a potential leading to the Taub–NUT metric is presented in detail.

5.1 Introduction

5.1.1 $N = 1$ *matter*

In the $N = 1$ case the basic matter multiplet (two spins 0 and one spin 1/2 on shell) is described by a complex unconstrained superfield defined in the $N = 1$ chiral superspace:*

$$\mathbb{C}^{4|2} = (\zeta_L^M) \equiv (x_L^m, \theta_L^\alpha) ;$$

$$\Phi(\zeta_L) = \phi(x_L) + \theta^\alpha \psi_\alpha(x_L) + \theta\theta F(x_L) . \tag{5.1}$$

Here $\phi(x_L)$, $\psi_\alpha(x_L)$ are the physical fields whereas $F(x_L)$ is an auxiliary field. The most general action for $N = 1$ rigid matter which yields second-order field equations for the bosons and first-order ones for the fermions has the following

* In order to avoid confusion, from now on we use the indices $m, n, \ldots = 0, 1, 2, 3$ for the Minkowski space coordinates and reserve the indices $a, b, \ldots$ for the target space coordinates.

form

$$S = \frac{1}{\xi^2}\Big[\int d^4x\, d^2\theta\, d^2\bar{\theta}\, K(\Phi^a(\zeta_L), \bar{\Phi}^a(\zeta_R)) \\ + \int d^4x_L\, d^2\theta\, P(\Phi^a(\zeta_L)) + \text{c.c.}\Big], \tag{5.2}$$

$$[\Phi] = 0, \qquad [\xi] = -1, \qquad \zeta_R = \overline{\zeta_L}\,.$$

Here K is a real and P a holomorphic arbitrary function. The target space index $a = 1, 2, \ldots$ labels the chiral superfields. The dimensionful constant ξ is introduced to make Φ^a dimensionless, thus allowing one to consider arbitrary functions of Φ^a. For example, the Wess–Zumino model of Chapter 2 corresponds to the choice

$$K = \Phi\bar{\Phi}, \qquad P = \frac{m}{2}(\Phi)^2 + \frac{g}{3}(\Phi)^3. \tag{5.3}$$

Another example is the action of an $N = 1$ supersymmetric sigma model. It is obtained by setting the potential term $P = 0$, but keeping a general $K(\Phi, \bar{\Phi})$. In this case, inserting the superfield expansion (5.1) into (5.2) and eliminating the auxiliary fields F^a, $\bar{F}^a$, one obtains the following action:

$$S = \frac{1}{\xi^2}\int d^4x \big(g_{a\bar{b}}\, \partial^m\phi^a\, \partial_m\bar{\phi}^b - \frac{i}{2} g_{a\bar{b}}\psi^a\sigma^m\nabla_m\bar{\psi}^b + R_{\bar{a}c\bar{b}d}\, \bar{\psi}^a\bar{\psi}^b\psi^c\psi^d\big)\,. \tag{5.4}$$

The first term in (5.4) is a sigma model with metric

$$g_{a\bar{b}} = \frac{\partial^2 K}{\partial\Phi^a\,\partial\bar{\Phi}^b}\bigg|_{\theta=0} \tag{5.5}$$

on a target manifold parametrized by the complex fields ϕ^a, $\bar{\phi}^a$. Metrics of the type (5.5) are called Kähler metrics and K is the Kähler potential (see Chapter 11 for a detailed discussion of sigma model geometry). The covariant derivative ∇_m in (5.4) involves a Christoffel symbol constructed from $g_{a\bar{b}}$, and $R_{\bar{a}c\bar{b}d}$ is the Riemann tensor.

The action (5.2) is invariant under general holomorphic changes of variables (i.e., reparametrizations of the Kähler manifold):

$$\delta\Phi^a \equiv \Phi^{a\prime} - \Phi^a = \lambda^a(\Phi)\,, \qquad \delta\bar{\Phi}^a = \bar{\lambda}^a(\bar{\Phi})\,, \tag{5.6}$$

provided K and P transform as scalars:

$$\delta K \equiv K'(\Phi', \bar{\Phi}') - K(\Phi, \bar{\Phi}) = 0\,, \qquad \delta P = 0\,. \tag{5.7}$$

In addition, this action does not change if the Kähler potential is shifted by holomorphic and antiholomorphic functions (the so-called Kähler invariance):

$$\delta K(\Phi, \bar{\Phi}) = \Lambda(\Phi) + \bar{\Lambda}(\bar{\Phi})\,. \tag{5.8}$$

The reason is that the full superspace integral in (5.2) vanishes if the integrand is (anti)chiral.

We should mention here that there exist other off-shell representations of $N = 1$ matter, which are described by constrained superfields. All of them can be reduced to certain subclasses of the general action (5.2) by duality transformations (see Chapter 6).

5.1.2 $N = 2$ matter multiplets on shell

On shell a massless $N = 2$ scalar multiplet (hypermultiplet) includes four real scalars (or two complex ones) and two spinors (see Section 2.4.3). Depending on the $SU(2)$ assignment of these fields one distinguishes two forms of the hypermultiplet.

(i) The Fayet–Sohnius (FS) hypermultiplet (or q^+ hypermultiplet in what follows) has an $SU(2)_A$ doublet* of scalars and a pair of isosinglet Dirac spinors:

$$f^i(x)\,, \qquad \psi^\alpha(x)\,, \qquad \bar\kappa^{\dot\alpha}(x)\,. \tag{5.9}$$

In Section 2.4.3 we have shown that the sum of the free actions of these fields, eq. (2.66), is invariant under the $N = 2$ supersymmetry transformations (2.67). They form an algebra which closes only modulo the free field equations:

$$\Box f^i = \partial\!\!\!/ \bar\kappa = \partial\!\!\!/ \psi = 0\,. \tag{5.10}$$

(ii) The real form of the hypermultiplet (or ω hypermultiplet henceforth) has the scalars in a $(\underline{1}+\underline{3})$ of $SU(2)_A$ and an $SU(2)_A$ doublet of Weyl fermions:

$$\omega(x)\,, \qquad \omega^{(ij)}(x)\,, \qquad \psi^i_\alpha(x)\,;$$
$$\bar\omega = \omega\,, \qquad \overline{\omega^{(ij)}} = \epsilon_{ik}\epsilon_{jl}\omega^{(kl)}\,. \tag{5.11}$$

The relevant free action and $N = 2$ supersymmetry transformations are

$$S^{\text{free}}_\omega = \int d^4x \left(\frac{1}{2}\partial^m\omega\partial_m\omega + \frac{1}{2}\partial^m\omega^{(ij)}\partial_m\omega_{(ij)} - \frac{i}{2}\psi^i\partial\!\!\!/\bar\psi_i\right); \tag{5.12}$$

$$\begin{aligned}
\delta\omega &= -\frac{1}{\sqrt{2}}(\epsilon^i\psi_i - \bar\epsilon^i\bar\psi_i)\,, \\
\delta\omega^{(ij)} &= -\epsilon^{(i}\psi^{j)} + \bar\epsilon^{(i}\bar\psi^{j)}\,, \\
\delta\psi^i_\alpha &= 2i(\sigma^m\bar\epsilon_j)_\alpha\left(\partial_m\omega^{(ij)} + \frac{1}{\sqrt{2}}\epsilon^{ij}\partial_m\omega\right).
\end{aligned} \tag{5.13}$$

---

* $SU(2)_A$ is the automorphism group of the supersymmetry algebra (2.17) (for $N = 2$). As we shall see later on, the free hypermultiplet actions possess some extra $SU(2)$ symmetries.

Once again, the algebra of the transformations (5.13) closes only on shell.

These two forms of the same $N = 2$ multiplet correspond to different choices of the Clifford vacuum. It carries helicity 1/2 in both cases, but has isospin 0 in the first case and 1/2 in the second case (see Section 2.3).

As explained in Section 2.3.4, in the absence of central charge the $N = 2$ matter multiplet is necessarily massless. If we add to the action (2.66) a mass term for f, ψ and κ,

$$-\int d^4x \left[m^2 f^i \bar{f}_i + \frac{m}{2}\left(\psi\kappa + \bar{\psi}\bar{\kappa}\right)\right], \tag{5.14}$$

then this sum is invariant under modified supersymmetry transformations (cf. (2.67)):

$$\begin{aligned} \delta f^i &= -(\epsilon^i \psi + \bar{\epsilon}^i \bar{\kappa})\,, \\ \delta \psi_\alpha &= -2i(\sigma^m \bar{\epsilon}^i)_\alpha \partial_m f_i - 2m\epsilon^i_\alpha f_i\,, \\ \delta \kappa_\alpha &= -2i(\sigma^m \bar{\epsilon}^i)_\alpha \partial_m \bar{f}_i + 2m\epsilon^i_\alpha \bar{f}_i\,. \end{aligned} \tag{5.15}$$

On shell they form the algebra of $N = 2$ supersymmetry with a non-vanishing central charge, equal to the mass m.

5.1.3 Relationship between q^+ and ω hypermultiplets

At the level of physical fields the difference between the two forms (5.9) and (5.11) of the hypermultiplet is merely conventional and can be attributed to the freedom in defining $SU(2)_A$. We mean that instead of $SU(2)_A$ we can always choose another automorphism group for the $N = 2$ superalgebra, the diagonal subgroup in the product $SU(2)_A \times SU(2)'$ where $SU(2)'$ is some extra $SU(2)$ commuting with supersymmetry. For example, for the set of fields (5.9) $SU(2)'$ is the Pauli–Gürsey group $SU(2)_{PG}$:*

$$f^i_a \equiv (f^i, \bar{f}^i)\,, \qquad \psi_{\alpha a} \equiv (\psi_\alpha, \kappa_\alpha)\,, \qquad \bar{\psi}_{\dot{\alpha} a} \equiv (\bar{\kappa}_{\dot{\alpha}}, -\bar{\psi}_{\dot{\alpha}})\,; \tag{5.16}$$

$$\delta_{PG} f^i_a = \alpha^b_a f^i_b\,, \qquad \delta_{PG}\psi_{\alpha a} = \alpha^b_a \psi_{\alpha b}\,. \tag{5.17}$$

It commutes with $SU(2)_A$ and the $N = 2$ superalgebra. Now, with respect to the diagonal in $SU(2)_A \times SU(2)_{PG}$ the fields (5.16) are reorganized into a singlet, a triplet and a doublet as in (5.11). After that the action (2.66) coincides with (5.12). We see that on shell the q^+ hypermultiplet action is the same as the ω hypermultiplet one.

* The extended $SU(2) \times SU(2)$ invariance of the free action of an isodoublet of Dirac fermions was first noticed by Pauli and Gürsey.

5.1.4 Off-shell $N = 2$ matter before harmonic superspace

As we just saw, on shell the difference between the hypermultiplets (5.9) and (5.11) is merely conventional. However, it becomes very important when trying to extend the $N = 2$ scalar multiplet off shell. There have been several attempts to formulate the real hypermultiplet (5.11) off shell. The first supermultiplet of this kind, the tensor one, was found by Wess [W6] and de Wit and van Holten [D19]. As in the $N = 1$ case, it contains a divergenceless vector field, so the scalar component $\omega(x)$ is described off shell by a gauge antisymmetric tensor (notoph) [O5]. As usual, this is an obstacle for minimal super-Yang–Mills couplings. Later on Howe, Stelle and Townsend [H15] discovered the relaxed hypermultiplet. Unlike the tensor one, it does not contain conserved vectors and the isosinglet field $\omega(x)$ is among its off-shell components. This makes possible the minimal coupling of the relaxed hypermultiplet to $N = 2$ Yang–Mills. Some 'further relaxed' hypermultiplets with larger (but finite) numbers of auxiliary fields were constructed by Yamron and Siegel [Y1].

A common feature of the above representations is that they are described by constrained superfields in ordinary $N = 2$ superspace. Their general actions always possess at least one isometry in the physical boson sector (see Chapter 6 for details). On the other hand, there exist geometric arguments that the most general $N = 2$ matter action should not have any isometries (see [E6] and the discussion in Chapters 6 and 11). Thus one concludes that the constrained $N = 2$ multiplets are not adequate for describing the most general off-shell $N = 2$ supersymmetric matter action.

One could instead try to extend the FS complex hypermultiplet (5.9) off shell. There are various reasons why this is preferable (for instance, the FS hypermultiplet can be put in a complex representation of a Yang–Mills group without doubling the number of physical degrees of freedom). Sohnius [S12] proposed a version of this multiplet with central charge. Though being off shell in four dimensions, it satisfies a five-dimensional equation of motion (the fifth dimension corresponds to the central charge). This does not allow for non-trivial self-interactions.

Without central charges the situation is even worse. Indeed, as explained in Section 2.4.3, there exists a 'no-go' theorem [H18, S21] stating that the problem has no solution in a framework with finite sets of auxiliary fields.

The puzzle found its natural resolution in 1984 in the framework of $N = 2$ harmonic superspace. There the FS hypermultiplet is described off shell by an unconstrained analytic superfield $q^+(\zeta, u)$, which contains an *infinite* number of auxiliary fields. All the previously known off-shell $N = 2$ matter multiplets have *finite* sets of auxiliary fields and are represented by constrained analytic $N = 2$ superfields [G10, G11]. The latter are related to the q^+ hypermultiplet by $N = 2$ duality transformations, which convert their general actions into particular subclasses of the general q^+ action. It should be stressed that the superfield q^+ admits off-shell self-couplings which are not realizable in terms of constrained superfields and which in general possess no symmetries besides $N=2$ supersymmetry itself.

After these introductory remarks we proceed to the harmonic superspace description of $N=2$ matter.

5.2 Free off-shell hypermultiplet

5.2.1 *The Fayet–Sohnius hypermultiplet constraints as analyticity conditions*

We begin by reviewing the formulation of the on-shell FS hypermultiplet in ordinary superspace. This leads us to the idea of using harmonic variables in order to interpret the hypermultiplet constraints and extend the theory off shell. Thus the concepts of harmonic superspace and Grassmann analyticity arise naturally.

The set of fields (5.9) can be embedded into a complex $N=2$ superfield $q^i(x,\theta,\bar\theta) = f^i(x) + \cdots$ of dimension 1 subject to the following constraints [S12]:

$$D^{(i}_\alpha q^{j)} = \bar D^{(i}_{\dot\alpha} q^{j)} = 0\,. \tag{5.18}$$

They follow from the requirement that the isotriplet of fermions of physical dimension 3/2 be absent from the multiplet, according to (5.9).

Unfortunately, the constraints (5.18) turn out too strong. They eliminate all the components of $q^i(X)$ but the physical fields and put the latter on shell:

$$\begin{aligned}(5.18) \quad \Rightarrow \quad & q^i(X) = f^i(x) + \theta^{i\alpha}\psi_\alpha(x) + \bar\theta^i_{\dot\alpha}\bar\kappa^{\dot\alpha}(x) + \text{derivative terms;}\\ & \partial\!\!\!/\bar\kappa = \partial\!\!\!/\psi = \Box f^i = 0\,.\end{aligned} \tag{5.19}$$

Equation (5.19) has been obtained from (5.18) using the $N=2$ covariant derivative algebra (3.23) and the definitions:

$$f^i(x) = q^i|_{\theta=0}\,, \qquad \psi_\alpha(x) = \frac{1}{2}D^i_\alpha q_i|_{\theta=0}\,, \qquad \bar\kappa^{\dot\alpha}(x) = \frac{1}{2}\bar D^{\dot\alpha}_i q^i|_{\theta=0}\,. \tag{5.20}$$

Attempting to extend this theory off shell, one might try to relax the constraints (5.18) and find an action, from which they follow as variational equations. Unfortunately, according to the no-go theorem of Section 2.4.3, this is not possible in the framework of ordinary $N=2$ superspace. It is harmonic superspace that provides the key to the problem.

Let us contract the $SU(2)_{\rm A}$ indices in (5.18) with the harmonics u^+_i, u^+_j:

$$D^+_\alpha q^+ = \bar D^+_{\dot\alpha} q^+ = 0\,, \tag{5.21}$$

where

$$D^+_\alpha = D^i_\alpha u^+_i\,, \qquad \bar D^+_{\dot\alpha} = \bar D^i_{\dot\alpha} u^+_i\,, \qquad q^+(X) = q^i(X)u^+_i\,. \tag{5.22}$$

Since u_i^+, u_j^+ are arbitrary commuting variables, eqs. (5.21), (5.22) are equivalent to (5.18).

Further, the linear dependence (5.22) of q^+ on the harmonics u^+ can be equivalently formulated as a differential constraint on an *arbitrary* harmonic function of $U(1)$ charge +1, $q^+ = q^+(X, u)$:

$$D^{++}q^+(X, u) = 0\,, \tag{5.23}$$

where $D^{++} = u^{+i}\partial/\partial u^{-i}$ is the harmonic derivative in the central basis of $\mathbb{HR}^{4+2|8}$ (3.73). Indeed, according to (4.20), (4.22) the solution of eq. (5.23) is a linear function of u^+.

So, we have succeeded in equivalently reformulating the constraint (5.18) on the ordinary $N = 2$ superfield $q^i(X)$ as a set of two constraints:

$$D_\alpha^+ q^+(X, u) = \bar{D}_{\dot{\alpha}}^+ q^+(X, u) = 0\,, \tag{5.24}$$

$$D^{++}q^+(X, u) = 0 \tag{5.25}$$

on the superfield $q^+(X, u)$ defined in $N = 2$ harmonic superspace. Note that these three constraints are consistent by virtue of the (anti)commutativity of the derivatives D_α^+, $\bar{D}_{\dot{\alpha}}^+$, D^{++}, eq. (3.87).

It is very important to realize that both constraints are *analyticity conditions*. As we explained in Section 3.3, eqs. (5.24) are *Grassmann analyticity conditions* (3.55). The second constraint (5.25) is nothing but an analyticity condition on the sphere S^2. Indeed, in the central basis (3.73) and in the stereographic parametrization (4.3) the operator D^{++} takes the form (4.18) of a 'covariant' derivative with respect to $\bar{t}$. As shown in Chapter 4, the solution of an equation of the type (5.25) is given in terms of a holomorphic function on S^2.

It must be stressed that the two analyticity conditions *together* amount to the original on-shell constraints (5.18). Each one of them, taken separately, is just a kinematical constraint and can be explicitly solved in the appropriate basis. As we just explained, the S^2 analyticity condition (5.25) is solved in the central basis. Thus, in this basis the dynamics is contained in the Grassmann analyticity conditions (5.24). Alternatively, the analytic basis (3.74) is suitable for solving the Grassmann analyticity conditions. In this basis D_α^+ and $\bar{D}_{\dot{\alpha}}^+$ are short (3.53), and these conditions imply

$$(5.24) \quad \Rightarrow \quad q^+ = q^+(\zeta, u^\pm)\,. \tag{5.26}$$

Thus we obtain a Grassmann analytic superfield $q^+(\zeta, u^\pm)$. This time the dynamical equation is the remaining S^2 analyticity constraint (5.25). In the analytic basis the latter reads (see (3.84))

$$D^{++}q^+(\zeta, u^\pm) = 0\,, \qquad \text{where} \qquad D^{++} = \partial^{++} - 2i\theta^+\sigma^m\bar{\theta}^+\partial_m\,. \tag{5.27}$$

Recall that the analytic superfield $q^+(\zeta, u)$ contains an infinite tower of fields emerging from its harmonic expansion. The rôle of eq. (5.27) is to eliminate all fields but a finite number of physical fields. In addition, eq. (5.27) should put the physical fields on shell. Let us have a closer look at this.

Substituting the θ decomposition of $q^+(\zeta, u)$:

$$\begin{aligned} q^+(\zeta,u) &= F^+(x,u) + \theta^{+\alpha}\psi_\alpha(x,u) + \bar\theta^+_{\dot\alpha}\bar\kappa^{\dot\alpha}(x,u) \\ &+(\theta^+)^2 M^-(x,u) + (\bar\theta^+)^2 N^-(x,u) + i\theta^+\sigma^m\bar\theta^+ A^-_m(x,u) \\ &+(\theta^+)^2\bar\theta^+_{\dot\alpha}\bar\chi^{(-2)\dot\alpha}(x,u) + (\bar\theta^+)^2\theta^{+\alpha}\lambda^{(-2)}_\alpha(x,u) \\ &+(\theta^+)^2(\bar\theta^+)^2 P^{(-3)}(x,u) \end{aligned} \tag{5.28}$$

into (5.27), one obtains the following equations for the coefficients in (5.28):

$$\partial^{++}F^+(x,u) = 0\,, \tag{5.29}$$

$$\partial^{++}\psi_\alpha(x,u) = \partial^{++}\bar\kappa^{\dot\alpha}(x,u) = 0\,, \tag{5.30}$$

$$\partial^{++}M^-(x,u) = \partial^{++}N^-(x,u) = 0\,, \tag{5.31}$$

$$\partial^{++}A^-_m(x,u) - 2\partial_m F^+(x,u) = 0\,, \tag{5.32}$$

$$\partial^{++}\lambda^{(-2)}_\alpha(x,u) + i(\partial\!\!\!/\bar\kappa)_\alpha(x,u) = 0\,, \tag{5.33}$$

$$\partial^{++}\bar\chi^{(-2)}_{\dot\alpha}(x,u) + i(\partial\!\!\!/\psi)_{\dot\alpha}(x,u) = 0\,, \tag{5.34}$$

$$\partial^{++}P^{(-3)}(x,u) + \partial^m A^-_m(x,u) = 0\,. \tag{5.35}$$

Equations (5.29)–(5.32) are purely kinematical. They eliminate the infinite sets of auxiliary fields in the harmonic expansions (see (4.20),(4.22)):

$$\begin{aligned} (5.29) &\Rightarrow F^+(x,u) = f^i(x)u^+_i\,, \\ (5.30) &\Rightarrow \psi_\alpha(x,u) = \psi_\alpha(x), \qquad \bar\kappa^{\dot\alpha}(x,u) = \bar\kappa^{\dot\alpha}(x), \\ (5.31) &\Rightarrow M^-(x,u) = N^-(x,u) = 0\,, \\ (5.32) &\Rightarrow A^-_m(x,u) = 2\partial_m f^i(x)u^-_i\,, \end{aligned} \tag{5.36}$$

leaving only the physical fields $f^i(x)$, $\psi_\alpha(x)$, $\bar\kappa^{\dot\alpha}(x)$. Equations (5.33)–(5.35) not only eliminate the rest of the auxiliary fields, they also put the remaining physical fields (5.36) on shell:

$$\begin{aligned} (5.33) &\Rightarrow \lambda^{(-2)}_\alpha = 0\,, \qquad \partial\!\!\!/\bar\kappa = 0\,, \\ (5.34) &\Rightarrow \bar\chi^{(-2)} = 0\,, \qquad \partial\!\!\!/\psi = 0\,, \\ (5.35) &\Rightarrow P^{(-3)} = 0\,, \qquad \partial^m A^-_m = 2\Box f^i u^-_i = 0 \quad\Rightarrow\quad \Box f^i(x) = 0\,. \end{aligned} \tag{5.37}$$

So, on shell the Grassmann analytic superfield q^+ is reduced to

$$\begin{aligned} q^+(\zeta,u) &= f^i(x)u^+_i + \theta^{+\alpha}\psi_\alpha(x) + \bar\theta^+_{\dot\alpha}\bar\kappa^{\dot\alpha}(x) \\ &+2i\theta^+\sigma^m\bar\theta^+\partial_m f^i(x)u^-_i\,, \end{aligned} \tag{5.38}$$

where

$$\Box f^i(x) = \partial\!\!\!/\psi = \partial\!\!\!/\bar\kappa = 0\,.$$

It is not hard to check that the $N = 2$ supersymmetry transformations of the on-shell components in (5.38) coincide with (2.67). To avoid confusion and for future reference, let us note that the short form of q^+ in (5.38) comes out at the step of the elimination of the auxiliary fields by eqs. (5.29)–(5.32) and the kinematical part of eqs. (5.33)–(5.35), before employing the dynamical part of the latter.

An alternative way to derive the component equations of motion (5.29)–(5.35) (without going through the explicit θ-expansion) consists of replacing the superfield eq. (5.27) by the equivalent:

$$(D^{--})^2 q^+ = 0\,. \tag{5.39}$$

The easiest way to see this is to go to the central basis where both (5.27) and (5.39) simply mean that q^+ is a linear function of u^+. Then one starts hitting eq. (5.39) with the spinor derivatives $D^+_{\alpha,\dot\alpha}$, uses their algebra and the analyticity of q^+ to obtain the various components. For instance, hitting it with one D^+_α gives

$$D^{--}(D^-_\alpha q^+) = 0\,.$$

If one defines $\psi_\alpha(x,u) = -D^-_\alpha q^+|_{\theta=0}$, then this equation means that $\psi_\alpha(x,u)$ does not depend on the harmonics. The same follows from eq. (5.30). Further, two derivatives $D^{+\alpha}D^+_\alpha = (D^+)^2$ give

$$(D^-)^2 q^+ = 0 \quad \Rightarrow \quad M^-(x,u) \equiv \frac{1}{2}(D^-)^2 q^+|_{\theta=0} = 0\,,$$

which coincides with the content of the first of eqs. (5.31). Similarly, hitting eq. (5.39) with all four derivatives $(D^+)^2(\bar D^+)^2$ immediately implies the field equation $\Box q^+ = 0$, etc.

5.2.2 *Free off-shell q^+ action*

So, we have convinced ourselves that the FS hypermultiplet constraints, rewritten in harmonic superspace, are equivalent to eq. (5.27) for the analytic superfield $q^+(\zeta,u)$. Now it is clear how one can immediately go off shell. One should treat $q^+(\zeta,u)$ as an *unconstrained $N = 2$ Grassmann analytic superfield* and find an action from which (5.27) will follow as an Euler–Lagrange equation. Such an action is easy to find, but let us first make a digression to explain the integration rules in harmonic superspace.

In the full harmonic superspace $\mathbb{HR}^{4+2|8}$ (3.73) (or (3.74)) the integral is simply

$$\begin{aligned}\int du\, d^{12}X \equiv \int du\, d^4x\, d^8\theta &= \int du\, d^4x_A\, d^4\theta^+\, d^4\theta^- \\ &= \int du\, d^4x_A (D^-)^4 (D^+)^4, \end{aligned} \tag{5.40}$$

where

$$(D^{\pm})^4 = \frac{1}{16}(D^{\pm\alpha}D^{\pm}_{\alpha})(\bar{D}^{\pm}_{\dot{\alpha}}\bar{D}^{\pm\dot{\alpha}}) = \frac{1}{16}(D^{\pm})^2(\bar{D}^{\pm})^2 . \tag{5.41}$$

This reflects the fact that $\mathbb{HR}^{4+2|8} = \mathbb{R}^{4|8} \times S^2$. The situation is more subtle in the analytic superspace $\mathbb{HA}^{4+2|4}$ (3.72). There the Grassmann measure $d^4\theta^+$ has harmonic $U(1)$ charge -4, since the Berezin integral is equivalent to differentiation. So, we define the integral over $\mathbb{HA}^{4+2|4}$ as follows:

$$\int du\, d\zeta^{(-4)} \equiv \int du\, d^4x_A\, d^4\theta^+ = \int du\, d^4x_A (D^-)^4 . \tag{5.42}$$

Note that the analytic superspace integration measure (5.42) is real with respect to the $\widetilde{\ }$ conjugation.

The Grassmann integration measures are normalized so that

$$\int d^8\theta\, \theta^8 = 1 , \qquad \int d^4\theta^+ (\theta^+)^4 = 1 , \tag{5.43}$$

where

$$\theta^8 = (\theta^+)^4(\theta^-)^4 , \qquad (\theta^{\pm})^4 = (\theta^{\pm})^2(\bar{\theta}^{\pm})^2 .$$

Now we recall that the harmonic integral has the property that it vanishes if the integrand is charged (see (4.10)). This implies that the integrand in the integral over $\mathbb{HA}^{4+2|4}$ must be an analytic quantity of $U(1)$ charge $+4$ in order to compensate the charge of the Grassmann measure:

$$I = \int du\, d\zeta^{(-4)} L^{+4}(\zeta, u) . \tag{5.44}$$

An important property of this integral is that it vanishes if the integrand is a total harmonic derivative D^{++}:

$$I' = \int du\, d\zeta^{(-4)} D^{++}L^{++}(\zeta, u) = \int du\, d^4x_A (D^-)^4 D^{++}L^{++}(\zeta, u) = 0 . \tag{5.45}$$

One can see this in the following way. Owing to the integral $\int d^4x$, the covariant derivatives D^- and D^{++} in (5.45) can be replaced by the partial ones:

$$I' = \int du\, d^4x_A (\partial^-)^4 \partial^{++} L^{++}(\zeta, u) , \tag{5.46}$$

where $\partial^- \equiv \partial/\partial\theta^+$. In the analytic basis the partial derivatives ∂^- and ∂^{++} commute since $\theta^{\pm}$ and the harmonics $u^{\pm}$ are independent variables. Therefore, one can pull ∂^{++} up to the harmonic integral and use (4.17) to show that $I' = 0$.

Let us now come back to the issue of writing down an off-shell action from which the equation of motion (5.27) follows. It has a very simple form

$$S_q^{\text{free}} = -\int du\, d\zeta^{(-4)} \tilde{q}^+ D^{++} q^+ , \tag{5.47}$$

where the $\sim$ conjugation is defined in Section 3.7. Varying (5.47) with respect to the unconstrained superfield q^+ and its conjugate $\tilde{q}^+$ and using the property (5.45) yield the equation of motion (5.27) (and its conjugate). This action is manifestly supersymmetric and real in the sense of ordinary complex conjugation:

$$\bar{S} = S\,. \tag{5.48}$$

This property follows from the fact that $\widetilde{(\tilde{q}^+)} = -q^+$ (see (3.104)).

All the higher-isospin fields in the harmonic expansion of $q^+(\zeta, u)$ become auxiliary fields. Indeed, doing the θ integral in (5.47) one easily sees that these fields have no kinetic terms. Varying with respect to them yields just the kinematical part of eqs. (5.29)–(5.35).* As we noticed earlier, the elimination of the auxiliary fields by these equations brings q^+ into the short form (5.38), but with the physical fields f^i, ψ, κ still not subject to the equations of motion. Substituting this short q^+ back into (5.47) and integrating over harmonics and θ, we recover just the on-shell action (2.66) for the physical fields.

So, the infinite number of auxiliary fields was the price to pay for circumventing the no-go theorem [H18, S21] and for constructing an off-shell theory of the FS hypermultiplet. An analogous but more complicated mechanism will work in the case of the $N = 3$ off-shell super-Yang–Mills theory (see Chapter 12).

The on-shell component-field action (2.66) is invariant under two different $SU(2)$ groups, the automorphism group $SU(2)_{\rm A}$ acting on the fields according to their internal symmetry indices, and the Pauli–Gürsey group $SU(2)_{\rm PG}$ (5.17). Both of them are realized in terms of the off-shell superfield q^+ in a natural way. The group $SU(2)_{\rm A}$ rotates the harmonics:

$$\begin{aligned} SU(2)_{\rm A}: \quad \delta^*_{\rm A} q^+(\zeta, u) &\equiv q^{+'}(\zeta, u) - q^+(\zeta, u) \\ &= A_i^j \left(u^+_{(j} \frac{\partial}{\partial u^+_{i)}} + u^-_{(j} \frac{\partial}{\partial u^-_{i)}} \right) q^+(\zeta, u)\,. \end{aligned} \tag{5.49}$$

The Pauli–Gürsey group $SU(2)_{\rm PG}$ (5.17) acts on the doublet composed of $q^+, \tilde{q}^+$:

$$q^+_a = \left(q^+, -\tilde{q}^+\right), \qquad \widetilde{q^+_a} \equiv q^{+a} = \epsilon^{ab} q^+_b\,; \tag{5.50}$$

$$SU(2)_{\rm PG}: \qquad \delta_{\rm PG} q^+_a = \alpha_a^b q^+_b\,. \tag{5.51}$$

Clearly, $SU(2)_{\rm PG}$ (5.51) commutes with $SU(2)_{\rm A}$ and supersymmetry. The invariance of the action (5.47) under $SU(2)_{\rm PG}$ becomes manifest after rewriting it as follows:

$$S_q^{\rm free} = \frac{1}{2} \int du\, d\zeta^{(-4)} q^+_a D^{++} q^{+a}\,. \tag{5.52}$$

* The detailed component form of the action will be given in the most general case of a self-interacting q^+ in Chapter 11.

The free action (5.52) is easily generalized for any number n of hypermultiplets q^+_a, $(a = 1, \dots, 2n)$. Then the Pauli–Gürsey group $SU(2)_{\mathrm{PG}}$ is extended to $Sp(n)$ and the reality condition (5.50) becomes

$$\widetilde{q^+_a} \equiv q^{+a} = \Omega^{ab} q^+_b \,, \tag{5.53}$$

where $\Omega^{ab} = -\Omega^{ba}$ is the invariant tensor of the group $Sp(n)$. Note that this corresponds to the following condition for the physical bosonic fields $f^{ai}(x)$ $(q^{+a} = f^{ai}(x)u^+_i + \cdots)$

$$\overline{(f^{ai})} = \Omega_{ab}\epsilon_{ik} f^{bk} \,. \tag{5.54}$$

5.2.3 Relationship between q^+ and ω hypermultiplets off shell

The correspondence between (5.9) and (5.11) mentioned in Section 5.1.3 can be extended off shell. To this end let us make the change of variables:

$$q^+_a = u^+_a \omega + u^-_a f^{++} \,, \qquad \tilde{\omega} = \omega \,, \qquad \tilde{f}^{++} = f^{++} \tag{5.55}$$

$$\Leftrightarrow \qquad \omega = -u^{-a} q^+_a \,, \qquad f^{++} = u^{+a} q^+_a \,.$$

Substituting (5.55) into (5.52) gives

$$S^{\mathrm{free}}_{\omega, f} = \int du\, d\zeta^{(-4)} \frac{1}{2}(f^{++} f^{++} + 2 f^{++} D^{++} \omega) \,. \tag{5.56}$$

The analytic superfield $f^{++}(\zeta, u)$ in (5.56) does not propagate and hence its variation yields an algebraic equation:

$$f^{++} = -D^{++} \omega \,. \tag{5.57}$$

Inserting this expression back into (5.56) one finds

$$S^{\mathrm{free}}_{\omega} = -\frac{1}{2} \int du\, d\zeta^{(-4)} (D^{++}\omega)^2 \,. \tag{5.58}$$

Varying (5.58) with respect to ω produces the equation of motion:

$$(D^{++})^2 \omega = 0 \,. \tag{5.59}$$

We leave it to the reader to check that eq. (5.59) reduces the field content of ω to that in (5.11):

$$\begin{aligned} \omega(\zeta, u) \;=\; & \frac{1}{\sqrt{2}} \omega(x) + \omega^{(ij)}(x) u^+_i u^-_j + \theta^{+\alpha} \psi^i_\alpha(x) u^-_i \\ & + \bar{\theta}^+_{\dot{\alpha}} \bar{\psi}^{\dot{\alpha}}_i(x) u^{-i} + 2i\theta^+ \sigma^m \bar{\theta}^+ \partial_m \omega^{(ij)}(x) u^-_i u^-_j \,. \end{aligned} \tag{5.60}$$

Further, the fields in (5.60) satisfy the standard free equations of motion corresponding to the component action (5.12).

The conclusion is that the free q^+ action (5.52) can be reduced to the free action (5.58) of the real analytic superfield $\omega(\zeta, u)$. This relationship is rather similar to that between the first- and second-order formulations of the ordinary scalar field action. Thus, the q^+ free action is the first-order form of the ω free action. The two multiplets coincide on shell (up to different $SU(2)_{\mathrm{A}}$ assignments of the physical fields), but differ off shell in the structure of the infinite sets of auxiliary fields. This relationship can be extended to the interacting case in a straightforward way.

Finally, we mention that in the central basis (3.73) eq. (5.59) means that ω has a short harmonic decomposition,

$$(\partial^{++})^2\omega = 0 \quad \Rightarrow \quad \omega = \omega(X) + \omega^{(ij)}(X)u_i^+u_j^- \,, \tag{5.61}$$

just as in the case of q^+. All the non-trivial dynamical information is encoded in the analyticity condition. In the central basis the latter becomes a set of constraints on the ordinary $N = 2$ superfields $\omega(X)$ and $\omega^{(ij)}(X)$:

$$D^+_\alpha\omega = \bar{D}^+_{\dot\alpha}\omega = 0 \quad \Rightarrow$$

$$D^i_\alpha\omega(X) = \frac{1}{3}D_{\alpha k}\omega^{(ki)}(X)\,, \qquad D^{(i}_\alpha\omega^{jk)} = 0 \qquad \text{and c.c.} \tag{5.62}$$

These equations constitute an equivalent description of the free ω hypermultiplet. They eliminate all the components from $\omega(X)$ and $\omega^{(ij)}(X)$ except the physical ones and put the latter on shell.

5.2.4 *Massive q^+ hypermultiplet*

As explained in Section 2.3.4, the only way to have a massive $N = 2$ matter multiplet is to let it have a central charge equal to the mass (the second central charge can be chosen to vanish). In other words, we demand

$$Zq^+ = mq^+\,. \tag{5.63}$$

In Section 3.6 we saw that Z is realized as a shift in the fifth (auxiliary) coordinate x^5, $Z = -i\partial/\partial x^5$ (see (3.99), (3.100)). Then the solution of (5.63) is

$$q^+(\zeta, x^5, u) = e^{imx^5}q^+(\zeta, u)\,. \tag{5.64}$$

Next we substitute (5.64) and the expression (3.98) for D^{++} in the presence of central charge,

$$D^{++}_{\mathrm{cc}} = D^{++} + i[(\theta^+)^2 - (\bar\theta^+)^2]\frac{\partial}{\partial x^5}\,, \tag{5.65}$$

into the action (5.47) for q^+:

$$S_q^{\rm cc} = -\int du\, d\zeta^{(-4)} \tilde{q}^+(\zeta, x^5, u) D_{\rm cc}^{++} q^+(\zeta, x^5, u)$$
$$= -\int du\, d\zeta^{(-4)} \tilde{q}^+(\zeta, u)[D^{++} - m((\theta^+)^2 - (\bar{\theta}^+)^2)]q^+(\zeta, u)\,. \quad (5.66)$$

Note that the integral in (5.66) goes over d^4x only, because the integrand does not depend on x^5. This is not accidental. In fact, we have identified the central charge generator Z with the generator of the $U(1)$ subgroup of the Pauli–Gürsey group $SU(2)_{\rm PG}$ (5.51),

$$\delta q^+ = i\alpha q^+, \qquad \delta\tilde{q}^+ = -i\alpha\tilde{q}^+. \quad (5.67)$$

This $U(1)$ commutes both with supersymmetry and $SU(2)_{\rm A}$, as required from a central charge. This trick is an example of the Scherk–Schwarz dimensional reduction method [S3] (see Section 5.3).

Another peculiarity of the massive action is the explicit θ's in it. This, of course, is not consistent with ordinary supersymmetry. However, the action is invariant under $N = 2$ supersymmetry with central charge (5.15):

$$q^{+\prime}(\zeta', x^{5\prime}, u') = q^+(\zeta, x^5, u) \quad \Rightarrow$$
$$\delta q^+ = q^{+\prime}(\zeta', u') - q^+(\zeta, u) = 2m(\theta^+\epsilon^i u_i^- - \bar{\theta}^+\bar{\epsilon}^i u_i^-)q^+(\zeta, u)\,. \quad (5.68)$$

In terms of physical component fields (upon elimination of the auxiliary fields) (5.68) reduces to (5.15).

5.2.5 Invariances of the free hypermultiplet actions

So far we have seen that the free action (5.52) exhibits two $SU(2)$ invariances, the automorphism $SU(2)_{\rm A}$ (5.49) and the Pauli–Gürsey $SU(2)_{\rm PG}$ (5.51). In addition, it is also invariant under the $N = 2$ superconformal group $SU(2,2|2)$. The realization of the latter in harmonic superspace will be given in Chapter 9, here we only indicate the realization of its subgroup $SU(2)_{\rm C}$ on q^+:

$$SU(2)_{\rm C}: \qquad \delta^*_{\rm C} q^+_a = \delta^*_{\rm A} q^+_a - \lambda^{ij} u^-_i u^-_j D^{++} q^+_a\,, \quad (5.69)$$

where $\delta^*_{\rm A} q^+_a$ was defined in (5.49) (the parameters A^{ij} are replaced by λ^{ij}).

It should be pointed out that the transformations (5.69) are defined off shell, so they do not depend on the dynamics of q^+_a. In the free case $SU(2)_{\rm C}$ cannot be distinguished from $SU(2)_{\rm A}$ (5.49) on shell. Take, for example, the transformation of the physical field $f^{ai}(x)$:

$$q^{+a} = f^{ai}(x)u^+_i + f^{a(lmj)}u^+_l u^+_m u^-_j + \cdots\,,$$
$$\delta^*_{\rm C} f^{ai}(x) = -\lambda^{(ik)} f^a_k(x) - \frac{1}{2}\lambda^{(lm)} f^a{}_{(lm}{}^{i)}(x)\,. \quad (5.70)$$

On shell the auxiliary field $f^{a\ \ i)}_{(lm}(x)$ vanishes, so the transformation law (5.70) coincides with the $SU(2)_{\rm A}$ one. However, for a self-interacting q^+_a (see Section 5.3) these $SU(2)$ groups do not in general coincide even on shell (the reason is that $f^{a\ \ i)}_{(lm}(x)$ may become a non-linear function of the physical fields).

Notice that $SU(2)_{\rm C}$ (5.69) commutes with $SU(2)_{\rm PG}$, but not with $SU(2)_{\rm A}$:

$$[\delta^*_{\rm A}, \delta^*_{\rm C}] \sim \delta^*_{\rm C}\,, \qquad [\delta^*_{\rm C}, \delta_{\rm PG}] = 0\,. \tag{5.71}$$

One can combine the variations $\delta^*_{\rm A}$ and $\delta^*_{\rm C}$:

$$(\delta^*_{\rm A} - \delta^*_{\rm C})q^+_a = \lambda^{--}D^{++}q^+_a\,, \tag{5.72}$$

and obtain a new $SU(2)$ group that commutes with $\delta^*_{\rm C}$:*

$$[(\delta^*_{\rm A} - \delta^*_{\rm C}), (\delta^*_{\rm A} - \delta^*_{\rm C})] \sim (\delta^*_{\rm A} - \delta^*_{\rm C})\,, \qquad [(\delta^*_{\rm A} - \delta^*_{\rm C}), \delta^*_{\rm C}] = 0\,. \tag{5.73}$$

It is important to know the variety of $SU(2)$ symmetries of the free q^{+a} action, because they (or their linear combinations) generate the isometry groups of many interesting self-interactions of q^+ (see Section 5.3).

Now we turn our attention to the ω hypermultiplet. The action (5.58) exhibits the same set of $SU(2)$ symmetries as (5.52) does, but their realization is different. Moreover, some of these symmetries now close only modulo the equation of motion for ω. The reason is that we had to use the field equation (5.57) when switching over to the second-order form (5.58). Nevertheless, one is still able to choose proper combinations of those symmetries which are realized from the very beginning in terms of ω only, so they close off shell. For instance, the automorphism and Pauli–Gürsey $SU(2)$ groups (5.49), (5.51) are realized on f^{++} and ω as follows:

$$\begin{aligned} \delta_{\rm A}\omega &= \omega'(\zeta, u') - \omega(\zeta, u) = (A^{+-})\omega + (A^{--})f^{++}\,, \\ \delta_{\rm A} f^{++} &= f^{++\prime}(\zeta, u') - f^{++}(\zeta, u) = -(A^{+-})f^{++} - (A^{++})\omega\,, \end{aligned} \tag{5.74}$$

$$\begin{aligned} \delta_{\rm PG}\omega &= \omega'(\zeta, u) - \omega(\zeta, u) = (\alpha^{+-})\omega + (\alpha^{--})f^{++}\,, \\ \delta_{\rm PG} f^{++} &= f^{++\prime}(\zeta, u) - f^{++}(\zeta, u) = -(\alpha^{+-})f^{++} - (\alpha^{++})\omega\,, \end{aligned} \tag{5.75}$$

with

$$A^{+-} = A^{(ik)}u^+_i u^-_k\,, \qquad A^{\pm\pm} = A^{(ik)}u^\pm_i u^\pm_k\,,$$

etc. This is an equivalent form of the q^+ transformations (5.49), (5.51). One can show that eq. (5.57) is not covariant under either (5.74) or (5.75). For example,

$$\delta_{\rm A}(f^{++} + D^{++}\omega) = (A^{+-})(f^{++} + D^{++}\omega) + (A^{--})D^{++}f^{++}\,. \tag{5.76}$$

---

* This is a manifestation of the following simple fact. Let G_I and G_{II} be two isomorphic groups which form a semi-direct product: $[T^\alpha_I, T^\beta_I] = c^{\alpha\beta\gamma}T^\gamma_I$, $[T^\alpha_{II}, T^\beta_{II}] = c^{\alpha\beta\gamma}T^\gamma_{II}$, $[T^\alpha_I, T^\beta_{II}] = c^{\alpha\beta\gamma}T^\gamma_{II}$. They can be rearranged into a tensor product as follows: $\hat T^\alpha_I = T^\alpha_I - T^\alpha_{II}$, $\hat T^\alpha_{II} = T^\alpha_{II} \;\Rightarrow\; [\hat T_I, \hat T_I] \sim \hat T_I$, $[\hat T_I, \hat T_{II}] = 0$.

To achieve covariance and hence closure of these transformations on ω, one must demand $D^{++} f^{++} = 0$, which is just the equation of motion (5.59). On the other hand, the diagonal in the product $SU(2)_{\rm A} \times SU(2)_{\rm PG}$ (corresponding to the identification $A^{(ab)} = -\alpha^{(ab)}$) transforms ω and f^{++} independently of each other:

$$\hat{\delta}_{\rm A} \equiv \delta_{\rm A} + \delta_{\rm PG}\,, \qquad \hat{\delta}_{\rm A}\omega = \hat{\delta}_{\rm A} f^{++} = 0\,. \tag{5.77}$$

So, it can naturally serve as the automorphism group of the $N = 2$ superalgebra in the ω representation (recall that $SU(2)_{\rm PG}$ commutes with $N = 2$ supersymmetry). Equation (5.57) is clearly covariant with respect to (5.77), so (5.77) is well defined off the ω mass shell.

Another off-shell $SU(2)$ which closes on ω alone is the subgroup $SU(2)_{\rm C}$ of the $N = 2$ superconformal group $SU(2,2|2)$. Its realization on q^+_a was given in (5.69). In terms of ω and f^{++} it becomes

$$\begin{aligned}
\delta^*_{\rm C}\omega &= \lambda^{+-}\omega - \lambda^{--}D^{++}\omega + (\delta^*_{\rm A} + \delta_{\rm PG})\omega \\
&= \lambda^{+-}\omega - \lambda^{--}D^{++}\omega + \lambda^j_i \left(u^+_{(j}\frac{\partial}{\partial u^+_{i)}} + u^-_{(j}\frac{\partial}{\partial u^-_{i)}} \right)\omega\,, \\
\delta^*_{\rm C} f^{++} &= -\lambda^{+-} f^{++} - \lambda^{++}\omega - \lambda^{--}D^{++} f^{++} \\
&\quad + \lambda^j_i \left(u^+_{(j}\frac{\partial}{\partial u^+_{i)}} + u^-_{(j}\frac{\partial}{\partial u^-_{i)}} \right) f^{++}\,.
\end{aligned} \tag{5.78}$$

It is easy to check covariance of eq. (5.57) with respect to this $SU(2)$:

$$\begin{aligned}
\delta^*_{\rm C}(f^{++} + D^{++}\omega) &= -\lambda^{+-}(f^{++} + D^{++}\omega) - \lambda^{--}D^{++}(f^{++} + D^{++}\omega) \\
&\quad + \lambda^j_i \left(u^+_{(j}\frac{\partial}{\partial u^+_{i)}} + u^-_{(j}\frac{\partial}{\partial u^-_{i)}} \right)(f^{++} + D^{++}\omega)\,.
\end{aligned} \tag{5.79}$$

Further, using the fact that the variation $(\delta^*_{\rm A} - \delta^*_{\rm C})$ generates an $SU(2)$ group (recall (5.73)) and commutes with $\delta_{\rm PG}$, one concludes that the variations

$$\hat{\delta}^* = \delta^*_{\rm A} - \delta^*_{\rm C} + \delta_{\rm PG}, \qquad \hat{\delta}^*\omega = \lambda^{--}D^{++}\omega - \lambda^{+-}\omega \tag{5.80}$$

also constitute an $SU(2)$ group which closes off shell. It is convenient to choose the $SU(2)$ groups generated by (5.77) and (5.80) as the basic internal $SU(2)$ symmetries in the ω representation. Note that they form a semi-direct product, $[\hat{\delta}_{\rm A}, \hat{\delta}] \sim \hat{\delta}$. The commuting set is formed by $\hat{\delta}\,,\ \delta_{\rm C} = \hat{\delta}_{\rm A} - \hat{\delta}$, in accord with the footnote before eq. (5.73).

From the discussion above it follows that q^+_a admits three independent off-shell $SU(2)$ symmetries ($SU(2)_{\rm C}$, $SU(2)_{\rm PG}$ and $SU(2)_{\rm A}$), whereas ω admits only two such symmetries, (5.77) and (5.80) (or (5.77) and (5.78)). A third one, e.g., $\delta_{\rm A}$ or $\delta_{\rm PG}$, can also be realized on ω, but in general it closes only on shell.

Finally, we point out one more invariance of the free action (5.52):

$$\delta q_a^+ = c_a^i u_i^+ , \tag{5.81}$$

where c_a^i is a real constant 2×2 matrix.

5.3 Hypermultiplet self-couplings

In this section we give an overview of the hypermultiplet self-interactions. We write down the most general self-coupling for q^+ hypermultiplets.* In terms of component fields it gives rise to a supersymmetric sigma model of hyper-Kähler type. We postpone the proof that the general self-couplings of q^+ indeed yield the most general hyper-Kähler sigma model in the bosonic sector until Chapter 11. Here we give an example to illustrate how, starting from a very simple monomial q^+ self-interaction, one ends up with a non-trivial hyper-Kähler metric. The same example demonstrates how the introduction of central charge gives rise to potential terms in the component field action. We also discuss the symmetries of the general self-interactions, a surprising analogy with the Hamiltonian mechanics of the point particle and give more examples of q^+ self-couplings resulting in some interesting hyper-Kähler metrics.

5.3.1 *General action for q^+ hypermultiplets*

The most general action for n hypermultiplets q_a^+ $(a = 1, \ldots, 2n)$ is

$$S = \frac{1}{\xi^2} \int du \, d\zeta^{(-4)} \left[L_a^+(q^+, u) D^{++} q^{+a} + L^{+4}(q^+, u) \right] . \tag{5.82}$$

This form can be justified in the following way. First of all, it is easy to see that the presence of spinor derivatives leads to higher-order space-time derivatives for the physical fields. Further, the harmonic derivative D^{--} is ruled out because it breaks analyticity (and thus would require extra spinor derivatives). Finally, terms of the type $(D^{++})^n q^+$, $n > 1$, cannot be ruled out on general grounds, but in Chapter 11 we shall see that the most general $N = 2$ supersymmetric sigma model corresponds to the action (5.82).

We emphasize that the action (5.82) is consistent with $N = 2$ supersymmetry. At the same time, it is not, in general, invariant under the automorphism group $SU(2)_A$ (since it may involve explicit harmonic dependence), nor under any of the other symmetries of the free action.

* There is no need to study the ω self-couplings separately. Indeed, one can pass to the first-order form of the ω action (see (5.56)) by substituting $(D^{++}\omega)^2 \to -(f^{++} f^{++} + 2 f^{++} D^{++})\omega$. Then one combines f^{++} and ω into a q^+ (see (5.55)) and thus rewrites the action for ω in terms of q^+.

The physical scalar fields (the lowest-order components in the θ expansion of q^+) are dimensionless, so after elimination of the auxiliary fields they must appear in the action in the sigma model form $\xi^{-2}\int d^4x g(f)(\partial_m f)^2$ (as we shall see in Chapter 10, they generically parametrize a hyper-Kähler manifold and the interaction Lagrangian in (5.82) is the basic geometric object of hyper-Kähler geometry, the hyper-Kähler potential). Thus the action (5.82) gives rise to an $N = 2$ supersymmetric sigma model. The introduction of potential terms is only possible by breaking or modifying $N = 2$ supersymmetry, for instance, by bringing in central charge [A4].

The action (5.82) is written down in terms of the unconstrained superfields q^+. Therefore, the straightforward variation of q^+ produces the equation of motion:

$$D^{++}q^{+a} = \frac{1}{2}L^{ab}(\partial^{++}L^+_b - \partial_{b+}L^{+4}) . \tag{5.83}$$

Here $\partial_{a+} \equiv \partial/\partial q^{+a}$ and the harmonic derivative ∂^{++} acts only on the explicit harmonic argument of $L(q, u)$. Further,

$$L^{ab}L_{bc} = \delta^a_c , \qquad L_{ab} \equiv \partial_{[a+}L^+_{b]} , \tag{5.84}$$

Clearly, one should demand that the matrix L_{ab} be non-degenerate.

5.3.2 An example of a q^+ self-coupling: The Taub–NUT sigma model

Usually, the main technical difficulty in evaluating the physical field action is the elimination of the infinite set of auxiliary fields. Essentially, this amounts to solving non-linear differential equations on S^2. In general, this is a highly non-trivial task. However, in certain cases the symmetry of the problem may substantially help. Here we discuss in detail an example of a q^+ self-interaction [G17], which is invariant under $SU(2)_{\rm A}$, as well as the $U(1)$ subgroup of the Pauli–Gürsey group $SU(2)_{\rm PG}$. The action involves a single q^+ hypermultiplet with a very simple quartic interaction term:

$$S = -\int du\, d\zeta^{(-4)} \left[\tilde{q}^+ D^{++} q^+ + \frac{\lambda}{2}(q^+)^2(\tilde{q}^+)^2\right] . \tag{5.85}$$

The group $U(1)$ is the one we used when introducing mass into the free q^+ action (see eq. (5.67)).

The equation of motion following from (5.85) is

$$D^{++}q^+ + \lambda(q^+\tilde{q}^+)q^+ = 0 \tag{5.86}$$

(and its $\tilde{}$ conjugate). Quite naturally, the $U(1)$ invariance (5.67) yields the conserved Noether current

$$\Lambda^{++} = iq^+\tilde{q}^+ , \qquad D^{++}\Lambda^{++} = 0 . \tag{5.87}$$

Here we are only interested in the bosonic part of the component field action, so we omit the fermion terms in the θ decomposition (5.28). Substituting the boson terms into (5.86), we get the equations of motion in (x, u) space:

$$\partial^{++}F^{+} + \lambda(F^{+}\tilde{F}^{+})F^{+} = 0, \quad (5.88)$$

$$\partial^{++}A^{-}_{m} - 2\partial_{m}F^{+} + 2\lambda(F^{+}\tilde{F}^{+})A^{-}_{m} + \lambda(F^{+})^{2}\tilde{A}^{-}_{m} = 0, \quad (5.89)$$

$$\partial^{++}M^{-} + 2\lambda(F^{+}\tilde{F}^{+})M^{-} + \lambda(F^{+})^{2}\tilde{N}^{-} = 0, \quad (5.90)$$

$$\partial^{++}N^{-} + 2\lambda(F^{+}\tilde{F}^{+})N^{-} + \lambda(F^{+})^{2}\tilde{M}^{-} = 0, \quad (5.91)$$

$$\begin{aligned} &\partial^{++}P^{(-3)} + \partial^{m}A^{-}_{m} \\ &+2\lambda(F^{+}\tilde{F}^{+})P^{(-3)} + \lambda(F^{+})^{2}\tilde{P}^{(-3)} - \frac{\lambda}{2}A^{-m}A^{-}_{m}\tilde{F}^{+} \\ &-\lambda A^{-m}\tilde{A}^{-}_{m}F^{+} + 2\lambda\tilde{F}^{+}M^{-}N^{-} + 2\lambda F^{+}(M^{-}\tilde{M}^{-} + N^{-}\tilde{N}^{-}) = 0. \end{aligned} \quad (5.92)$$

All of these equations, except for (5.92), are kinematical and serve to eliminate the infinite towers of auxiliary fields. The last equation is dynamical, and hence will not be used in what follows.

Now we substitute the decomposition (5.28) into the action (5.85), integrate over $\theta^{+}, \bar{\theta}^{+}$ and use the kinematical equations (5.88)–(5.91). The contributions proportional to M^{-}, N^{-} and $P^{(-3)}$ drop out, and the bosonic action is reduced to

$$S^{\text{bosonic}} = \frac{1}{2}\int du\, d^{4}x\, (\tilde{A}^{-}_{m}\partial^{m}F^{+} - A^{-}_{m}\partial^{m}\tilde{F}^{+}), \quad (5.93)$$

where $F^{+}(x, u)$ and $A^{-}_{m}(x, u)$ obey eqs. (5.88), (5.89).

As mentioned earlier, the $U(1)$ invariance (5.67) greatly simplifies solving eq. (5.88). Indeed, the conservation law (5.87) implies

$$\partial^{++}(F^{+}\tilde{F}^{+}) = 0 \quad \Rightarrow \quad F^{+}(x, u)\tilde{F}^{+}(x, u) = c^{(ij)}(x)u^{+}_{i}u^{+}_{j}. \quad (5.94)$$

From (3.104) it follows $\widetilde{(F^{+}\tilde{F}^{+})} = -F^{+}\tilde{F}^{+}$, hence

$$\overline{c^{(ij)}} = -\epsilon_{il}\epsilon_{jn}c^{(ln)}. \quad (5.95)$$

This suggests the following change of variables:

$$\begin{aligned} F^{+}(x, u) &= f^{+}(x, u)\exp(-\lambda c^{+-}), \\ c^{+-}(x, u) &\equiv c^{(ij)}(x)u^{+}_{i}u^{-}_{j} = -\widetilde{c^{+-}}(x, u), \end{aligned} \quad (5.96)$$

which reduces (5.88) to the linear homogeneous equation

$$\partial^{++}f^{+}(x, u) = 0 \quad \Rightarrow \quad f^{+}(x, u) = f^{i}(x)u^{+}_{i}. \quad (5.97)$$

Taking into account that

$$F^{+}\tilde{F}^{+} = f^{+}\tilde{f}^{+} \quad \Rightarrow \quad c^{(ij)}(x) = -f^{(i}(x)\bar{f}^{j)}(x), \quad (5.98)$$

where $\bar{f}^i = \epsilon^{ij}\bar{f}_j$, $\bar{f}_j \equiv \overline{(f^j)}$, we obtain the general solution of (5.88) in the form

$$F^+(x,u) = f^i(x)u_i^+ \exp\left(\lambda f^{(j}(x)\bar{f}^{k)}(x)u_j^+u_k^-\right). \tag{5.99}$$

So, all the components in the harmonic expansion of $F^+(x,u)$ are expressed in terms of $f^i(x)$. The latter are the physical bosonic fields. At the same time, they are the four integration constants of the first-order harmonic differential equation (5.88).

The remaining equation (5.89)* is simplified by the substitution

$$A_m^-(x,u) = B_m^-(x,u)e^{-\lambda c^{+-}}.$$

Equation (5.89) implies that the harmonic expansion of B_m^- contains only linear (u^-) and trilinear ($u^-u^-u^+$) terms. Finally,

$$\begin{aligned} A_m^- &= e^{-\lambda c^{+-}}\Bigg[2\partial_m f^i u_i^- + \frac{\lambda f^i u_i^-}{1+\lambda f\bar{f}}(f^j\partial_m\bar{f}_j - \bar{f}_j\partial_m f^j) \\ &\qquad + 2\lambda f^i u_i^+\partial_m\left(f^{(k}\bar{f}^{j)}u_k^-u_j^-\right)\Bigg]. \end{aligned} \tag{5.100}$$

Let us emphasize once again that this simple form of the solution is due to the $U(1)\times SU(2)_A$ invariance of the action (5.85).

Finally, we insert (5.99) and (5.100) into the action (5.93) and perform the harmonic integral using the reduction identities (4.15) and integration rules (4.11),(4.12) for harmonics. The result is the following sigma model action:

$$S^{\text{bosonic}} = \int du\, d^4x\, \left(h^i_j\partial_m f^j\partial^m\bar{f}_i + g_{ij}\partial_m f^i\partial^m f^j + \bar{g}^{ij}\partial_m\bar{f}_i\partial^m\bar{f}_j\right), \tag{5.101}$$

where

$$\begin{aligned} h^i_j &= \delta^i_j(1+\lambda f\bar{f}) - \frac{\lambda(2+\lambda f\bar{f})}{2(1+\lambda f\bar{f})}f^i\bar{f}_j, \qquad f\bar{f} \equiv f^i\bar{f}_i, \\ g_{ij} &= \frac{\lambda(2+\lambda f\bar{f})}{4(1+\lambda f\bar{f})}\bar{f}_i\bar{f}_j, \qquad \bar{g}^{ij} = \frac{\lambda(2+\lambda f\bar{f})}{4(1+\lambda f\bar{f})}f^i f^j. \end{aligned} \tag{5.102}$$

It is remarkable that the simple monomial $N=2$ hypermultiplet interaction (5.85) entails a complicated non-polynomial Lagrangian for the physical bosons. Note the manifest $U(2)$ invariance of (5.101) and (5.102), which reflects the $U(2)$ invariance of the original action.

---

* In fact, the solution of this equation can be derived from the solution of (5.88), as explained in Chapter 11.

The metric (5.102) turns out to be one of the well-known hyper-Kähler metrics, the so-called Taub–NUT metric [E6]. The actual identification requires a suitable change of coordinates:

$$\begin{aligned} f^1 &= \rho \cos \frac{\theta}{2} \exp \frac{i}{2}(\psi + \phi) \,, \\ f^2 &= \rho \sin \frac{\theta}{2} \exp \frac{i}{2}(\psi - \phi) \,, \qquad f\bar{f} = \rho^2 \,. \end{aligned} \tag{5.103}$$

Assuming that $\lambda > 0$, we substitute

$$\rho^2 = 2(r - \mu)\mu \,, \qquad r \geq \mu = \frac{1}{2\sqrt{\lambda}} \,. \tag{5.104}$$

All this results in the following form of the invariant interval:

$$\begin{aligned} ds^2 &= \frac{1}{2}\frac{r + \mu}{r - \mu} dr^2 + \frac{1}{2}(r^2 - \mu^2)(d\theta^2 + \sin^2\theta \, d\phi^2) \\ &\quad + 2\mu^2 \frac{r - \mu}{r + \mu}(d\psi + \cos\theta \, d\phi)^2 \,, \end{aligned} \tag{5.105}$$

which is the standard Taub–NUT metric [E6, H4].

For future use, it is worthwhile to give one more form of this metric demonstrating that it belongs to the multicenter class of four-dimensional hyper-Kähler metrics [G32]. After passing to the new variables

$$\begin{aligned} x^1 &= \hat{r} \sin\theta \cos\phi \,, \qquad & x^2 &= \hat{r} \sin\theta \sin\phi \,, \qquad & x^3 &= \hat{r}\cos\theta \,, \\ \tau &= \tfrac{1}{2}(\psi + \phi) \,, & \hat{r} &\equiv \rho^2 , \end{aligned} \tag{5.106}$$

the distance (5.105) can be rewritten as

$$ds^2 = V_{\text{TN}}^{-1}(\hat{r}) \left(d\tau + \vec{V} \cdot d\vec{x} \right)^2 + V_{\text{TN}}(\hat{r}) \, d\vec{x} \cdot d\vec{x} \,, \tag{5.107}$$

where

$$V_{\text{TN}}(\hat{r}) = \frac{1}{2}\left(\frac{1}{\hat{r}} + \lambda \right) , \tag{5.108}$$

$$\begin{aligned} V^1 &= -\frac{1}{2\hat{r}} \frac{\cos\theta - 1}{\sin\theta} \sin\phi \,, \\ V^2 &= \frac{1}{2\hat{r}} \frac{\cos\theta - 1}{\sin\theta} \cos\phi \,, \\ V^3 &= 0 \end{aligned} \tag{5.109}$$

are a particular solution of the general multicenter class conditions [G32]

$$\Delta V = 0 \,, \qquad \vec{\nabla} V = \vec{\nabla} \wedge \vec{V} \,, \qquad (\Delta \equiv \vec{\nabla} \cdot \vec{\nabla}) \,. \tag{5.110}$$

We use the example of the Taub–NUT sigma model to explain how one can introduce a potential term into an $N = 2$ sigma model. The idea is the same as for the mass term in the free case (see Section 5.2.4). We assume that the superfield q^+ also depends on the central charge coordinate x^5 as shown in (5.64). This dependence has the form of a $U(1)$ transformation, and our action (5.85) is $U(1)$ invariant. Then the only modification to (5.85) will be a mass-like term, as in (5.66). This yields a modification of the component-field equations (5.88)–(5.92). The net result is the following potential term added to the sigma model action (5.101):

$$\begin{aligned} S^{\text{potential}} &= -m^2 \int du\, d^4x \left(h^i_j f^j \bar{f}_i - g_{ij} f^i f^j - \bar{g}^{ij} \bar{f}_i \bar{f}_j\right) \\ &= -m^2 \int du\, d^4x \, \frac{f\bar{f}}{1+\lambda f\bar{f}} . \end{aligned} \tag{5.111}$$

Its origin is easy to trace back to the derivative terms in (5.101), assuming that they involve ∂_5 as well and that the fields $f, \bar{f}$ depend on x^5 as indicated in (5.64). The mechanism illustrated here can be generalized to making use of any number of $U(1)$ isometries of an $N = 2$ sigma model action for building up potential terms (see Section 5.3.3).

5.3.3 Symmetries of the general hypermultiplet action

Like the $N = 1$ matter action (5.2), the general q^+ action (5.82) is invariant under reparametrizations of the manifold $\{q^{+a}\}$:

$$\delta q^{+a} = \lambda^{+a}(q^+, u), \tag{5.112}$$

provided L^+_a and L^{+4} transform as follows:

$$\delta L^+_a = -\partial_{a+}\lambda^{+b} L^+_b, \qquad \delta L^{+4} = -\partial^{++}\lambda^{+a} L^+_a \tag{5.113}$$

(here the partial derivative ∂^{++} acts on the explicit harmonics in $\lambda^{+a}(q^+, u)$). Notice that not only the action, but even the Lagrangian in (5.82) is invariant under the transformations (5.112), (5.113).

Using this freedom, one can always fix a gauge, in which L^+_a equals its flat limit:

$$\left(L^+_a\right)_{\text{fixedgauge}} = q^+_a, \qquad L_{ab} = -\Omega_{ab} \tag{5.114}$$

(L_{ab} was defined in (5.84)). Then the action (5.82) is simplified:

$$S = \frac{1}{\xi^2} \int du\, d\zeta^{(-4)} \left[q^+_a D^{++} q^{+a} + L^{+4}(q^+, u)\right], \tag{5.115}$$

although it loses its reparametrization covariance.

The general action (5.82) possesses another invariance [B2], which is the analog of Kähler invariance (5.8). This time the target space coordinates q^+ do not transform, while the Lagrangian is changed by a total harmonic derivative:

$$\delta q^+ = 0\,, \qquad \delta L^+_a = \partial_{a+}\lambda^{++}\,, \qquad \delta L^{+4} = \partial^{++}\lambda^{++} \tag{5.116}$$

$$\Rightarrow \quad \delta(L^+_a D^{++} q^{+a} + L^{+4}) = D^{++}\lambda^{++}\,. \tag{5.117}$$

Here $\lambda^{++}(q^+, u)$ is an arbitrary real function of q^+ and $u^\pm$.

The above transformations leave the action, considered as a functional of q^+, L^+ and L^{+4}, invariant in the following sense:

$$S'[q', L^{+'}, L^{+4'}] = S[q, L^+, L^{+4}]\,. \tag{5.118}$$

Strictly speaking, these are not symmetries, since in general they alter the form of the action (i.e., L^{+a} and L^{+4} change). One may wonder whether there exists a subset of the transformations (5.112), (5.113) and (5.116), such that L^{+a} and L^{+4} do not change under it. Such symmetries

$$\delta q^{+a} = \epsilon\, \lambda^{+a}(q, u)\,, \tag{5.119}$$

ϵ being a group parameter and $\lambda^{+a}(q,u)$ a *Killing vector*, are called *isometries*, because they give rise to isometries of the corresponding bosonic sigma model metrics. The conditions for the existence of isometries are

$$\begin{aligned} \delta^* L^+_a &\equiv L^{+'}_a(q,u) - L^+_a(q,u) \\ &= -\epsilon\left(\lambda^{+b}\partial_{b+}L^+_a + \partial_{a+}\lambda^{+b}L^+_b - \partial_{a+}\lambda^{++}\right) = 0\,, \end{aligned} \tag{5.120}$$

$$\delta^* L^{+4} = -\epsilon\left(\lambda^{+a}\partial_{a+}L^{+4} + L^+_a\partial^{++}\lambda^{+a} - \partial^{++}\lambda^{++}\right) = 0\,. \tag{5.121}$$

Introducing the notation

$$\Lambda^{++} = \lambda^{++} - L^+_a\lambda^{+a}\,, \tag{5.122}$$

and using (5.84), one can solve (5.120) for λ^{+a}:

$$\lambda^{+a} = -\frac{1}{2}L^{ab}\partial_{b+}\Lambda^{++}\,. \tag{5.123}$$

We see that the Killing vector λ^{+a} (5.119) is a secondary object, it is expressed in terms of the Killing potential Λ^{++}.* The latter must satisfy the equation [B2]

$$\partial^{++}\Lambda^{++} = \frac{1}{2}L^{ab}\partial_{b+}\Lambda^{++}(\partial^{++}L^+_a - \partial_{a+}L^{+4})\,, \tag{5.124}$$

following from (5.121)–(5.123). Note the important property

$$\delta^*\Lambda^{++} = -\epsilon\,\lambda^{+a}\partial_{a+}\Lambda^{++} = 0 \tag{5.125}$$

* The same phenomenon takes place in the case of $N = 1$ Kähler sigma models [B3, H23].

which follows from eq. (5.123). We have already seen an example of a Killing potential in the Taub–NUT case (Section 5.3.2), it is given by (5.87) and corresponds to $U(1)_{\rm PG}$ symmetry as the isometry (eq. (5.124) is reduced to $\partial^{++}\Lambda^{++} = 0$ in this case).

Remarkably enough, in the generic case, similarly to the Taub–NUT example, the Killing potential turns out to be the conserved Noether current associated with the isometry generated by the Killing vector (5.119). To see this, one performs a local variation

$$\delta q^{+a} = \epsilon(\zeta, u)\lambda^{+a}(q, u) \tag{5.126}$$

in the action (5.82). Using (5.120)–(5.122), one obtains

$$\delta S = -\int du\, d\zeta^{(-4)}\Lambda^{++}D^{++}\epsilon\,. \tag{5.127}$$

Since on shell $\delta S = 0$, one gets the conservation law (cf. (5.87))

$$D^{++}\Lambda^{++} = 0\,. \tag{5.128}$$

It can be directly verified by employing the equation of motion (5.83) in the Killing equation (5.124).

It should be pointed out that the whole discussion of this section refers to the so-called triholomorphic isometries, i.e., those commuting with supersymmetry.* Just for them the Killing vectors have the form (5.119), i.e., bear dependence only on q^{+a} and explicit harmonics, and not on harmonic derivatives of q^{+a}. In all the examples we are considering (see also Section 5.3.5) the isometries of the corresponding hyper-Kähler metrics are represented, in the q^+ language, by the Killing vectors of this type.

We conclude this subsection with a generalization of the mechanism for introduction of potential terms, illustrated by the example of the Taub–NUT sigma model in Section 5.3.2. Suppose that we are given a q^+ action possessing an isometry. Then we assign to q^+ a dependence on x^5, such that $\partial/\partial x^5$ can be identified with the Killing vector of the isometry:

$$\frac{\partial}{\partial x^5}q^{+a} = m\lambda^{+a}(q, u)\,. \tag{5.129}$$

Substituting $q^+(\zeta, x^5, u)$ and (5.65) into the action (5.82), one obtains

$$S = \frac{1}{\xi^2}\int du\, d\zeta^{(-4)}\left[L^+_a D^{++}q^{+a} + L^{+4} + im[(\theta^+)^2 - (\bar\theta^+)^2]L^+_a\lambda^{+a}\right]. \tag{5.130}$$

Although the integral in (5.130) goes over d^4x, the action does not depend on x^5, because of the isometry. The extra term in (5.130) is easily checked to be invariant under the isometry transformations as a consequence of eqs. (5.120), (5.123) and (5.125).

* Sometimes such isometries are called ‘translational’.

5.3.4 Analogy with Hamiltonian mechanics

An intriguing analogy exists between the harmonic superspace description of q^+ hypermultiplets and Hamiltonian mechanics of a point particle [G25].

In the symplectic notation (see, e.g., [D27]), the action of a particle moving in n-dimensional space reads

$$S_{\text{particle}} = \int dt \left[\frac{1}{2} q_a \frac{d}{dt} q^a - H(q,t) \right].. \tag{5.131}$$

Here

$$q_a = \begin{pmatrix} p_k \\ x^k \end{pmatrix}, \qquad q^a = \Omega^{ab} q_b = \begin{pmatrix} -x^k \\ p_k \end{pmatrix}, \tag{5.132}$$

$x^k,\ p_k\ (k = 1, 2, \dots n)$ are the coordinates of the particle and their canonically conjugated momenta, $H(q,t)$ is the Hamiltonian and Ω^{ab} is an antisymmetric $Sp(n)$ metric of the same kind as in (5.53), (5.114).

Comparing (5.131) with the general q^+ action (5.82) in the gauge (5.114), one observes that the phase space coordinates $q_a(t)$ are analogous to the hypermultiplet superfields $q_a^+(\zeta, u)$, the Hamiltonian $H(q,t)$ to the hyper-Kähler potential $L^{+4}(q^+, u)$ and the time dependence in the classical mechanics is similar to the harmonic dependence in the theory of q^+ hypermultiplets.

This surprising analogy spreads even further. Like in standard Hamiltonian mechanics, one can define the Poisson brackets (this time carrying harmonic $U(1)$ charge -2)

$$\{F(q^+, u), G(q^+, u)\}^{--} = \frac{1}{2} \Omega^{ab} \partial_{a+} F \partial_{b+} G\,. \tag{5.133}$$

Then, using these brackets, one can rewrite the q^+ equation of motion (5.83), the conservation law (5.128), etc., in the following suggestive Hamiltonian-like way (in the gauge (5.114))

$$\begin{aligned} D^{++} q^{+a} &= \{L^{+4}, q^{+a}\}^{--}\,, \\ \partial^{++} \Lambda^{++} + \{L^{+4}, \Lambda^{++}\}^{--} &= 0\,. \end{aligned} \tag{5.134}$$

The last equation can be compared with the conservation law for the particle

$$\frac{\partial}{\partial t} f + \{H, f\} = 0\,.$$

Finally, the invariances of the general q^+ action discussed in the preceding subsection also have their counterparts in Hamiltonian mechanics. For instance, the subgroup of (5.112), (5.113), (5.116), (5.117) which preserves the gauge (5.114) is analogous to the so-called group of canonical transformations:

$$\delta q^a = \Omega^{ab} \partial_b \Lambda(q,t)\,, \qquad \delta H = H'(q',t) - H(q,t) = -\frac{\partial}{\partial t} \Lambda(q,t)\,. \tag{5.135}$$

The possible implications of this unexpected analogy are as yet unclear to us. The apparently strange similarity between the one-dimensional evolution parameter t in the Hamiltonian mechanics and the two-dimensional harmonic sphere S^2 in the theory of q^+ hypermultiplets can perhaps be justified by the following reasoning. Though the harmonics u^+, u^- and the derivatives $\partial^{++}, \partial^{--}$ are not real with respect to the ordinary complex conjugation which takes them into each other, they are separately real with respect to the involution $\widetilde{\ }$ and in this sense are one-dimensional.

5.3.5 *More examples of q^+ self-couplings: The Eguchi–Hanson sigma model and all that*

In Section 5.3.2 we presented the simplest (quartic) self-interaction of one q^+ which gives rise to the well-known Taub–NUT metric. Here we consider a few other examples of q^+ actions leading to interesting hyper-Kähler metrics.

We start with the Eguchi–Hanson (EH) metric [E7, E8]. The corresponding hyper-Kähler manifold is also four-dimensional. It appeared as the bosonic target manifold in the first example of the component $N = 2$ sigma model constructed by Curtright and Freedman [C13].

The EH metric, like the Taub–NUT metric, is known to possess a $U(2)$ isometry. However, as opposed to the Taub–NUT case, it is the $SU(2)$ part of the EH isometry group which is triholomorphic (i.e., commuting with supersymmetry). So, in the corresponding q^+ action the $SU(2)_{\rm A}$ symmetry should be broken down to $U(1)_{\rm A}$, while yet another $SU(2)$ commuting with supersymmetry should remain unbroken. Such an action is easy to construct [G22]. It looks most transparent in the ω representation:

$$S_{\rm EH} = -\frac{1}{4\gamma^2}\int d\zeta^{(-4)}\left[(D^{++}\omega)^2 - \frac{(\xi^{++})^2}{\omega^2}\right], \qquad \xi^{++} = \xi^{ik}u_i^+u_k^+ = \widetilde{\xi^{++}}\,. \tag{5.136}$$

It is straightforward to check that the 'potential' in (5.136) is shifted by a full harmonic derivative under the $\hat{SU}(2)$ group transformations (5.80). Moreover, it is the *unique* ω self-interaction with this invariance. The constant vector ξ^{ik} in (5.136) breaks $SU(2)_{\rm A}$ defined by eq. (5.77) to just $U(1)_{\rm A}$. Thus the full invariance group of (5.136) is $U(2) = \hat{SU}(2)\times U(1)_{\rm C}$ where $U(1)_{\rm C}$ is generated by the difference of the $U(1)_{\rm A}$ generator and the third generator of $\hat{SU}(2)$.

Using the q^+–ω correspondence (5.55), we can rewrite the action (5.136) in terms of q^{+a}:

$$S_{\rm EH} = \frac{1}{4\gamma^2}\int d\zeta^{(-4)}\left[q_a^+D^{++}q^{+a} + \frac{(\xi^{++})^2}{(u^{-a}q_a^+)^2}\right]. \tag{5.137}$$

The $\hat{SU}(2)$ symmetry of (5.137) is realized by the transformations (recall eqs.

(5.73), (5.51), (5.72))

$$\hat{\delta} q^+_a = \lambda^{--} D^{++} q^+_a - \lambda^b_a q^+_b \,. \tag{5.138}$$

Modulo a 'trivial' $SU(2)$ variation vanishing on shell, (5.138) can be rewritten as

$$\hat{\delta} q^+_a = -\lambda^b_a q^+_b + \lambda^{--} u^-_a \frac{(\xi^{++})^2}{(u^- q^+)^3} \,. \tag{5.139}$$

It is easy to check that these transformations indeed generate $SU(2)$ and leave (5.137) invariant. They are more convenient to deal with because they contain no harmonic derivatives on q^+_a and so fall into the general class of the triholomorphic isometries (5.119). The corresponding Killing potential reads

$$\Lambda^{++}_{(ab)} = -q^+_a q^+_b - u^-_a u^-_b \frac{(\xi^{++})^2}{(u^- q^+)^2} \,. \tag{5.140}$$

To prove that the above actions indeed yield the EH sigma model in the bosonic sector, one could apply the same techniques as in the Taub–NUT example. However, in the present case it is easier to pass to an equivalent representation of these actions by making use of the $N = 2$ version of the so-called quotient construction. It was first applied at the component level in [C13] and then worked out in the general case in terms of $N = 1$ superfields in [H8, L1]. The basic idea is to gauge some isometries of the flat hyper-Kähler manifold of appropriate dimension with the help of a non-propagating gauge multiplet. After a proper gauge-fixing, the remaining components of this multiplet are either eliminated by their algebraic equations of motion or become Lagrange multipliers for the constraints on the original bosonic fields. As a result, a non-trivial hyper-Kähler sigma model action arises for the remaining independent fields.

Following [G22], we apply this method to the sum of two free q^+ actions:

$$S_{\text{free}} = -\frac{1}{2\gamma^2} \int du\, d\zeta^{(-4)}\, \tilde{q}^+_A D^{++} q^+_A \,, \qquad A = 1, 2 \,.$$

Along with the general invariances of the free q^+ actions (see Section 5.2.5), this action is invariant under global $O(2)$ transformations with a real parameter λ:

$$\delta q^+_A = \lambda\, \epsilon_{AB} q^+_B \,, \qquad \epsilon_{12} = -\epsilon_{21} = 1 \,. \tag{5.141}$$

Let us gauge this $O(2)$ symmetry by promoting λ to an arbitrary analytic function $\lambda(\zeta, u)$, $\tilde{\lambda} = \lambda$, and introducing an Abelian analytic gauge superfield $V^{++}(\zeta, u)$ (see Chapter 7):

$$S_{\text{gauged}} = -\frac{1}{2\gamma^2} \int du\, d\zeta^{(-4)} \left[\tilde{q}^+_A D^{++} q^+_A + V^{++}(\epsilon_{AB} \tilde{q}^+_A q^+_B + \xi^{++})\right], \tag{5.142}$$

$$\delta q^+_A = \lambda\, \epsilon_{AB} q^+_B \,, \qquad \delta V^{++} = -D^{++} \lambda \,. \tag{5.143}$$

We have also added a 'Fayet–Iliopoulos' term proportional to $\xi^{++} = \xi^{ik}u_i^+u_k^+$ which is gauge-invariant on its own owing to the property $D^{++}\xi^{++} = 0$. The action (5.142) also respects, besides the $O(2)$ gauge invariance, the global automorphism $U(1)_{\rm A}$, the $SU(2)_{\rm PG}$ (for each q^+ separately) and the $\hat{SU}(2)$ symmetries (5.138):

$$\hat{\delta} q^+_{a\,A} = \lambda^{--} D^{++} q^+_{a\,A} - \lambda_a^b\, q^+_{b\,A}\,, \qquad \hat{\delta} V^{++} = D^{++}(\lambda^{--} V^{++})\,, \tag{5.144}$$

where we used the doublet notation $q_a^+ = (q^+, -\tilde{q}^+)$ (see eq. (5.50)).

Switching to the ω, f^{++} representation of the action (5.142) according to the general relations (5.55) and eliminating the superfields f_A^{++} by their algebraic equations of motion, we obtain the ω-form of (5.142) as follows:

$$S_{\rm gauged} = -\frac{1}{4\gamma^2} \int du\, d\zeta^{(-4)} \left[\mathcal{D}^{++}\omega \mathcal{D}^{++}\bar{\omega} + 2\,\xi^{++} V^{++}\right], \tag{5.145}$$

where

$$\omega = \omega_1 + i\,\omega_2\,, \qquad \mathcal{D}^{++}\omega = D^{++}\omega + i\,V^{++}\omega\,, \qquad \delta\omega = -i\,\lambda\,\omega\,. \tag{5.146}$$

Further, fixing the $O(2)$ gauge freedom by the gauge condition

$$\omega = \bar{\omega} \tag{5.147}$$

(ω is assumed to have a non-vanishing constant part) and eliminating V^{++} by its algebraic equation of motion

$$V^{++} = -\frac{\xi^{++}}{\omega^2}\,, \tag{5.148}$$

we arrive at the action (5.136). Thus we have demonstrated that (5.136) amounts to choosing a special gauge in the q^+ action (5.142) and then rewriting it in the ω-form.

Now it is easy to prove that the physical boson target metric in (5.136) (or in (5.137)) is just the EH metric, without explicitly evaluating it. Let us make another gauge choice in the 'parent' action (5.142), namely the Wess–Zumino gauge (see Chapter 7). In it V^{++}, with all fermion fields omitted, is reduced to

$$V^{++}_{\rm WZ} = (\theta^+)^2 M(x) + (\bar{\theta}^+)^2 \bar{M}(x) + i\theta^+\sigma^m\bar{\theta}^+ V_m(x) + (\theta^+)^2(\bar{\theta}^+)^2 D^{ik}(x) u_i^- u_k^-\,. \tag{5.149}$$

Substituting this gauge-fixed V^{++} into the action (5.142), one easily finds that:

(i) The purely harmonic part of the equation of motion for q^+_{aA} turns out to be the same as in the free case,

$$\partial^{++} F^+_{a\,B}(x,u) = 0 \quad \Rightarrow \quad F^+_{a\,B}(x,u) = f^i{}_{a\,B}(x) u_i^+\,. \tag{5.150}$$

(ii) The terms with the auxiliary fields $M(x), \bar{M}(x)$ drop out.

(iii) The field $V_m(x)$ is expressed through $f^i{}_{aB}(x)$ as the solution to an algebraic equation of motion.

(iv) The field $D^{ik}(x)$ appears as the Lagrange multiplier for a non-linear constraint on the physical fields $f^i{}_{aB}(x)$.

The final form of the bosonic action is as follows:

$$S^{\text{bosonic}}_{\text{EH}} = \frac{1}{2\gamma^2} \int d^4x \left[\nabla^m f^{ia}{}_A \nabla_m f_{ia\,A} + \frac{1}{6} D_{ij} (\epsilon_{AB} f^{ia}{}_A f^j{}_{aB} + 2\xi^{ij}) \right], \tag{5.151}$$

where

$$\nabla^m f^{ia}{}_A = \partial^m f^{ia}{}_A - \epsilon_{AB} V^m f^{ia}{}_B\,, \qquad V^m = \frac{\epsilon_{AB} f^{ia}{}_A \partial^m f_{ia\,B}}{f^{ia}{}_C f_{ia\,C}}\,. \tag{5.152}$$

The action (5.151) is just the bosonic part of the action of ref. [C13]. There it was proved that the target metric for the four independent bosonic fields is just the EH metric.* Note that the group $\hat{SU}(2)$ is reduced to $SU(2)_{\text{PG}}$ on the fields f^{ia}, as follows from the transformation law (5.144) and eqs. (5.150) which eliminate an infinite set of bosonic auxiliary fields. Thus the invariance (isometry) group of the resulting EH bosonic metric is $SU(2)_{\text{PG}} \times U(1)_{\text{A}}$.

In [G22] some more sophisticated examples of the quotient construction in harmonic superspace were also considered. They were shown to lead to important classes of hyper-Kähler metrics appearing in the bosonic sector of the q^+ actions, in particular, Eguchi–Hanson multi-instantons [G32, G33] and $4n$-dimensional Calabi metrics [C1]. The discussion of these models is outside the scope of the present book. Instead, we illustrate the quotient method on another simple example, namely, we show how the Taub–NUT q^+ action can be reproduced in this way.

In this case one again starts from the sum of two free actions but one then gauges a different global $U(1)$ symmetry, as compared to the EH case. It is convenient to take one of these actions in the q^+ form and the second one in the ω form:

$$S_{\text{free}} = -\int du\, d\zeta^{(-4)}\, \tilde{q}^+ D^{++} q^+ - \frac{1}{2} \int du\, d\zeta^{(-4)}\, (D^{++}\omega)^2. \tag{5.153}$$

Let us consider the following particular $U(1)$ symmetry:

$$\delta q^+ = ig\alpha\, q^+\,, \qquad \delta\tilde{q}^+ = -ig\alpha\, \tilde{q}^+\,, \qquad \delta\omega = \alpha\,, \tag{5.154}$$

* To see that the dimension of the physical boson manifold equals four, one should take into account that the constraint in (5.151) eliminates three out of the original eight bosonic degrees of freedom, while one more degree of freedom is gauged away by the $O(2)$ gauge invariance which is still respected by (5.151).

where α is the group parameter and g is a $U(1)$ 'charge'. Gauging this $U(1)$, $\alpha \to \alpha(\zeta, u)$, amounts to the following modification of (5.153):

$$S_{\text{gauged}} = - \int du\, d\zeta^{(-4)}\, \tilde{q}^{+}(D^{++} + igV^{++})q^{+} - \frac{1}{2} \int du\, d\zeta^{(-4)}\, (D^{++}\omega + V^{++})^{2}, \tag{5.155}$$

where V^{++} undergoes the gauge transformation $\delta V^{++} = -D^{++}\alpha$. Choosing the gauge

$$\omega = 0, \tag{5.156}$$

and eliminating in this gauge V^{++} by its algebraic equation of motion

$$V^{++} = -ig\, \tilde{q}^{+} q^{+}, \tag{5.157}$$

we arrive at the familiar Taub–NUT action (5.85):

$$S_{\text{TN}} = - \int du\, d\zeta^{(-4)} \left[\tilde{q}^{+} D^{++} q^{+} + \frac{\lambda}{2} (\tilde{q}^{+} q^{+})^{2} \right], \qquad \lambda \equiv g^{2}. \tag{5.158}$$

We leave it to the reader to make an alternative calculation of the bosonic Taub–NUT metric by choosing the Wess–Zumino gauge (5.149) in the corresponding 'parent' action (5.155) and to see how this construction works while using the q^{+} representation of the second term in (5.153) (in this case the shift of ω in (5.154) is realized as the q^{+} shift (5.81)). Note that one can also add a Fayet–Iliopoulos term to (5.155), but it may be absorbed into a field redefinition.

It seems that any hyper-Kähler metrics which can be obtained within the quotient construction (e.g., those derived on a purely geometric ground in the recent paper [G36]) can be alternatively reproduced from the q^{+} actions by applying the above $N = 2$ supersymmetric quotient construction to the adequate number of free q^{+} actions with the appropriate rigid symmetries gauged.

In refs. [G34, O10] some direct generalizations of the Taub–NUT action (5.85) with unbroken $U(1)_{\text{PG}}$ invariance were considered. It turned out that this invariance, just like in the Taub–NUT case, makes it possible to solve the corresponding harmonic differential equations in a comparatively simple way. The simplest modification of (5.85) is to add the so-called 'dipolar' term to the Taub–NUT interaction [O10]:

$$L^{+4}_{\text{TN}} \Rightarrow \frac{\lambda}{2} (\tilde{q}^{+} q^{+})^{2} + \frac{1}{3} \xi^{--} (\tilde{q}^{+} q^{+})^{3}, \tag{5.159}$$

with $\xi^{--} = \xi^{ik} u_i^{-} u_k^{-} = -\widetilde{(\xi^{--})}$. It was shown in [O10] that the corresponding bosonic metric is again of the multicenter form (5.107), with the potential V being a 'dipolar perturbation' of the one-center, i.e., Taub–NUT potential V_{TN} (5.108):

$$V(\hat{r}, \theta) = \frac{1}{2\hat{r}} + \frac{\lambda}{2} - \frac{1}{3} \xi^{12} \hat{r} \cos\theta,$$

where the $SU(2)_A$ frame has been fixed so as to leave only one component in the $SU(2)_A$ breaking vector ξ^{ik}. Note that the full invariance group of (5.159) and hence the isometry of the bosonic metric is $U(1)_A \times U(1)_{PG}$.

As a generalization of this observation, in [G34] the q^+ self-interaction leading to the most general four-dimensional multicenter hyper-Kähler metric was constructed:

$$L^{+4}_{\text{multicenter}} = \sum_{l=0}^{\infty} \frac{1}{l+2} \xi^{-2l} (\tilde{q}^+ q^+)^{l+2}, \qquad \xi^{-2l} = \xi^{i_1 \ldots i_{2l}} u^-_{i_1} \ldots u^-_{i_{2l}} . \tag{5.160}$$

In the bosonic sector arises the metric which has the generic form (5.107), with the potential $V(\hat{r}, \theta, \phi)$ being the *most general* solution to the Laplace equation in (5.110). The coefficients ξ in (5.160) are in one-to-one correspondence with the coefficients in the expansion of this V in terms of spherical harmonics. Once again, the $U(1)_{PG}$ invariance of (5.160) plays a crucial rôle in achieving this result. As for $SU(2)_A$, it is completely broken by the self-interaction (5.160), so the only isometry of the metric in the general case is $U(1)_{PG}$. This property is not accidental: In [G35, G34, H7] it has been proved that any four-dimensional hyper-Kähler metric with a triholomorphic $U(1)$ isometry falls into the multicenter class (or into its appropriate generalization [H8, P1] in the case of $4n$-dimensional manifolds with n-commuting isometries, the so-called 'toric' hyper-Kähler manifolds [G36]). This theorem supplies a powerful criterion in searching for new non-trivial hyper-Kähler metrics in the harmonic superspace approach. Indeed, in order to find a q^+ action which cannot be reduced to the multicenter form (5.160), one should fully break $SU(2)_{PG}$ symmetry in such a candidate action. The simplest example of this sort is provided by the self-interaction

$$L^{+4} = \lambda (q^+)^4 + \bar{\lambda} (\tilde{q}^+)^4 , \tag{5.161}$$

in which $SU(2)_{PG}$ is fully broken.* On the other hand, it preserves $SU(2)_A$ which does not commute with supersymmetry and so corresponds to non-triholomorphic isometries of the bosonic metric. The only known four-dimensional hyper-Kähler metric with such an $SU(2)$ isometry group is the Atiyah–Hitchin metric [A8]. So, the Lagrangian (5.161) is expected to yield this metric. No explicit proof of this conjecture has so far been given, as the relevant harmonic equation is very difficult to solve because of the lack of $U(1)_{PG}$ invariance which greatly helped in the multicenter case.

Another merit of the above theorem in the context of the harmonic approach is the possibility to identify any $U(1)_{PG}$ invariant q^+ (or ω) Lagrangian, even one not having the form (5.160), as belonging to the multicenter class and hence admitting an equivalence transformation to (5.160) (to find such a

* Any other $SU(2)_{PG}$ non-invariant quartic self-interaction of q^+ can be brought into the form (5.161) by means of an $SU(2)_{PG}$ rotation.

field redefinition on its own can be a complicated problem). For instance, it immediately follows that the EH Lagrangian, being $U(1)_{\rm PG}$ invariant, should be representable in this form, like the Tab-NUT one. Such a representation was found in [G34]:

$$S_{\rm EH} \propto -\int du\, d\zeta^{(-4)} \left[\tilde{q}^+ D^{++} q^+ + \frac{1}{2}\mu^{1/2} \left(\frac{\tilde{q}^+ q^+}{1+\sqrt{1+\eta^{--}\tilde{q}^+ q^+}} \right)^2 \right], \tag{5.162}$$

where

$$\eta^{--} = \eta^{ik} u_i^- u_k^- = -\widetilde{(\eta^{--})}, \qquad \mu = \frac{1}{2}\bar{\eta}^{ik}\eta_{ik}.$$

Note that the EH action in the form (5.162) naturally arises after performing an $N = 2$ duality transformation in a particular action of a single $N = 2$ tensor multiplet (see Section 6.6.1 of the next chapter). It would be of interest to find the field redefinition relating (5.162) to the EH action in the form (5.137). Anticipating the content of Chapter 6, we also notice that the concept of an $N = 2$ duality transformation allows one to give a simple proof of the above theorem (see Section 6.6.4).

So much for the q^+ actions. In conclusion, we briefly summarize the most characteristic features of the hyper-Kähler $N = 2$ sigma models in the harmonic superspace approach and formulate some problems yet to be solved.

- The isometries of the bosonic hyper-Kähler metrics have, as a rule, a transparent appearance in the q^+ language as symmetries of the q^+ actions.
- The computation of the bosonic metric for a given q^+ action is essentially simplified when such isometries are present. All the examples which have so far been explicitly worked out correspond to at least one unbroken $U(1)_{\rm PG}$ symmetry (or any other $U(1)$ symmetry which commutes with $N = 2$ supersymmetry and is reduced to $U(1)_{\rm PG}$ on shell).
- When trying to use the q^+ actions approach as a tool for the explicit construction of new hyper-Kähler metrics (which are not reducible, e.g., to the multicenter ones) one should consider those actions which have no Pauli–Gürsey type symmetries.
- The same hyper-Kähler metric can be derived from several apparently different q^+ actions. How to consistently divide all possible q^+ actions into equivalence classes with respect to field redefinitions is as yet unsolved.
- It is still an open question how the non-trivial global properties of hyper-Kähler manifolds are encoded in the q^+ actions. Being locally represented by the same metrics, these manifolds can radically differ in their global,

topological characteristics. This subtlety is directly related, e.g., to the problem of describing the unique compact four-dimensional hyper-Kähler manifold, the famous K3 manifold, within the q^+ actions approach.

The discussion of $N = 2$ hyper-Kähler sigma models and potentials given in this chapter will be continued in Chapter 11. There we shall concentrate on the geometric aspects of these theories.

6

$N = 2$ matter multiplets with a finite number of auxiliary fields. $N = 2$ duality transformations

In this chapter we demonstrate that all known off-shell $N = 2$ matter multiplets with finite sets of auxiliary fields are described by Grassmann analytic superfields with properly chosen harmonic constraints. We define the $N = 2$ superfield duality transformation and use it to show explicitly that the general self-couplings of these constrained $N = 2$ matter superfields are reduced to particular classes of the q^+ hypermultiplet self-coupling.

6.1 Introductory remarks

As was mentioned in Section 5.1.4, before the invention of harmonic superspace $N = 2$ off-shell matter had been described by constrained superfields in ordinary $N = 2$ superspace. These are the $N = 2$ tensor multiplet [D19, S6, W6], non-linear multiplet [D16, D17], relaxed hypermultiplet [H15] and its generalizations [Y1], etc. Since such superfields have finite sets of components, they are sometimes easier to deal with than the unconstrained Grassmann analytic superfields, e.g., when constructing physical component actions. However, they are certainly not adequate for describing the most general $N = 2$ matter action and, respectively, the most general set of hyper-Kähler metrics. The natural object representing $N = 2$ off-shell matter is the Grassmann analytic superfield q^+. Only with its help one can establish a one-to-one correspondence between $N = 2$ matter and hyper-Kähler manifolds. Nevertheless, to make contact with earlier studies, we explain how ordinary $N = 2$ matter superfields can be rewritten in harmonic superspace and how their actions are related to the general off-shell q^+ action (5.82). In the present chapter we show that the actions of all known constrained matter $N = 2$ superfields (without central charges) are reduced to particular subclasses of the generic action (5.82) by duality transformations [G11].

We start by recalling the similar situation in the case of $N = 1$ matter. The on-shell $N = 1$ matter multiplet comprising two spins 0 and one spin

1/2 also admits different off-shell representations. It can be described by a chiral superfield $\Phi(x_L,\theta)$, or by a tensor superfield $G(x,\theta,\bar\theta)$, $G=\bar G$, $D^\alpha D_\alpha G \equiv D^2 G = 0$, or by a complex linear superfield $L(x,\theta,\bar\theta)$, $D^2 L = 0$, etc. (see, e.g., the Appendix in [Y1]). These off-shell descriptions differ in the auxiliary fields and sometimes in the physical field representations. It is well known that the chiral superfield description is the most geometric one, since it makes explicit the underlying Kähler structure of $N=1$ matter. The actions of the other off-shell $N=1$ matter multiplets are reduced to some special classes of the general chiral superfield action (5.2) by means of duality transformations [H11, L1]. Take, for instance, the general $N=1$ tensor multiplet action

$$S_G = \int d^4x\, d^4\theta f(G)\,, \tag{6.1}$$

$$D^2 G = \bar D^2 G = 0\,, \tag{6.2}$$

where $f(G)$ is an arbitrary function. One can implement the constraint (6.2) with the help of a Lagrange multiplier:

$$S_{G,X} = \int d^4x\, d^4\theta\, \big[f(G) + (D^2X + \bar D^2 \bar X)G\big]\,. \tag{6.3}$$

Varying with respect to X we come back to eqs. (6.1), (6.2). Instead, varying with respect to G we find an algebraic equation:

$$f'(G) = -(D^2X + \bar D^2\bar X) \equiv -(\Phi + \bar\Phi) \equiv -Y\,,$$

where $\Phi = \bar D^2 \bar X$ ($\bar\Phi = D^2 X$) is a left (right)-handed chiral superfield. This equation can always be solved for G,

$$G = G(Y)\,, \tag{6.4}$$

provided f is non-degenerate ($f'(G) \neq 0$). Substituting (6.4) into (6.3) gives

$$S_\Phi = \int d^4x\, d^4\theta\, \{f\,[G(Y)] + Y\, G(Y)\}\,, \qquad Y = \Phi + \bar\Phi\,. \tag{6.5}$$

Thus we see that the general self-interactions of the $N=1$ tensor multiplet (6.1) are equivalent to a restricted class of chiral multiplet self-interactions (5.2) with $P(\Phi) = 0$ and

$$K(\Phi,\bar\Phi) = f\,[G(Y)] + Y\, G(Y)\,. \tag{6.6}$$

Note that this particular $K(\Phi,\bar\Phi)$ depends only on $Y = \Phi + \bar\Phi$. For this reason it is invariant under the following shifts:

$$\Phi \to \Phi + ia\,, \qquad \bar\Phi \to \bar\Phi - ia\,, \qquad a = \bar a = \text{const}, \tag{6.7}$$

while $K(\Phi,\bar\Phi)$ in (5.2) has no such symmetries in the general case.

A similar situation takes place for other off-shell representations of $N = 1$ matter. The conclusion is that the most general description of $N = 1$ matter is achieved in terms of chiral superfields, although in certain cases one can have alternative descriptions in terms of some other constrained $N = 1$ superfields.

Now we shall see how all this generalizes to the case of $N = 2$ matter. We begin by presenting the harmonic superspace formulation of the off-shell $N = 2$ matter multiplets with finite sets of auxiliary fields.

6.2 $N = 2$ tensor multiplet

The harmonic superspace description of the $N = 2$ tensor multiplet was considered in [G10, G16]. Here we recall some of its basic features, starting from the formulation of this multiplet in ordinary $N = 2$ superspace $(x^m, \theta^\alpha_i, \bar\theta^{\dot\alpha i})$.

The $N = 2$ tensor multiplet consists of an $SU(2)$ triplet of physical scalars L^{ij}, a doublet of Weyl fermions $\psi^i_\alpha(x)$, an antisymmetric gauge tensor $E_{mn}(x)$ ('notoph' [O5]) with field strength $V^m = \frac{1}{2}\epsilon^{mnkl}\partial_n E_{kl}$ and a complex auxiliary scalar $M(x)$. The field strength of this multiplet is represented in ordinary superspace by a real triplet superfield $L^{(ij)}(X)$ [W6] satisfying the following constraint:

$$D^{(i}_\alpha L^{jk)} = \bar D^{(i}_{\dot\alpha} L^{jk)} = 0 . \tag{6.8}$$

Multiplying (6.8) by $u^+_i u^+_j u^+_k$, one equivalently represents it as the following set of harmonic superspace constraints:

$$D^+_\alpha L^{++} = \bar D^+_{\dot\alpha} L^{++} = 0 , \tag{6.9}$$

$$D^{++} L^{++} = 0 \Leftrightarrow L^{++} = L^{(ij)}(X) u^+_i u^+_j . \tag{6.10}$$

One immediately recognizes (6.9) as the familiar $N = 2$ Grassmann analyticity condition. So, it is natural to switch to the analytic basis of $N = 2$ harmonic superspace where the $N = 2$ tensor multiplet is represented by the analytic superfield $L^{++}(\zeta, u)$ with the additional harmonic constraint:

$$D^{++} L^{++}(\zeta, u) = 0 . \tag{6.11}$$

It is a simple exercise to solve (6.11) using the explicit form of D^{++} in the analytic basis (3.84). One obtains

$$\begin{aligned} L^{++}(\zeta, u) \;=\; & L^{ij}(x_A) u^+_i u^+_j + 2[\theta^{+\alpha}\psi^i_\alpha(x_A) - \bar\theta^+_{\dot\alpha}\bar\psi^{\dot\alpha i}(x_A)] u^+_i \\ & + (\theta^+)^2 M(x_A) + (\bar\theta^+)^2 \bar M(x_A) \\ & + 2i\theta^+\sigma^m\bar\theta^+ [V_m(x_A) + \partial_m L^{ij}(x_A) u^+_i u^-_j] \\ & + 2i[(\bar\theta^+)^2\theta^{+\alpha}\partial_{\alpha\dot\alpha}\bar\psi^{\dot\alpha i}(x_A) + (\theta^+)^2\bar\theta^+_{\dot\alpha}\partial^{\alpha\dot\alpha}\psi^i_\alpha(x_A)] u^-_i \\ & - (\theta^+)^4 \Box L^{ij}(x_A) u^-_i u^-_j , \end{aligned} \tag{6.12}$$

where

$$\partial^m V_m = 0 \,. \tag{6.13}$$

This exactly corresponds to the component content of $L^{ij}(X)$ mentioned above. Note that condition (6.11) eliminates all supermultiplets in L^{++} except the one with the lowest superisospin 0.

It is not hard to guess what the general form of the relevant action must be. The free action of L^{++} is given by

$$S_{\text{free}} = \frac{1}{2} \int du\, d\zeta^{(-4)}\, (L^{++})^2 \,, \qquad D^{++}L^{++} = 0 \,. \tag{6.14}$$

This gives rise to the following component action:

$$S_{\text{free}} = \frac{1}{2} \int d^4x \left(\frac{1}{2} \partial^m L^{ij} \partial_m L_{ij} - V^m V_m - i\psi^i_\alpha \partial^{\alpha\dot\alpha} \bar\psi_{\dot\alpha i} + M\bar M \right), \tag{6.15}$$

where

$$\partial^m V_m = 0 \,.$$

Introducing a dimensionful constant γ, ($[\gamma] = -1$) one can make L^{++} dimensionless: $L^{++} \to \gamma L^{++}$. Then the most general action for L^{++} can be written in the form

$$S_L^{\text{gen}} = \frac{1}{\gamma^2} \int du\, d\zeta^{(-4)} F^{+4}(u^\pm, L^{++}) \,, \qquad D^{++}L^{++} = 0 \,, \tag{6.16}$$

where F^{+4} is an arbitrary function of the field strength L^{++} and the harmonics $u^\pm$ carrying $U(1)$ charge +4.

An important example is the conformally invariant action of the improved tensor multiplet. The definition (6.11) is superconformally invariant whereas the free action (6.14) is not. However, one can construct an action of the type (6.16) which is conformally invariant. The supermultiplet with such an action is called the improved tensor multiplet. It can be used as a compensator for $N = 2$ conformal supergravity [D17] (see Chapter 10). The improved tensor multiplet was first described in terms of component fields [D17], then in terms of $N = 1$ and constrained $N = 2$ superfields [L1, K6] and finally in terms of $N = 2$ harmonic superfields [G10]. In the latter case the improved tensor multiplet action reads

$$S_{\text{impr}} = \frac{1}{\gamma^2} \int du\, d\zeta^{(-4)} \left(\frac{\ell^{++}}{1 + \sqrt{1 + \ell^{++}c^{--}}} \right)^2 \equiv \frac{1}{\gamma^2} \int du\, d\zeta^{(-4)}\, (g^{++})^2 \,, \tag{6.17}$$

where

$$\ell^{++} \equiv L^{++} - c^{++} \,, \qquad c_{ij} = \text{const}, \qquad c^{\pm\pm} = c^{ij} u^\pm_i u^\pm_j \,, \qquad c^{ij} c_{ij} = 2 \,. \tag{6.18}$$

The conformal properties of this action will be discussed in Chapter 9.

In Section 6.6.1 we shall show that the action (6.17) actually describes trivial dynamics. It can be reduced to a free q^+ action by a duality transformation. At the same time, the sum of (6.14) and (6.17) yields a non-trivial self-interaction [G18], which becomes just the Taub–NUT action (5.85) when translated into the q^+ language.

It is worthwhile mentioning that a minimal Yang–Mills coupling is not possible for the $N = 2$ tensor multiplet. The reason is that this multiplet includes a gauge antisymmetric tensor field which cannot be coupled to the Yang–Mills field.

6.3 The relaxed hypermultiplet

Howe, Stelle and Townsend (HST) [H15] invented an $N = 2$ matter multiplet which can have minimal Yang–Mills couplings, unlike the tensor multiplet. They called it the relaxed hypermultiplet. In harmonic superspace it is described by the real analytic superfields $L^{++}(\zeta, u)$ and $V(\zeta, u)$ [G11]. One of them, $L^{++}(\zeta, u)$, satisfies a relaxed form of the constraint (6.11):

$$(D^{++})^2 L^{++} = 0\,, \tag{6.19}$$

while the other, V, is defined up to a gauge transformation:

$$V' = V + D^{++}\Lambda^{--}\,, \tag{6.20}$$

where Λ^{--} is an analytic gauge parameter. The free action

$$S^{\text{free}}_{\text{HST}} = \frac{1}{\gamma^2}\int du\, d\zeta^{(-4)}\ \left[(L^{++})^2 + VD^{++}L^{++}\right] \tag{6.21}$$

is compatible with the gauge invariance (6.20) because of the constraint (6.19).

The general form of the self-coupling is uniquely determined by dimensionality, analyticity and gauge invariance arguments:

$$S_{\text{HST}} = \frac{1}{\gamma^2}\int du\, d\zeta^{(-4)}\ \left[F^{+4}(L^{++}, D^{++}L^{++}, u) + VD^{++}L^{++}\right], \tag{6.22}$$

where $F^{+4}(L^{++}, D^{++}L^{++}, u)$ is an arbitrary function of charge +4. Note that varying with respect to V in (6.22) one obtains $D^{++}L^{++} = 0$ and thus recovers the general action for the tensor multiplet (6.19) with the constraint (6.11). So, as far as the on-shell self-interactions are concerned, the relaxed hypermultiplet and the tensor multiplet are equivalent. The advantage of the relaxed hypermultiplet is that it can have minimal Yang–Mills couplings. To turn this coupling on, one takes n copies of the superfields L^{++} and V, puts them into a real n-dimensional representation of the Yang–Mills group and, finally, covariantizes the constraint (6.19) and the gauge transformation (6.20) in an obvious way (see Chapter 7). Note that the reality condition is a strong restriction on the possible choice of representations.

We have seen how simple the harmonic superspace description of the relaxed hypermultiplet is. To show the exact correspondence to the original description, one should use the central basis. There (6.19) implies

$$L^{++} = u_i^+ u_j^+ L^{ij}(X) + 5u^+_{(i} u^+_j u^+_k u^-_{l)} L^{ijkl}(X) . \tag{6.23}$$

From the analyticity of L^{++} it follows that

$$\begin{aligned} D^{(i}_{\alpha(\dot\alpha)} L^{jk)} &= D_{\alpha(\dot\alpha)l} L^{ijkl} , \\ D^{(i}_{\alpha(\dot\alpha)} L^{jklm)} &= 0 . \end{aligned} \tag{6.24}$$

This set of constrained $N = 2$ superfields was first introduced in [H15]. Finally, the scalar superfield $V(X)$ of HST [H15] is given by

$$V(X) = \int du \; V(\zeta(X, u), u) . \tag{6.25}$$

Obviously, $V(X)$ is invariant under (6.20) and solves the following constraints [H15]:

$$D^i_\alpha D_{\beta i} V = \bar D_{\dot\alpha i} \bar D^i_{\dot\beta} V = [D^i_\alpha, \bar D_{\dot\beta i}] V = 0 . \tag{6.26}$$

One can show [G11] that in the central basis the action (6.22) coincides with the original one.

6.4 Further relaxed hypermultiplets

Now it is clear that one can go on in relaxing the constraints. At the next step one obtains a multiplet described by the same real analytic superfields L^{++}, V. However, L^{++} satisfies a more relaxed form of the constraint (6.19):

$$(D^{++})^3 L^{++} = 0 \tag{6.27}$$

and V is defined up to the gauge transformation:

$$V' = V + (D^{++})^2 \Lambda^{(-4)} , \tag{6.28}$$

where $\Lambda^{(-4)}(\zeta, u)$ is a real analytic gauge parameter. The free gauge-invariant action coincides with the 'first-order multiplet' action of Yamron and Siegel [Y1]. In analytic superspace it again has the form (6.21). The general self-coupling is given by

$$S_{\rm YS} = \frac{1}{\gamma^2} \int du\, d\zeta^{(-4)} \left[F^{+4}(L^{++}, D^{++}L^{++}, (D^{++})^2 L^{++}, u) + V D^{++} L^{++} \right] . \tag{6.29}$$

Varying with respect to V, one obtains the constraint (6.11) and so goes back to the tensor multiplet action (6.16).* Thus, as far as the self-interactions are concerned, the tensor multiplet, the relaxed hypermultiplet and the further relaxed hypermultiplet are equivalent to each other after elimination of some auxiliary fields.

One can construct new multiplets by further relaxing the constraint (6.11). After the n-th step the constraint becomes

$$(D^{++})^n L^{++} = 0 \tag{6.30}$$

and the scalar analytic Lagrange multiplier $V_n(\zeta, u)$ undergoes gauge shifts:

$$V_n' = V_n + (D^{++})^{n-1} \Lambda^{-2(n-1)} . \tag{6.31}$$

The general action is

$$S_n = \frac{1}{\gamma^2} \int du\, d\zeta^{(-4)} \left[F^{+4}(L^{++}, D^{++}L^{++}, \ldots, (D^{++})^{n-1}L^{++}, u) \right.$$
$$\left. + V_n D^{++} L^{++} \right] . \tag{6.32}$$

The superspin 0 superfield L^{++} carries superisospins $0, 1, \ldots, n-1$ and describes $(8+8)(1+2+\cdots+2n-1) = 8n^2+8n^2$ degrees of freedom. Owing to the gauge invariance the superspin 0 superfield V_n contains only superisospins $1, 2, \ldots, n-1$ and hence contributes $(8+8)(n^2-1)$ additional components. Altogether the theory describes $8(2n^2-1)+8(2n^2-1)$ field degrees of freedom. In particular, these are the $8+8$ degrees of freedom of the tensor multiplet ($n=1$), the $56+56$ of the relaxed hypermultiplet ($n=2$), the $136+136$ of the further relaxed hypermultiplet ($n=3$), etc. Despite the rapidly growing number of components, for any finite n one obtains theories with equivalent self-couplings. Indeed, the Lagrange multiplier V_n always yields the constraint $D^{++}L^{++} = 0$, which leads to the same self-interacting tensor multiplet. A radically new situation arises at '$n = \infty$', when L^{++} becomes an unconstrained analytic superfield and V does not undergo any gauge transformations. Then one can make F^{+4} depend on V as well. After that, combining L^{++} and V into a single superfield q^+ as in (5.55), one actually arrives at the general $N=2$ matter action (5.82). This argument clearly shows the privileged rôle of analytic superfields with their infinite sets of components in realizing the most general $N=2$ matter self-couplings.

We would like to note that the relaxed multiplet actions given above can easily be extended to non-zero central charges by the method of Section 5.3.3. In

* One might add the term $V^{--}(D^{++})^2L^{++}$ to (6.29) and thus equally reproduce the relaxed hypermultiplet action (6.22) and the constraint (6.19) by varying with respect to V^{--}. However, this modification is insignificant, since the new action can be reduced to (6.29) by a shift in the Lagrange multiplier $V \to V + D^{++}V^{--}$.

particular, the actions (6.22), (6.29), (6.32) have the obvious invariance $V_n \to V_n + \text{const}$. One can identify its generator (times a mass parameter) with the central charge and thus produce mass terms coinciding with those in [Y1].

6.5 Non-linear multiplet

The non-linear multiplet was introduced in [D16, D17] in the context of local $N = 2$ supersymmetry as a compensator giving rise to the historically first off-shell version of $N = 2$ Einstein supergravity. Later on it was used in rigid $N = 2$ supersymmetry to construct a class of $N = 2$ sigma models [H8, K7, L1].

In ordinary $N = 2$ superspace this multiplet is represented by a 2×2 matrix superfield $L_{ai}(X)$ subject to the pseudoreality condition

$$\overline{L_{ai}} = \epsilon^{ab}\epsilon^{ij}L_{bj} \tag{6.33}$$

and to the constraints

$$L^{a(i}(X)D^l_\alpha L^{j)}_a(X) = L^{a(i}(X)\bar{D}^l_{\dot\alpha} L^{j)}_a(X) = 0\,. \tag{6.34}$$

Equations (6.33) and (6.34) are covariant with respect to two independent $SU(2)$ groups acting on the indices a and i:

$$SU(2)_{\text{ext}}: \quad \delta L^{ai} \simeq L^{ai'}(X,u) - L^{ai}(X,u) = \lambda^a_b L^{bi}\,; \tag{6.35}$$
$$SU(2)_{\text{A}}: \quad \delta L^{ai} \simeq L^{ai'}(X,u') - L^{ai}(X,u) = A^i_j L^{aj}\,. \tag{6.36}$$

The definition (6.33), (6.34) allows for arbitrary real rescalings of L^{ai}, so, without loss of generality one can choose

$$\det L = 1 \quad \Rightarrow \quad L^a_i L^j_a = -\delta^i_j\,, \qquad L^a_i L^i_b = -\delta^a_b\,. \tag{6.37}$$

Now we rewrite the constraints (6.34), (6.37) and (6.33) in harmonic superspace. Contracting all $SU(2)$ indices with harmonics and using the completeness relation (3.79) one gets the following equivalent set of constraints:

$$\begin{aligned} L^{-+}D^+_\alpha L^{++} - L^{++}D^+_\alpha L^{-+} &= 0\,, \\ L^{-+}\bar{D}^+_{\dot\alpha} L^{++} - L^{++}\bar{D}^+_{\dot\alpha} L^{-+} &= 0\,, \end{aligned} \tag{6.38}$$
$$L^{-+}L^{+-} - L^{++}L^{--} = -1\,, \tag{6.39}$$

where

$$\begin{aligned} L^{\pm\pm} &= L^{ai}u^\pm_a u^\pm_i = L^{(ai)}u^\pm_a u^\pm_i = \widetilde{L^{\pm\pm}}\,, \\ L^{\pm\mp} &= L^{ai}u^\pm_a u^\mp_i = \pm 1 + \cdots\,. \end{aligned} \tag{6.40}$$

Next we define

$$N^{++} = \frac{L^{++}}{L^{-+}} \tag{6.41}$$

and observe that eqs. (6.38) imply that N^{++} is analytic:

$$D^+_\alpha N^{++} = \bar{D}^+_{\dot\alpha} N^{++} = 0\,. \qquad (6.42)$$

It is straightforward to obtain

$$D^{++}N^{++} + (N^{++})^2 = 0\,. \qquad (6.43)$$

Equations (6.42) and (6.43) define the non-linear $N = 2$ multiplet in harmonic superspace. At the end of this subsection we show that (6.42) (6.43) are equivalent to (6.34), (6.37), but first we write down the most general action for N^{++}.

As usual, in the analytic basis (6.42) simply means that

$$N^{++} = N^{++}(\zeta, u)\,, \qquad (6.44)$$

whereas (6.43) is a constraint eliminating the infinite tower of fields appearing in the harmonic expansion of N^{++}. Equation (6.43) implies that any higher-order harmonic derivative of N^{++} is reduced to some power of N^{++}. Thus the most general action for k independent superfields N^{++}_A $(A = 1, \ldots, k)$ is

$$S_N = \frac{1}{\gamma^2}\int du\, d\zeta^{(-4)}\; F^{+4}_N(N^{++}_1, \ldots, N^{++}_k, u)\,; \qquad (6.45)$$

$$D^{++}N^{++}_A + (N^{++}_A)^2 = 0\,. \qquad (6.46)$$

The action is quite simple for a single N^{++}:

$$S_N = \frac{1}{\gamma^2}\int du\, d\zeta^{(-4)}\; N^{++}(\zeta, u) b^{++}(u^\pm)\,, \qquad (6.47)$$

where b^{++} is an arbitrary function of the harmonics. Any other power of N^{++} in the action (6.47) could be reduced to a term linear in N^{++} with the help of the constraint (6.43). In the case of several N^{++} one can make the action linear in any given N^{++}, e.g.,

$$S_N = \frac{1}{\gamma^2}\int du\, d\zeta^{(-4)}\; N^{++}_1(\zeta, u) b^{++}(N^{++}_2, \ldots, N^{++}_k, u^\pm)\,, \qquad (6.48)$$

It is useful to present the transformation laws of N^{++} under the two $SU(2)$ groups (6.35), (6.36). A simple central basis calculation gives

$$\begin{aligned} SU(2)_{\rm ext}: \quad \delta N^{++} &\simeq N^{++'}(\zeta, u) - N^{++}(\zeta, u) \\ &= -\lambda^{++} + 2\lambda^{-+}N^{++} - \lambda^{--}(N^{++})^2\,; \qquad (6.49) \\ SU(2)_{\rm A}: \quad \delta N^{++} &\simeq N^{++'}(\zeta, u') - N^{++}(\zeta, u) \\ &= A^{++} - 2A^{-+}N^{++} + A^{--}(N^{++})^2\,, \qquad (6.50) \end{aligned}$$

where $\lambda^{\pm\pm} = \lambda^{(ab)} u_a^\pm u_b^\pm$, etc. The difference between these two realizations is that in the first case only the form of N^{++} is changed whereas in the second case the harmonics are transformed as well. Thus the conventional $SU(2)'_{\rm A}$ which rotates the harmonics is the diagonal in the product of the two $SU(2)$ groups above. It is defined by the identification

$$\lambda^{(ab)} = A^{(ab)}\,, \tag{6.51}$$

whence

$$SU(2)'_{\rm A}: \qquad \delta N^{++} \simeq N^{++\prime}(\zeta, u') - N^{++}(\zeta, u) = 0\,. \tag{6.52}$$

Note that the defining constraint (6.43) is manifestly covariant under both $SU(2)'_{\rm A}$ and $SU(2)_{\rm ext}$:

$$\delta\left[D^{++}N^{++} + (N^{++})^2\right] = 2(\lambda^{-+} - \lambda^{--}N^{++})\left[D^{++}N^{++} + (N^{++})^2\right]. \tag{6.53}$$

In the actions (6.47), (6.48), $SU(2)_{\rm A}$ is explicitly broken because of the harmonic dependence of the coefficient $b^{++}(u^\pm)$. Nevertheless, there still exists an action invariant under $SU(2)_{\rm ext}$. Indeed, if one substitutes

$$b^{++} = b^{ij}u_i^+u_j^+ \quad \Leftrightarrow \quad D^{++}b^{++} = 0\,, \tag{6.54}$$

into the Lagrangian (6.47), it will transform by a full harmonic derivative under (6.49):

$$\delta\left[N^{++}(\zeta, u)b^{++}(u^+)\right] = -D^{++}\left[\lambda^{+-}b^{++} - \lambda^{--}N^{++}b^{++}\right]. \tag{6.55}$$

We note that the action for N^{++} explicitly involves only three out of the four physical scalars of the on-shell matter multiplet. The fourth scalar, as in the case of the $N = 2$ tensor multiplet, is represented by the constrained vector field $V_m(x)$ ($N^{++} = \theta^+\sigma^m\bar\theta^+ V_m(x) + \cdots$). This constraint follows from (6.43):

$$\partial^m V_m + V^m V_m = 0 \tag{6.56}$$

(discarding contributions from other fields). Unlike the tensor multiplet constraint $\partial^m V_m = 0$, which has the solution $V_m = \epsilon_{mnkl}\partial^n f^{kl}$, eq. (6.56) cannot be solved explicitly. It seems that the only reasonable way to deal with (6.56) is to implement it into the action with a scalar Lagrange multiplier. The latter will become the fourth bosonic degree of freedom upon elimination of V_m. This naturally happens in the framework of $N = 2$ duality transformations (see Section 6.6).

Finally, we can obtain the conventional description of the non-linear multiplet in ordinary $N = 2$ superspace starting from its harmonic superspace description. Unlike the $N = 2$ tensor multiplet L^{++}, the relation is not so direct because the defining constraint (6.43) remains non-linear in the central basis of harmonic superspace. Therefore, the harmonic dependence of N^{++} is non-trivial in the central basis. Nevertheless, one can make a change of variables and recover the original definition of the non-linear multiplet. The change goes as follows (now we switch back to the central basis):

$$L^{++} = \frac{N^{++}}{\sqrt{1-N}}\,, \qquad L^{-+} = \frac{1}{\sqrt{1-N}}\,,$$

$$L^{--} = \frac{1}{2(1-N)^{3/2}}N^{--}, \qquad L^{+-} = \frac{-2(1-N)^2 + N^{++}N^{--}}{2(1-N)^{3/2}}\,, \tag{6.57}$$

where

$$N \equiv \partial^{--}N^{++}, \qquad N^{--} \equiv (\partial^{--})^2 N^{++}\,. \tag{6.58}$$

As a consequence of the main constraint (6.43) the superfields (6.58) satisfy the following constraints:

$$\partial^{++}N = 2N^{++}(1-N), \qquad \partial^{++}N^{--} = 2N(1-N) - 2N^{++}N^{--}\,. \tag{6.59}$$

Then it is not hard to check that the superfields (6.57) satisfy the constraints (6.38), (6.39), which are just a reformulation of the original definition (6.34), (6.37).

We would like to point out that all the higher-order derivatives ∂^{--} of N^{++} are expressed in terms of the three basic superfields N^{++}, N and N^{--}:

$$(\partial^{--})^{n+1}N^{++} = (\partial^{--})^n N = (-1)^{n+1}2^{-n}(n+1)!(1-N)^{1-n}(N^{--})^n, \quad n \geq 1. \tag{6.60}$$

The existence of these three independent superfields is not accidental. In Chapter 9 we shall see that the object N^{++} obeying the harmonic constraint (6.46) naturally appears in the context of a harmonic extension $S^3 \times \tilde{S}^2$ of the three-sphere S^3. There N^{++}, N and N^{--} play the rôle of coordinates. Further, the group $SU(2)_{\text{ext}}$ becomes the group of motion of S^3.

This concludes our discussion of the non-linear multiplet. The component sigma model action emerging from the general self-interaction of the non-linear multiplet and the related hyper-Kähler metrics have been studied in great detail in [H8, K7]. Our aim here was to demonstrate that this multiplet has a concise and transparent description in $N = 2$ analytic superspace.

6.6 $N = 2$ duality transformations

In this section we show that all the self-interactions of the tensor, relaxed, further relaxed and non-linear multiplets are classically equivalent to some restricted classes of q^+ (or ω) self-interactions.

6.6.1 Transforming the tensor multiplet

The $N = 2$ duality transformations are direct generalizations of the $N = 1$ ones [H11, L1, R3] discussed in Section 6.1 (eqs. (6.5), (6.6)). In the simplest case of the tensor multiplet the superfield L^{++} in (6.14), (6.16) is constrained by eq. (6.11). One can implement this constraint in the action with the help of a Lagrange multiplier ω:

$$S = \frac{1}{\gamma^2} \int du\, d\zeta^{(-4)} \left[F^{+4}(u^{\pm}, L^{++}) - \omega D^{++} L^{++} \right]. \tag{6.61}$$

Now both L^{++} and ω are unconstrained superfields. Varying with respect to ω we recover the original constraint (6.11). Making the change of variables (see (5.55))

$$\begin{aligned} L^{++} &= u^{+i} q_i^+ , \qquad \omega = -u^{-i} q_i^+ , \\ q_i^+ &= u_i^+ \omega + u_i^- L^{++} , \end{aligned} \tag{6.62}$$

we obtain a q^+ hypermultiplet action:

$$S^{\text{dual}} = \frac{1}{\gamma^2} \int du\, d\zeta^{(-4)} \left[F^{+4}(u^+ q^+, u) - \frac{1}{2}(u^+ q^+)^2 + \frac{1}{2} q_i^+ D^{++} q^{+i} \right]. \tag{6.63}$$

Note that the action dual to the free tensor multiplet action (6.14) with the Lagrangian $F^{+4} = \frac{1}{2}(L^{++})^2$ is the free action (5.52) for q^+.

The action (6.63) is clearly less general than (5.82). Indeed, it has the following isometry:

$$\delta q_i^+ = \text{const}\, u_i^+ , \tag{6.64}$$

which amounts to the shift

$$\delta\omega = \text{const}$$

in the L^{++}, ω form (6.61) of this action. Hence the corresponding hyper-Kähler metric has at least one Killing vector. Actually, upon elimination of the infinite set of auxiliary fields eq. (6.63) produces the well-known general Ansatz for four-dimensional multicenter hyper-Kähler metrics [G32]. We shall explicitly show this in Section 11.3.8, proceeding from the purely geometric description of hyper-Kähler manifolds. Note that $L^{++} = u^{+i} q_i^+$ coincides (up to a numerical factor) with the Killing potential for the isometry (6.64). In Section 6.6.4 we shall see that this is not an accident.

The above consideration generalizes in an obvious way to n interacting tensor multiplets. The corresponding dual q^+ action necessarily possesses n independent commuting $U(1)$ symmetries which are realized as constant shifts of n Lagrange multipliers ω, or, equivalently, as the shifts (6.64) of the relevant superfields q^+. Thus the related $4n$-dimensional hyper-Kähler metrics possess n commuting Killing vectors. This is in agreement with the general statement

[L1] that any $4n$-dimensional hyper-Kähler metric having at least n commuting triholomorphic (i.e., commuting also with supersymmetry) isometries can be constructed starting from some self-interactions of n tensor multiplets. A proof of this theorem within the harmonic approach will be given in Section 6.6.4.

The conformally invariant action of the improved tensor multiplet (6.17) can also be rewritten as a q^+ action by means of (6.62). There one finds a fake interaction term. Indeed, one can combine L^{++} and ω into a single q^+ by a more sophisticated change of variables [G10, G18]:

$$\hat{q}_i^+ = (c_{ij} + i\epsilon_{ij})(u^{+j} - iu^{-j}g^{++})e^{-i\frac{\omega}{2}} + (c_{ij} - i\epsilon_{ij})(u^{+j} + iu^{-j}g^{++})e^{i\frac{\omega}{2}}, \tag{6.65}$$

$$g^{++} \equiv \frac{\ell^{++}}{1 + \sqrt{1 + \ell^{++}c^{--}}}. \tag{6.66}$$

Thus one finds

$$\frac{1}{2}\hat{q}_i^+ D^{++}\hat{q}^{+i} = 2(g^{++})^2 + D^{++}L^{++}\omega, \qquad L^{++} = \ell^{++} + c^{++}. \tag{6.67}$$

The term $(g^{++})^2$ is just the Lagrangian of the improved tensor multiplet (6.17). Thus we see that the latter is duality-equivalent to the free q^+ action.*

With the help of (6.67), (6.66) one can rewrite any L^{++} action in terms of $\hat{q}_i^+$. Remarkably, L^{++} itself has a very simple expression:

$$L^{++} = \frac{1}{4}c^{ij}\hat{q}_i^+\hat{q}_j^+, \qquad D^{++}L^{++}|_{\text{on shell}} = 0, \tag{6.68}$$

and again coincides (up to a numerical factor) with the Killing potential (Noether current) of the $U(1)$ isometry of the Pauli–Gürsey type:

$$\delta\hat{q}_i^+ = \alpha c_i^n\hat{q}_n^+, \qquad \delta L^{++} = \delta\ell^{++} = 0. \tag{6.69}$$

It is easy to verify that (6.69) is generated by the constant shift

$$\delta\omega = \alpha \tag{6.70}$$

in (6.65) and so is an equivalent form of the isometry (6.64). Thus we arrive at the following important conclusion. Any action for a single L^{++} is duality-equivalent, via the relations (6.67) and (6.68), to the following q^+ action:

$$S_{\text{dual}}^{L^{++}} = \frac{1}{\gamma^2}\int du\, d\zeta^{(-4)}\left[\frac{1}{2}\hat{q}_i^+ D^{++}\hat{q}^{+i} + F'^{+4}(c^{ij}\hat{q}_i^+\hat{q}_j^+, u)\right]. \tag{6.71}$$

* In the pioneering studies this equivalence has been proved in terms of components by complicated indirect arguments [D17]. The first direct proof was given by Lindtsröm and Roček [L1] who used $N = 1$ duality transformations.

The latter is none other than the action (5.160) corresponding to the general family of multicenter four-dimensional hyper-Kähler metrics (for complete correspondence one needs to choose a special $SU(2)_{\rm A}$ frame: $c^{11} = c^{22} = 0\,,\; c^{12} = i$). Thus we see that in the q^+ language the multicenter family can be described in two equivalent ways, either by the action (6.63) or by (6.71). Like in the case of (6.63), one can argue that the q^+ actions which, by the substitutions (6.65), (6.66), (6.68), are duality-equivalent to the general actions for n tensor multiplets necessarily possess at least n commuting triholomorphic isometries of the Pauli–Gürsey type (6.69). This is in agreement with the results of [L1] obtained by using $N = 1$ superfields. In Section 6.6.4 we shall also prove the inverse statement, namely that any $4n$-dimensional hyper-Kähler metric with at least n commuting Pauli–Gürsey type $U(1)$ isometries (realized as invariances of the corresponding q^+ action involving n hypermultiplets) can be equivalently derived from a self-interaction of n tensor multiplets. The $N = 2$ duality transformation plays a key rôle in this proof. Note that this particular class of $4n$-dimensional hyper-Kähler manifolds is called 'toric manifolds' (see, e.g., [G36]). The general Ansatz for the relevant metrics generalizing the multicenter Ansatz (5.107), (11.256) was given in [H8, L1, P1]. It can be derived in the present approach as well, by extending the $n = 1$ construction of Section 11.3.8 of Chapter 11 to a generic n.

It is instructive to see which L^{++} actions correspond to the textbook Taub–NUT and Eguchi–Hanson examples. Following the general argument above one finds that the Taub–NUT action is given by the sum of the simple and improved L^{++} actions (6.14) and (6.17):

$$S_{\rm TN} = \frac{1}{\gamma^2}\int d\zeta^{(-4)}\, du\, \big[\lambda (L^{++})^2 + (g^{++})^2\big]\,. \tag{6.72}$$

Indeed, performing the duality transformation (6.66) and using (6.67), (6.68), one arrives at the action (5.85) with an interaction term generated by the first term in (6.72).* The EH action in the form (5.162) emerges as a result of a duality transformation with constants c^{ij} in the action:

$$S_{\rm EH} = \frac{1}{\gamma^2}\int d\zeta^{(-4)}\, du \left[(g^{++})^2 + \gamma \left(\frac{L^{++}}{1 + \sqrt{1 + L^{++}\eta^{--}}} \right)^2 \right]. \tag{6.73}$$

The first term in (6.73) generates the free $\hat{q}^+$ action and the second one leads to the self-interaction in (5.162), after substituting $L^{++} \sim c^{ij} \hat{q}^+_i \hat{q}^+_j$ and properly choosing the constant $SU(2)$ vector c^{ij}.

* One can find an equivalent form of the Taub–NUT action by making a duality transformation in (6.72) according to (6.62). Then the first and second terms in (6.72) produce the kinetic and interaction terms for q^+. The $U(1)$ isometry of the Taub–NUT action in such a formulation is realized by shifts (6.64).

Finally, we note that the $U(2)$ symmetries of the Taub–NUT and EH actions have a transparent form in the original q^+ formulations (5.85), (5.137). On the other hand, they are not so obvious in the L^{++} formulation. For example, $SU(2)_A$ is realized in (6.72) as the $SU(2)$ subgroup of the $N = 2$ superconformal group (the latter as a whole is broken). Also, only the $U(1)$ subgroup of the $SU(2)$ factor of the full $U(2)$ isometry group of the EH action is manifest in (6.73) (alongside the $U(1)_A$ factor). It is realized by the transformations (6.69).

6.6.2 *Transforming the relaxed hypermultiplet*

Once again, we introduce the constraint (6.19) for the relaxed hypermultiplet into the action (6.21) with the help of a real analytic Lagrange multiplier V^{--}:

$$S = \frac{1}{\gamma^2} \int du\, d\zeta^{(-4)} \left[F^{+4}(L^{++}, D^{++}L^{++}, u) + V D^{++}L^{++} + V^{--}(D^{++})^2 L^{++} \right]. \tag{6.74}$$

Here V^{--} should transform under the gauge group (6.20) as follows:

$$V^{--\prime} = V^{--} + \lambda^{--} \tag{6.75}$$

in order to maintain the gauge invariance of the action. Now we can combine V and V^{--} into one (non-gauge) analytic superfield ω:

$$\omega = V - D^{++}V^{--}\,. \tag{6.76}$$

Making the same change of variables (6.62) as in (6.63) we obtain the dual form of the action S (6.74):

$$\begin{aligned} S^{\text{dual}} &= \frac{1}{\gamma^2} \int du\, d\zeta^{(-4)} \Big[F^{+4}(u^+q^+, u^{+i}D^{++}q_i^+, u) \\ &\qquad - \frac{1}{2}(u^+q^+)^2 + \frac{1}{2} q_i^+ D^{++} q^{+i} \Big]. \end{aligned} \tag{6.77}$$

Again, this action is not the most general one. Moreover, it is in fact guaranteed to be equivalent to the action (6.63) for the tensor multiplet on shell (upon eliminating the auxiliary fields). Further, it is invariant under the shift (6.64).*

The same procedure applies to the Yamron–Siegel multiplet (6.27)–(6.29) and to the further relaxed multiplets. All their self-interactions turn out equivalent to some particular self-couplings of q^+ hypermultiplets, which are invariant under the shift (6.64).

* This isometry has been revealed already at the level of the original relaxed hypermultiplet actions (6.21), (6.19), (6.32) (see the remark at the end of Section 6.4).

6.6.3 Transforming the non-linear multiplet

As before, we insert the constraint (6.46) into the action (6.45) with the help of a suitable Lagrange multiplier:

$$S_N = \frac{1}{\gamma^2}\int du\, d\zeta^{(-4)}\Big[F_N^{+4}(N_1^{++},\ldots,N_k^{++},u) + \frac{1}{2}\sum_A \omega_A\left[D^{++}N_A^{++} + (N_A^{++})^2\right]\Big]. \quad (6.78)$$

Making the field redefinition:

$$\tilde{\omega}_A = \sqrt{\omega_A}\,, \qquad \tilde{N}_A^{++} = \sqrt{\omega_A}N_A^{++}\,,$$
$$q_A^{+a} = -u^{+a}\tilde{\omega}_A + u^{-a}\tilde{N}_A^{++}\,, \quad (6.79)$$

one obtains the following general dual form of the N^{++} action:

$$S_N^{\rm dual} = \frac{1}{\gamma^2}\int du\, d\zeta^{(-4)}\Big[\frac{1}{2}\sum_A q_{Aa}^{+}D^{++}q_A^{+a} + F_N^{+4}\left(\frac{u^{+a}q_{a1}^{+}}{u^{-a}q_{a1}^{+}},\ldots,\frac{u^{+a}q_{ak}^{+}}{u^{-a}q_{ak}^{+}},u\right)\Big]. \quad (6.80)$$

In particular, for a single N^{++} one has, in accordance with (6.47),

$$S_N^{\rm dual} = \frac{1}{\gamma^2}\int du\, d\zeta^{(-4)}\left[\frac{1}{2}q_a^{+}D^{++}q^{+a} + \frac{u^{+a}q_a^{+}}{u^{-a}q_a^{+}}b^{++}(u^{\pm})\right]. \quad (6.81)$$

Note that in general neither (6.78) nor (6.81) have obvious isometries, unlike the dual action for the tensor multiplet.

It is interesting to see how the group $SU(2)_{\rm ext}$ (6.49) is realized in terms of q_a^{+}. To find this realization, one starts with the dual form of the $SU(2)$ invariant N^{++} action (the latter corresponds to the choice of $b^{++} = b^{ij}u_i^{+}u_j^{+}$ in (6.81)) and specifies the variation of $\tilde{\omega}$ so that this dual action is invariant. The resulting transformations are

$$SU(2)_{\rm ext}: \qquad \delta q_a^{+} = -\lambda_a^{b}q_b^{+} + q_a^{+}\frac{\lambda^{--}b^{++}}{(u^{-}q^{+})^2}\,. \quad (6.82)$$

One can check that their algebra closes for any b^{++}. Note the close resemblance between (6.82) and the $SU(2)$ group (5.139), which is the invariance group of the EH action (5.137). Indeed, the second term in (6.82) is reduced to $D^{++}q_a^{+}$ on shell, as implied by the equation of motion following from (6.81):

$$D^{++}q_a^{+} = q_a^{+}\frac{b^{++}}{(u^{-}q^{+})^2}\,. \quad (6.83)$$

Similarly, in the present case one can also implement the group $\hat{SU}(2)$ (5.138):

$$\hat{SU}(2): \qquad \delta q_a^+ = -\lambda_a^b q_b^+ + \lambda^{--} D^{++} q_a^+ . \tag{6.84}$$

On shell the transformations (6.84) and (6.82) coincide, but off shell they constitute two different $SU(2)$ groups. The action (6.81) is invariant under (6.84) provided $b^{++} = b^{ij} u_i^+ u_j^+$.

Finally, we give the expression of the general duality-transformed action for a single N^{++} (6.81) in terms of the superfield $\tilde{\omega}$. The field equation for $\tilde{N}^{++}$ following from (6.81) (with q_A^{+a} represented according to (6.79)) is

$$\tilde{N}^{++} = D^{++}\tilde{\omega} - \frac{b^{++}}{\tilde{\omega}}, \tag{6.85}$$

so one can eliminate $\tilde{N}^{++}$ and obtain

$$S^{\text{dual}}_{\omega,N} = -\frac{1}{\gamma^2} \int du\, d\zeta^{(-4)} \left[\frac{1}{2}(D^{++}\tilde{\omega})^2 + \frac{1}{2}\frac{(b^{++})^2}{\tilde{\omega}^2} + \ln\tilde{\omega}\, D^{++} b^{++} \right]. \tag{6.86}$$

Since in the $SU(2)$ invariant case $D^{++}b^{++} = 0$, the action (6.86) in this case takes the form which differs from the EH action (5.136) merely by the sign of the second term:

$$S^{\text{dual}}_{\omega,N} = -\frac{1}{\gamma^2} \int du\, d\zeta^{(-4)} \left[\frac{1}{2}(D^{++}\tilde{\omega})^2 + \frac{1}{2}\frac{(b^{++})^2}{\tilde{\omega}^2} \right]. \tag{6.87}$$

Clearly, it is manifestly invariant under the $SU(2)$ group (5.80) which is none other than the u_a^- projection of (6.84). It would be interesting to inquire which class of hyper-Kähler metrics is encoded by (6.86) for general b^{++}. In the general case the action (6.86) seems to possess no isometry.

6.6.4 General criterion for equivalence between hypermultiplet and tensor multiplet actions

So far we have shown that any known off-shell action for $N = 2$ matter multiplets with a finite number of auxiliary fields can be written down in terms of Grassmann analytic $N = 2$ superfields and, furthermore, can be duality-transformed to a subclass of the general q^+ hypermultiplet self-interactions. The inverse is not always true, so one would like to know under which conditions a given q^+ action admits a duality-equivalent representation in terms of $N = 2$ multiplets with finite sets of auxiliary fields. Here we formulate such a criterion for equivalence between the hypermultiplet and tensor multiplet actions.

The necessary and sufficient condition for an action of n q^+ hypermultiplets to be equivalent to some tensor multiplets action is the existence of n commuting $U(1)$ isometries, at least one isometry for each hypermultiplet. In Section

6.6.1 we have already seen that this condition is necessary, when considering the duality transformation from L^{++} to q^+. It remains to show that it is also sufficient. For simplicity we restrict ourselves to the case of a single q^+ (the proof can be easily extended to the case of several q^+ hypermultiplets).

Suppose that the Lagrangian

$$H^{+4} = q_a^+ D^{++} q^{+a} + L^{+4}(q^+, u) \tag{6.88}$$

has the $U(1)$ isometry

$$\delta q^{+a} = \epsilon \lambda^{+a}(q^+, u)\,; \qquad \lambda^{+a} = -\frac{1}{2}\frac{\partial \Lambda^{++}}{\partial q_a^+}\,,$$

$$\partial^{++}\Lambda^{++} = -\frac{1}{2}\frac{\partial L^{+4}}{\partial q^{+a}}\frac{\partial \Lambda^{++}}{\partial q_a^+}\,, \tag{6.89}$$

where $\lambda^{+a}(q^+, u)$ and $\Lambda^{++}(q^+, u)$ are, respectively, the corresponding Killing vector and potential (for their general definition see Section 5.3.3). We wish to show that there exists a change of variables

$$q^{+a} = q^{+a}(L^{++}, \omega, u)\,, \tag{6.90}$$

such that (6.88) takes the form (6.61) of a tensor multiplet action. The superfield L^{++} should satisfy the tensor multiplet constraint

$$D^{++}L^{++} = 0 \tag{6.91}$$

(which should arise as the equation of motion obtained by varying with respect to the superfield ω). An obvious candidate for L^{++} is the Killing potential Λ^{++} which, as we saw in Section 6.6.1, indeed coincides with L^{++} in the two important particular cases. In Section 5.3.3 we found that in the general case Λ^{++} satisfies the harmonic conservation law (5.128) as a consequence of the equation of motion for q^+. Using (6.89) and the equation of motion for q^+ following from (6.88), it is straightforward to see that

$$D^{++}\Lambda^{++}(q^+, a) = D^{++}q^{+a}\frac{\partial \Lambda^{++}}{\partial q^{+a}} + \partial^{++}\Lambda^{++} = 0\,. \tag{6.92}$$

Thus we can identify*

$$L^{++} = \Lambda^{++}(q^+, u)\,. \tag{6.93}$$

Now we can regard (6.93) as the equation expressing q^+ in terms of a given tensor multiplet superfield L^{++}. Note that Λ^{++} does not change under the transformation (6.89) with an arbitrary analytic superfield parameter $\omega(\zeta, u)$:

$$\delta\Lambda^{++} = \delta q^{+a}\frac{\partial \Lambda^{++}}{\partial q^{+a}} = -\frac{1}{2}\omega\frac{\partial \Lambda^{++}}{\partial q_a^+}\frac{\partial \Lambda^{++}}{\partial q^{+a}} = 0\,. \tag{6.94}$$

* The fact that Killing potentials for $N = 1, 2$ supersymmetric sigma models are identical to tensor (linear) multiplets was established in [H11] using $N = 1$ superfield language.

So, the solution of (6.93)

$$q^{+a} = q^{+a}(L^{++}, \omega, u) \tag{6.95}$$

depends on the parameter ω. Furthermore, eq. (6.89) implies

$$\frac{\partial}{\partial \omega} q^{+a} = \lambda^{+a}(q^{+}(L^{++}, \omega, u), u)\,. \tag{6.96}$$

This means that $q^{+}(L^{++}, \omega, u)$ can be obtained from $q^{+}(L^{++}, 0, u)$ by a finite transformation of the $U(1)$ isometry group with ω as a group parameter.

In summary, one can find the sought non-degenerate change of variables (6.90) as a solution of eqs. (6.93) and (6.96) for some given analytic superfields L^{++} and ω. It remains to show that substituting (6.90) into the q^{+} Lagrangian (6.88) makes it a tensor multiplet Lagrangian of the type (6.61). To this end, we use the transformation properties of the Lagrangian (6.88) under an infinitesimal $U(1)$ transformation (6.89) with the local parameter $\epsilon(\zeta, u)$:

$$\delta H^{+4}(q^{+}, u) = -\Lambda^{++} D^{++} \epsilon(\zeta, u) + \text{total harmonic derivative}\,. \tag{6.97}$$

Since $\delta \Lambda^{++} = 0$, eq. (6.97) actually yields, up to a total harmonic derivative, the *finite* form of a $U(1)$ transformation of H^{+4}. This clearly demonstrates that, up to a total harmonic derivative, $H^{+4}(q^{+}(L^{++}, \omega, u), u)$ contains the superfield ω only *linearly* (recall that ω appeared in (6.95), (6.96) just as a $U(1)$ group parameter). So,

$$H^{+4}(L^{++}, \omega, u) = H^{+4}(L^{++}, 0, u) + D^{++}L^{++}\omega + \text{total harmonic derivative}$$

$$\begin{aligned} \Rightarrow\ & q^{+}_{a}(L^{++}, 0, u) D^{++} q^{+a}(L^{++}, 0, u) + L^{+4}(q^{+}(L^{++}, 0, u), u) + D^{++}L^{++}\omega \\ =\ & F^{+4}(L^{++}, u) + D^{++}L^{++}\omega\,. \end{aligned} \tag{6.98}$$

Here

$$F^{+4}(L^{++}, u) \equiv q^{+}_{a}(L^{++}, 0, u) D^{++} q^{+a}(L^{++}, 0, u) + L^{+4}(q^{+}(L^{++}, 0, u), u) \tag{6.99}$$

is the self-interaction term for L^{++} according to (6.61). Note that if condition (6.91) holds, one can replace D^{++} in (6.99) by a partial derivative ∂^{++} acting only on the explicit harmonics in $q^{+}(L^{++}, 0, u)$. This completes the proof of the equivalence criterion.

It is straightforward to extend the proof to the case of several hypermultiplets. The only new condition is

$$\det \frac{\partial q^{iA+}(L^{++}, \omega, u)}{\partial (L^{++B}, \omega^{B})} \neq 0\,, \tag{6.100}$$

and it amounts to the assumption that the change of variables (6.90) is non-degenerate. This is certainly the case if the isometries (6.89) act on the manifold $\{q^{iA}\}$, $A = 1, \ldots, n$ effectively, i.e., in such a way that one cannot find a combination of q^{iA} invariant under all isometries.

Thus we have given, within the harmonic superspace approach, a proof of the already mentioned theorem [L1] stating that all $4n$-dimensional hyper-Kähler metrics with at least n commuting triholomorphic isometries ('toric' hyper-Kähler metrics) can be constructed starting from self-interactions of n tensor multiplets. As a corollary of this generic property for $n = 1$, we have also proved that any four-dimensional hyper-Kähler metric with at least one triholomorphic $U(1)$ isometry belongs to the multicenter class (in accord with the theorem of refs. [G34, G35, H7], see the discussion in Section 5.3.5). Indeed, as argued in Section 6.6.1, the general self-interaction of a single L^{++} is duality-equivalent to the multicenter subclass of general q^+ actions.

6.7 Conclusions

In this chapter we have shown that the entire variety of $N = 2$ matter off-shell representations is actually reduced to a single universal representation with an infinite number of auxiliary fields, the q^+ hypermultiplet. Its general self-interactions encompass all possible self-couplings of other multiplets.* In addition, there are new couplings which cannot in principle be described by finite-component multiplets. So, the q^+ hypermultiplet is the most adequate representation of $N = 2$ matter. It is the genuine analog of the $N = 1$ chiral superfield. In fact, it meets almost all the requirements for the 'ultimate' $N = 2$ matter multiplet as they were formulated in [Y1]. It has 'an internal $SU(2)$ symmetry in the free action, an internal $U(1)$ symmetry that couples to complex Yang–Mills, and no global symmetry in interactions corresponding to hyper-Kähler manifolds with no Killing vectors' (there are no reasons to expect any global symmetries in the general action (5.82), except for $N = 2$ supersymmetry). The only requirement of [Y1] that is not satisfied by q^+ is a finite set of auxiliary fields. However, this is just the crucial property of q^+ which makes it possible to realize the most general self-couplings. One may recall the no-go theorem of [H18, S21], which states that the complex form of the $N = 2$ scalar multiplet cannot exist off shell with a finite number of auxiliary fields (see Section 2.4.3).

As we shall see in Chapter 11, q^+ is also preferred on purely geometrical grounds. It is intimately related to the unconstrained formulations of hyper-Kähler and quaternionic geometries. This makes it the universal compensator multiplet for $N = 2$ supergravity (Chapter 10). It naturally gives rise to a

* Some multiplets of this type, alongside the tensor ones, were employed in [K7, L2, R3] for explicit construction of hyper-Kähler metrics.

new version of $N = 2$ Einstein supergravity with infinitely many auxiliary fields. Unlike all the other known versions, this one allows for the most general couplings to $N = 2$ matter, which, in turn, is also described by q^+ superfields.

7
Supersymmetric Yang–Mills theories

In this chapter we describe the formulation of $N = 2$ SYM theory in harmonic superspace. In the first part of the chapter we show how one can arrive at the idea of an analytic unconstrained prepotential V^{++} for $N = 2$ SYM starting from very simple considerations of minimal gauge coupling to $N = 2$ supersymmetric matter. As an illustration of this method we first briefly discuss the $N = 0$ and $N = 1$ theories. In the second part we develop a suitable geometric framework. There we begin with a set of constraints on the differential geometry in superspace and derive the prepotential as the solution to the constraints. All the curvature tensors can then be expressed in terms of the prepotential and an off-shell action can be constructed. Once again the case $N = 1$ serves as an example of the general approach.

7.1 Gauge fields from matter couplings

7.1.1 $N = 0$ *gauge fields*

One of the possible ways of introducing the notion of gauge field is to consider the action of a matter spinor field $\psi_\alpha(x)$ belonging to some representation of an internal symmetry group G:

$$S^{N=0}_{\text{matter}} = -\frac{i}{2}\int d^4x\ \psi\sigma^m\partial_m\bar{\psi}\,. \tag{7.1}$$

The idea is to try to make the internal symmetry local in space-time:

$$\psi'_\alpha(x) = e^{i\lambda(x)}\psi_\alpha(x)\,. \tag{7.2}$$

Obviously, a problem occurs when the derivative ∂_m in (7.1) acts on the gauge parameter $\lambda(x)$. In order to make the action (7.1) invariant one needs a compensating (gauge) field:

$$\partial_m\psi_\alpha \to (\partial_m + iA^r_m(x)t_r)\psi_\alpha \equiv \mathcal{D}_m\psi_\alpha\,, \tag{7.3}$$

where t_r are the generators of the representation of G to which ψ_α belongs. The field $A_m(x) \equiv A^r_m(x)t_r$ transforms inhomogeneously under the gauge group:

$$A'_m(x) = -ie^{i\lambda}\partial_m e^{-i\lambda} + e^{i\lambda}A_m e^{-i\lambda}\,, \tag{7.4}$$

so that the action (7.1) with the covariant derivative (7.3) becomes invariant.

The gauge-covariant object which can be built from the gauge field A_m is the curvature (or field strength) tensor:

$$[\mathcal{D}_m, \mathcal{D}_n] = iF_{mn} = i(\partial_m A_n - \partial_n A_m + i[A_m, A_n])\,, \qquad F'_{mn} = e^{i\lambda}F_{mn}e^{-i\lambda}\,. \tag{7.5}$$

With its help one can write down the gauge-invariant action for $N = 0$ Yang–Mills theory:

$$S^{N=0}_{\text{YM}} = -\frac{1}{4}\text{Tr}\int d^4x\; F^{mn}F_{mn}\,. \tag{7.6}$$

7.1.2 $N = 1$ SYM gauge prepotential

A suitable starting point now is the free $N = 1$ matter action (the kinetic part of the general action (5.2)):

$$S^{N=1}_{\text{matter}} = \int d^8X\; \bar{\Phi}(\zeta_R)\Phi(\zeta_L)\,. \tag{7.7}$$

Here the superfields Φ and $\bar{\Phi}$ are chiral and antichiral and transform under the group G with a rigid real parameter $\lambda = \bar{\lambda} = \lambda^r\, t_r$:

$$\Phi'(\zeta_L) = e^{i\lambda}\Phi(\zeta_L)\,, \qquad \bar{\Phi}'(\zeta_R) = \bar{\Phi}(\zeta_R)e^{-i\lambda}\,, \tag{7.8}$$

where t_r are generators of the representation of G to which Φ, $\bar{\Phi}$ belong. The next step is to try to make the parameter λ local in superspace, by analogy with the case $N = 0$ (7.2). However, since the matter superfields are (anti)chiral, one should preserve this property. To this end one complexifies the parameter λ ($\bar{\lambda} \neq \lambda$) and chooses it chiral:

$$\lambda = \lambda(\zeta_L)\,, \qquad \bar{\lambda} = \bar{\lambda}(\zeta_R)\,. \tag{7.9}$$

Then the action (7.7) ceases to be invariant, since $e^{-i\bar{\lambda}}e^{i\lambda} \neq 1$ any more. The problem can be solved by introducing a gauge superfield $V(x, \theta, \bar{\theta})$ which compensates for the transformations of the matter superfields:

$$S^{N=1}_{\text{matter}} = \int d^8X\; \bar{\Phi}\, e^V\, \Phi\,; \tag{7.10}$$

$$e^{V'} = e^{i\bar{\lambda}}e^V e^{-i\lambda}\,. \tag{7.11}$$

From (7.11) it is clear that the gauge superfield $V = V^r t_r$ can be chosen Hermitian, $V^r = \bar{V}^r$.

The question now is whether the gauge superfield $V(x, \theta, \bar{\theta})$ contains the usual gauge field $A_m(x)$ together with its superpartners described in Section 2.4.2. To check this one can use part of the gauge freedom (7.11) to gauge away those components of $V^r(x, \theta, \bar{\theta})$ which are pure gauges. This is most easily done in the Abelian case (or linearized approximation):

$$\delta V(x, \theta, \bar{\theta}) = i[\bar{\lambda}(x^m - i\theta\sigma^m\bar{\theta}, \bar{\theta}) - \lambda(x^m + i\theta\sigma^m\bar{\theta}, \theta)] . \tag{7.12}$$

Decomposing the superfields in (7.12) in $\theta, \bar{\theta}$ one sees that the first few terms in $V(x, \theta, \bar{\theta})$ are indeed pure gauges and can be eliminated. Then $V(x, \theta, \bar{\theta})$ gets the following form (called the Wess–Zumino gauge):

$$V_{\mathrm{WZ}}(x, \theta, \bar{\theta}) = 2\theta\sigma^m\bar{\theta}A_m(x) + 2i\bar{\theta}^2\theta^\alpha\psi_\alpha(x) - 2i\theta^2\bar{\theta}_{\dot{\alpha}}\bar{\psi}^{\dot{\alpha}}(x) + \theta^2\bar{\theta}^2 D(x) . \tag{7.13}$$

Here one recognizes the gauge field A_m, the photino field ψ_α, $\bar{\psi}_{\dot{\alpha}}$ and the auxiliary field D of Section 2.4.2. Moreover, in the gauge (7.13) the action (7.10) becomes polynomial in V (V is now nilpotent, $(V)^3 = 0$) and reduces to the usual minimal matter–gauge couplings for the component fields.

The conclusion is that the unconstrained superfield $V(x, \theta, \bar{\theta})$ does indeed carry the $N = 1$ SYM multiplet, so it can be called the prepotential of the theory (the potentials are the vector and spinor gauge connections $A_m(x, \theta, \bar{\theta})$, $A_\alpha(x, \theta, \bar{\theta})$ which will be expressed in terms of $V(x, \theta, \bar{\theta})$ in Section 7.2.2). The problem of constructing gauge invariants and an action for V is not trivial and will be dealt with in Section 7.2.2 in the context of the differential geometry formalism.

Note that the $N = 1$ SYM prepotential has a nice geometric meaning: It can be interpreted, in the spirit of the non-linear realizations theory, as a Goldstone superfield parametrizing the coset space G^c/G where G^c is a complexification of the gauge group G [I3, I4]. Indeed, the necessity to complexify the parameter λ in (7.8) just means that the group one actually gauges in $N = 1$ SYM theory is G^c with the double set of generators t_r, $i\,t_r$. In this sense the prepotential V is similar to the prepotential of $N = 1$ conformal supergravity, an axial-vector superfield $H^m(x, \theta, \bar{\theta})$, which is the imaginary part of the complexified bosonic coordinate of the $N = 1$ chiral superspace $\mathbb{C}^{4|2}$ [O6] (see Section 10.2).* As we shall see later, similar objects ('bridges') are also present in the geometric superfield formulations of $N = 2$ SYM and supergravity theories. But in these cases they turn out to be secondary, being expressible in terms of more fundamental geometric entities for which room is just provided by harmonic superspace.

* This prepotential also admits an interpretation as the Goldstone superfield [I9].

7.1.3 $N = 2$ SYM gauge prepotential

Once again we start with the simplest $N = 2$ matter action, that of the q^+ hypermultiplet (5.47):

$$S^{N=2}_{\text{matter}} = -\int du\, d\zeta^{(-4)}\, \tilde{q}^+ D^{++} q^+ . \tag{7.14}$$

This time the gauge group parameters should be made local in the analytic subspace (ζ, u) in order to preserve the analyticity of q^+:

$$q^{+\prime} = e^{i\lambda} q^+ , \qquad \lambda = \lambda(\zeta, u) = \widetilde{\lambda} . \tag{7.15}$$

Then the situation becomes very similar to that in the case $N = 0$. What is needed now is to covariantize the harmonic derivative:

$$D^{++} \to \mathcal{D}^{++} = D^{++} + iV^{++}(\zeta, u) , \qquad \widetilde{V^{++}} = V^{++} . \tag{7.16}$$

This gauge superfield V^{++} has the same charge and reality property as D^{++}, and is analytic in order to preserve the analyticity of the action (7.14):

$$S^{N=2}_{\text{matter}} = -\int du\, d\zeta^{(-4)}\, \tilde{q}^+ (D^{++} + iV^{++}) q^+ . \tag{7.17}$$

Finally, V^{++} should transform as follows:

$$V^{++\prime} = -ie^{i\lambda} D^{++} e^{-i\lambda} + e^{i\lambda} V^{++} e^{-i\lambda} , \tag{7.18}$$

so that the action (7.17) becomes gauge invariant.

Unlike the case $N = 1$ where the gauge superfield $V(x, \theta, \bar{\theta})$ contained only a few pure gauge degrees of freedom, the $N = 2$ candidate-prepotential $V^{++}(\zeta, u)$ contains infinitely many such degrees of freedom coming from its harmonic expansion. So, we have to make sure that the infinite set of components of V^{++} is just a little bit bigger than that of the gauge parameter λ, the difference being the $N = 2$ SYM multiplet. The latter has superspin 0 and superisospin 0 (see Section 2.3.2). In general, according to Section 3.4, the analytic superfield V^{++} contains multiplets with superspin 0 and an infinite sequence of superisospins:

$$V^{++} : \qquad I = 0, 1, 2, \ldots . \tag{7.19}$$

On the other hand, the analytic parameter λ contains superspin 0 and superisospins

$$\lambda : \qquad I = 1, 2, 3, \ldots . \tag{7.20}$$

The difference between (7.19) and (7.20) is precisely the desired superspin 0, superisospin 0 multiplet of $N = 2$ SYM theory.

The content of V^{++} can be made more explicit if one goes to the Wess–Zumino gauge. To this end one considers the Abelian version of the gauge transformation (7.18),

$$\delta V^{++} = -D^{++}\lambda \tag{7.21}$$

and expands V^{++} and λ both in θ^+ and $u^\pm$. Take, for example, the following two terms:

$$\begin{aligned} V^{++} &= v^{++}(x,u) - 2i\theta^+\sigma^m\bar\theta^+ A_m(x,u) + \cdots ; \\ \lambda &= \lambda(x,u) + i\theta^+\sigma^m\bar\theta^+\lambda_m^{--}(x,u) + \cdots . \end{aligned} \tag{7.22}$$

Inserting (7.22) into (7.21), and using the expression for D^{++} in the analytic basis from eqs. (3.84), one finds

$$\delta v^{++} = -\partial^{++}\lambda\,, \qquad \delta A_m = \frac{1}{2}\partial^{++}\lambda_m^{--} - \partial_m\lambda\,. \tag{7.23}$$

Comparing the harmonic expansions of $v^{++}(x,u)$ and $\lambda(x,u)$ (recall (3.63)) one sees that v^{++} can be entirely gauged away using the whole of $\lambda(x,u)$ but its lowest component $\lambda(x)$. Further, λ_m^{--} contains the same set of components as $A_m(x,u)$ with the exception of the lowest one $A_m(x)$. Thus, what remains from the gauge fields and parameters in (7.23) is just the ordinary gauge field:

$$\delta A_m(x) = -\partial_m\lambda(x)\,. \tag{7.24}$$

Proceeding in a similar manner one can easily show that the Wess–Zumino gauge for V^{++} is

$$\begin{aligned} V^{++}_{\rm WZ}(\zeta,u) = &-2i\theta^+\sigma^m\bar\theta^+A_m(x) - i\sqrt{2}(\theta^+)^2\bar\phi(x) + i\sqrt{2}(\bar\theta^+)^2\phi(x) \\ &+4(\bar\theta^+)^2\theta^{+\alpha}\psi^i_\alpha(x)u^-_i - 4(\theta^+)^2\bar\theta^+_{\dot\alpha}\bar\psi^{\dot\alpha i}(x)u^-_i \\ &+3(\theta^+)^2(\bar\theta^+)^2D^{ij}(x)u^-_iu^-_j\,, \end{aligned} \tag{7.25}$$

which is exactly the set of $N=2$ SYM off-shell fields (this definition of fields is chosen for future convenience). One can conclude that the analytic superfield V^{++} with the gauge transformation law (7.18) is indeed the prepotential for the $N=2$ SYM theory.

The construction of the full non-Abelian action for V^{++} will be the subject of Section 7.3. Here we show just the Abelian part of it. It is rather unusual, being non-local in the harmonic variables:

$$S^{N=2}_{\rm Maxwell} = \frac{1}{4}\int d^4x\, d^8\theta\, du_1\, du_2\, \frac{V^{++}(x,\theta,u_1)V^{++}(x,\theta,u_2)}{(u_1^+u_2^+)^2}\,. \tag{7.26}$$

Here $V^{++}(x,\theta,u_{1,2})$ are written as analytic superfields in the central basis:

$$V^{++}(x,\theta,u) = V^{++}(x^m - 2i\theta^{(i}\sigma^m\bar\theta^{j)}u^+_iu^-_j,\ \theta^i_\alpha u^+_i,\ \bar\theta^i_{\dot\alpha}u^+_i,\ u^\pm)\,, \tag{7.27}$$

and the integral is over the superspace $\mathbb{R}^{4|8} \times S^2 \times S^2$. The singular harmonic distribution $(u_1^+ u_2^+)^{-2}$ was defined in Chapter 4. To see that the action (7.26) is invariant under the Abelian gauge transformation (7.21) one makes use of (4.47):

$$\begin{aligned} \delta S &= -\frac{1}{2}\int d^{12}X\, du_1\, du_2\, \frac{D_1^{++}\lambda(1)V^{++}(2)}{(u_1^+u_2^+)^2} \\ &= \frac{1}{2}\int d^{12}X\, du_1\, du_2\, [D_1^{--}\delta^{(2,-2)}(u_1,u_2)]\lambda(1)V^{++}(2) \\ &= -\frac{1}{2}\int d^{12}X\, du\, [D^{--}\lambda(X,u)]V^{++}(X,u)\,. \end{aligned} \tag{7.28}$$

Next one splits the full superspace measure:

$$\int d^{12}X = \int d\zeta^{(-4)}\,(D^+)^4 \tag{7.29}$$

and rewrites δS as follows:

$$\delta S = -\frac{1}{2}\int du\, d\zeta^{(-4)}\,[(D^+)^4 D^{--}\lambda]V^{++} = 0\,, \tag{7.30}$$

since at least one of the spinor derivatives D^+ can reach the analytic parameter λ. Note that we had to integrate by parts with respect to D_1^{++} in (7.28), so the action is invariant up to a total derivative and the Lagrangian is not a tensor.

One can also substitute the Wess–Zumino gauge expansion (7.25) of V^{++} into (7.26) and obtain the kinetic part of the component action of $N=2$ SYM. It should be pointed out that the origin of the derivative terms is the coordinate shift defining the analytic basis (7.27) (just like in the $N=1$ matter action (7.7)). Take, for instance, the spinor field term:

$$\begin{aligned} V^{++}(X,u) &= \cdots + 4(\bar\theta^+)^2\theta^{+\alpha}\psi_\alpha^i(x)u_i^- - 2i(\bar\theta^+)^2(\theta^+)^2\bar\theta_{\dot\beta}^-(\tilde\sigma^m)^{\dot\beta\alpha}\partial_m\psi_\alpha^i(x)u_i^- \\ &\quad + (\tilde{}\ \text{conj})\,. \end{aligned} \tag{7.31}$$

Then one inserts this into (7.26) and collects the terms with eight θ's (needed for the θ integral), e.g.,

$$\begin{aligned} &(\theta_1^+)^2\theta_1^{+\alpha}\psi_\alpha^i u_{1i}^-(\bar\theta_2^+)^2(\theta_2^+)^2(\theta_2^-\sigma^m\partial_m\bar\psi^j)u_{2j}^- \\ &\quad = \frac{1}{2}(\theta)^8(u_1^+u_2^+)^3(\psi^i\sigma^m\partial_m\bar\psi^j)u_{1i}^-u_{2j}^-\,. \end{aligned} \tag{7.32}$$

The factor $(u_1^+u_2^+)^3$ in (7.32) cancels out the singular denominator in (7.26), after which the harmonic integrals can easily be done and, putting together the different pieces, one obtains the Dirac term:

$$-i\int d^4x\, \psi^i\sigma^m\partial_m\bar\psi_i\,. \tag{7.33}$$

Proceeding in a similar manner one finds the complete $N = 2$ Maxwell component action:

$$S^{N=2}_{\text{Maxwell}} = \int d^4x \left(-\frac{1}{4} F_{mn} F^{mn} + \partial_m \phi \partial^m \bar{\phi} - i \psi^i \sigma^m \partial_m \bar{\psi}_i + \frac{1}{4} D^{ij} D_{ij} \right). \tag{7.34}$$

Finally, we note that the harmonic superspace description of $N = 2$ SYM provides an answer to the question posed in [I4, I5]: What is the true $N = 2$ generalization of the complex group G^c which underlies the prepotential formulation of $N = 1$ SYM theory? Or, in other words, what is the analog of the complexification of the rigid group parameters? One can infer from (7.15) that this is the 'harmonization', i.e., extending these parameters to functions on the two-sphere S^2. Thus the group one actually gauges in the $N = 2$ case is a sort of infinite-dimensional Kac–Moody-like extension of the rigid group G, with the products of t_r and the harmonic monomials of any degree as the full set of generators.

7.2 Superspace differential geometry

In this section we develop a systematic approach to the gauge theories considered in Section 7.1. It applies to any supersymmetric gauge (and supergravity) theory. It consists in constructing the formalism of differential geometry for the corresponding superspace and imposing a set of constraints on the field strength (or torsion and curvature) tensors (see, e.g.,[C6, D23, G40, H9, S23, W7, W11]). These constraints reduce the excessively large number of components of the supertensors to the desired set of degrees of freedom of the physical theory. The constraints have a clear geometric meaning, the main ones having to do with the preservation of the appropriate notion of analyticity in the presence of gauge fields. The non-trivial part of the application of the differential geometry formalism is to find a solution to the constraints in terms of unconstrained objects. Thus we arrive once again at the prepotentials for SYM theories described in Section 7.1. The main advantage of this approach is that it gives a systematic scheme for constructing invariant tensors in terms of the prepotentials. As we shall see in Chapter 12, in the case of $N = 3$ SYM, which is radically different from the cases $N = 0, 1, 2$, the differential geometry formalism is the only available framework.

Although our main interest in this section is concentrated on the case $N = 2$, for pedagogical reasons we briefly sketch the differential geometry treatment of the $N = 1$ theory as well.

7.2.1 General framework

The starting point for the differential geometry formulation of a SYM theory [G40] is the real superspace $\mathbb{R}^{4|4N}$. As is usual for Yang–Mills theories, one introduces a gauge group with parameters $\tau^a(X)$ and considers matter superfields transforming as follows:

$$\Phi'(X) = e^{i\tau}\Phi(X)\,, \qquad \tau = \tau^a(X)t_a\,, \qquad \tau = \bar{\tau}\,, \tag{7.35}$$

where t_a are the generators of the YM group representation. In order to make the superspace derivatives $D_{\alpha i}$, $\bar{D}^i_{\dot{\alpha}}$, $\partial_{\alpha\dot{\alpha}}$ (3.22) covariant one needs gauge connections:

$$\begin{aligned} \mathcal{D}^i_\alpha &= D^i_\alpha + iA^i_\alpha(X)\,, \\ \bar{\mathcal{D}}_{\dot{\alpha}i} &= \bar{D}_{\dot{\alpha}i} + i\bar{A}_{\dot{\alpha}i}(X)\,, \\ \mathcal{D}_{\alpha\dot{\alpha}} &= \partial_{\alpha\dot{\alpha}} + iA_{\alpha\dot{\alpha}}(X) \end{aligned} \tag{7.36}$$

with the standard transformation law:

$$A'_A(X) = -ie^{i\tau(X)}\mathcal{D}_A e^{-i\tau(X)}\,, \qquad A = (\alpha i,\, \dot{\alpha}i,\, \alpha\dot{\alpha})\,. \tag{7.37}$$

(Anti)commuting the covariant derivatives one finds the gauge-covariant objects of the theory, the curvature tensors:

$$\begin{aligned} \left\{\mathcal{D}^i_\alpha, \mathcal{D}^j_\beta\right\} &= iF^{ij}_{\alpha\beta}\,, \qquad \{\bar{\mathcal{D}}_{\dot{\alpha}i}, \bar{\mathcal{D}}_{\dot{\beta}j}\} = iF_{\dot{\alpha}\dot{\beta}ij}\,, \\ \{\mathcal{D}^i_\alpha, \bar{\mathcal{D}}_{\dot{\beta}j}\} &= -2i\delta^i_j\mathcal{D}_{\alpha\dot{\beta}} + iF^{\,i}_{\alpha\dot{\beta}j}\,, \\ [\mathcal{D}^i_\alpha, \mathcal{D}_{\beta\dot{\beta}}] &= iF^{\,i}_{\alpha\beta\dot{\beta}}, \qquad [\bar{\mathcal{D}}_{\dot{\alpha}i}\,, \mathcal{D}_{\beta\dot{\beta}}] = iF_{\dot{\alpha}\beta\dot{\beta}i}\,, \\ [\mathcal{D}_{\alpha\dot{\alpha}}, \mathcal{D}_{\beta\dot{\beta}}] &= iF_{\alpha\dot{\alpha}\beta\dot{\beta}}\,. \end{aligned} \tag{7.38}$$

The vector covariant derivative $\mathcal{D}_{\alpha\dot{\beta}}$ appears in the third anticommutator in (7.38) as a consequence of the algebra (3.23) of the rigid-superspace covariant derivatives. The curvature tensors in (7.38) are subject to a number of Bianchi identities of the generic form

$$\begin{aligned} (-1)^{p(A)(p(B)+p(C))}\left[\mathcal{D}_A\left[\mathcal{D}_B, \mathcal{D}_C\right]_\pm\right]_\pm + \text{cycle}(ABC) &= 0\,, \\ (-1)^{p(A)(p(B)+p(C))}\mathcal{D}_A F_{BC} + \text{cycle}(ABC) &= 0\,. \end{aligned} \tag{7.39}$$

The main drawback of the above scheme is the presence of a huge number of ordinary fields in the form of components of the gauge superfields $A_A(X)$ or of the curvature superfields $F_{AB}(X)$. Compare, for example, the θ expansion of the $N = 1$ gauge superfields $A_\alpha(X)$, $\bar{A}_{\dot{\alpha}}(X)$, $A_{\alpha\dot{\alpha}}(X)$ with the single scalar prepotential $V(X)$ (7.13) which describes $N = 1$ SYM theory perfectly well. Clearly, in order to make the differential geometry formalism work, one has to impose a set of constraints on the curvature tensors in (7.38) which will reduce the field content of the theory to the desired one. Rather than presenting a general philosophy of how to do this, we consider the cases of interest ($N = 1$ and 2) one by one.

7.2.2 $N = 1$ SYM theory

The constraints which define $N = 1$ SYM theory have the following form

$$\begin{aligned} \{\mathcal{D}_\alpha, \mathcal{D}_\beta\} &= 0\,, && (7.40)\\ \{\bar{\mathcal{D}}_{\dot\alpha}, \bar{\mathcal{D}}_{\dot\beta}\} &= 0\,, && (7.41)\\ \{\mathcal{D}_\alpha, \bar{\mathcal{D}}_{\dot\beta}\} &= -2i\mathcal{D}_{\alpha\dot\beta}\,, && (7.42) \end{aligned}$$

or in other words:

$$F_{\alpha\beta} = F_{\dot\alpha\dot\beta} = F_{\alpha\dot\beta} = 0\,. \tag{7.43}$$

As a consequence of the Bianchi identities (7.39) these constraints lead to further restrictions on the curvature tensors:

$$\begin{aligned} F_{\alpha\beta\dot\beta} &= \epsilon_{\alpha\beta}\overline{W}_{\dot\beta}\,, \\ \mathcal{D}_\alpha \overline{W}_{\dot\beta} &= \bar{\mathcal{D}}_{\dot\alpha} W_\beta = 0\,, \\ \mathcal{D}^\alpha W_\alpha &= \bar{\mathcal{D}}_{\dot\alpha}\overline{W}^{\dot\alpha}\,. \end{aligned} \tag{7.44}$$

One way to show that eqs. (7.43) and (7.44) do indeed correspond to the $N = 1$ SYM theory is to make a detailed analysis of the component content of the curvature tensors and compare it with the gauge-independent fields in the $N = 1$ SYM multiplet. We do not do this here. Instead, we solve the above constraints and show how the prepotential $V(X)$ of Section 7.1.2 emerges as the single unconstrained object in the theory.

The key point leading to the solution of the constraints (7.40)–(7.42) is the observation that eq. (7.40) has the meaning of the integrability condition for the existence of antichiral superfields defined by

$$\mathcal{D}_\alpha \overline{\Phi} = 0 \tag{7.45}$$

in the presence of Yang–Mills fields. Correspondingly, the conjugated constraint (7.41) means integrability for chiral superfields:

$$\bar{\mathcal{D}}_{\dot\alpha}\Phi = 0\,. \tag{7.46}$$

Equation (7.42) is an example of the so-called conventional constraints, since it allows one to express the vector connection $A_{\alpha\dot\alpha}(X)$ in terms of the spinor ones $A_\alpha(X), \bar{A}_{\dot\alpha}(X)$.

The interpretation of eqs. (7.40), (7.41) suggests the way to solve them. According to Frohbenius' theorem, eq. (7.41) has the following general solution:

$$\bar{A}_{\dot\alpha}(X) = -ie^{ib}\bar{D}_{\dot\alpha}e^{-ib}\,, \tag{7.47}$$

where $b(X) = b^a(X)t_a$ and $b^a(X)$ are complex superfields and t_a are the generators of the adjoint representation of the Yang–Mills group. The object

$b(X)$ will be called a *gauge bridge* for reasons which will soon become clear. It is defined up to the following gauge transformations:

$$e^{ib'} = e^{i\tau} e^{ib} e^{-i\lambda}. \tag{7.48}$$

The τ factor in (7.48) is responsible for the gauge transformation property (7.37) of $\bar{A}_{\dot{\alpha}}$. The λ factor is the freedom of the solution (7.47), provided the parameter $\lambda = \lambda^a t_a$ satisfies the condition:

$$\bar{D}_{\dot{\alpha}} \lambda = 0\,, \tag{7.49}$$

i.e., it is a chiral superfield itself. This gauge freedom is usually called *pregauge freedom*, since it emerges in the process of solving the constraints.

The geometric meaning of the solution (7.47) and the bridge $b(X)$ becomes clear if we make the following gauge rotation:

$$\phi = e^{-ib}\Phi \quad \rightarrow \quad \phi' = e^{i\lambda}\phi\,. \tag{7.50}$$

We see that the new superfield ϕ transforms under the λ gauge group with chiral parameters instead of the original τ group with real parameters. Correspondingly, if the superfield $\Phi(X)$ is covariantly chiral (see (7.45)), the new one $\phi(X)$ is manifestly chiral:

$$\bar{\mathcal{D}}_{\dot{\alpha}} \Phi = 0 \quad \Rightarrow \quad \bar{D}_{\dot{\alpha}} \phi = 0 \quad \Rightarrow \quad \phi = \phi(\zeta_L)\,. \tag{7.51}$$

In other words, in the λ gauge frame the covariant derivative $\bar{\mathcal{D}}_{\dot{\alpha}}$ coincides with the rigid one $\bar{D}_{\dot{\alpha}}$ (but $\mathcal{D}_\alpha$ still has a non-trivial connection). Therefore, we call $b(X)$ a bridge from the old τ gauge frame to the new (chiral) λ gauge frame, in which chirality becomes manifest. Similarly, there exists a $\bar{\lambda}$ gauge frame in which antichirality is manifest (i.e., $\mathcal{D}_\alpha$ has no connection), and the bridge to that frame is $\bar{b}(X)$. Moreover, one can find a bridge from the λ frame directly to the $\bar{\lambda}$ one:

$$\begin{aligned} e^V &\equiv e^{-i\bar{b}} e^{ib}\,, \qquad \bar{V} = V\,; \\ e^{V'} &= e^{i\bar{\lambda}} e^V e^{-i\lambda}\,. \end{aligned} \tag{7.52}$$

Comparing (7.52) with (7.11) one realizes that the bridge (7.52) is nothing but the $N = 1$ SYM prepotential introduced in Section 7.1.2. Note that it transforms only under the λ gauge group and not under the τ one. This means that part of the complex bridge b is pure τ-gauge freedom. Indeed, one can fix a τ gauge in which

$$\mathrm{Re}\, b = 0 \quad \Rightarrow \quad V = -2\,\mathrm{Im}\, b\,. \tag{7.53}$$

At this point we can say that the constraints (7.40)–(7.42) have been solved in terms of the prepotential $b(X)$ (or $V(X)$ in the gauge (7.53)). The connection $\bar{A}_{\dot{\alpha}}(X)$ is given by (7.47), $A_\alpha(X)$ is its conjugate, $A_{\alpha\dot{\alpha}}(X)$ can be found from

(7.42). The curvature tensor $\overline{W}_{\dot\alpha}$ (7.44) has the following expression in terms of b:

$$\left(\overline{W}_{\dot\alpha}\right)_\tau = \frac{1}{2} e^{i\bar{b}} \left\{ D^\alpha D_\alpha \left[e^{-i\bar{b}} e^{ib} \left(\bar{D}_{\dot\alpha} e^{-ib} \right) e^{i\bar{b}} \right] \right\} e^{-i\bar{b}} , \tag{7.54}$$

and all of its properties stated in (7.44) can be checked directly. The tensor (7.54) is τ-gauge covariant, whereas

$$\left(\overline{W}_{\dot\alpha}\right)_{\bar\lambda} = e^{-i\bar{b}} \left(\overline{W}_{\dot\alpha}\right)_\tau e^{i\bar{b}} \tag{7.55}$$

transforms under the $\bar\lambda$ gauge group and is thus manifestly (not covariantly) antichiral (see (7.44)). Similar expressions, both in τ-gauge and λ-gauge covariant forms, can be found for the chiral tensor W_α.

Finally, the gauge-invariant action of $N = 1$ SYM theory can be written down as the real part of a chiral superspace integral:

$$S^{N=1}_{\text{SYM}} = \frac{1}{16} \int d\zeta_L \, \text{Tr}(W^\alpha W_\alpha) + \frac{1}{16} \int d\zeta_R \, \text{Tr}(\overline{W}_{\dot\alpha} \overline{W}^{\dot\alpha}) \tag{7.56}$$

(the numeric factor has been chosen in order to reproduce the component action (2.65) in the Wess–Zumino gauge (7.13)). Note that the presence of the trace under the integrals in (7.56) allows one to take the (anti)chiral superfields W ($\overline{W}$) in either the τ- or $\lambda(\bar\lambda)$-gauge covariant forms; in either case the trace is manifestly (anti)chiral and thus can be integrated over the (anti)chiral superspace.

7.2.3 $N = 2$ SYM theory

The constraints defining $N = 2$ SYM theory are

$$\begin{aligned}
\left\{ \mathcal{D}^i_\alpha, \mathcal{D}^j_\beta \right\} &= -2i\epsilon^{ij}\epsilon_{\alpha\beta}\overline{W} , &(7.57)\\
\left\{ \bar{\mathcal{D}}_{\dot\alpha i}, \bar{\mathcal{D}}_{\dot\beta j} \right\} &= -2i\epsilon_{ij}\epsilon_{\dot\alpha\dot\beta} W , &(7.58)\\
\left\{ \mathcal{D}^i_\alpha, \bar{\mathcal{D}}_{\dot\beta j} \right\} &= -2i\delta^i_j \mathcal{D}_{\alpha\dot\beta} . &(7.59)
\end{aligned}$$

In the case $N = 1$ we saw that the key point was the interpretation of the constraints (7.40), (7.41) as integrability conditions for chirality. The curvature term in (7.57) does not allow for the same interpretation in the case $N = 2$. However, the geometric picture becomes clear if one introduces $SU(2)/U(1)$ harmonic variables. Indeed, after multiplying the constraints (7.57)–(7.59) by $u^+_i u^+_j$ one obtains

$$\left\{ \mathcal{D}^+_\alpha, \mathcal{D}^+_\beta \right\} = \left\{ \bar{\mathcal{D}}^+_{\dot\alpha}, \bar{\mathcal{D}}^+_{\dot\beta} \right\} = \left\{ \mathcal{D}^+_\alpha, \bar{\mathcal{D}}^+_{\dot\beta} \right\} = 0 . \tag{7.60}$$

These are just the integrability conditions* for the existence of covariantly Grassmann analytic superfields:

$$\mathcal{D}^+_\alpha \Phi(X,u) = \bar{\mathcal{D}}^+_{\dot\alpha} \Phi(X,u) = 0\,. \tag{7.61}$$

This immediately suggests solving (7.60) in the form

$$A^+_\alpha = -ie^{-ib} D^+_\alpha e^{ib}\,, \qquad \bar{A}^+_{\dot\alpha} = -ie^{-ib} \bar{D}^+_{\dot\alpha} e^{ib}\,, \tag{7.62}$$

where $b = b(X,u)$ is a bridge with non-trivial harmonic dependence. Without loss of generality we can take it real:

$$\tilde{b}(X,u) = b(X,u)\,, \tag{7.63}$$

after which the two connections in (7.62) become related by $\widetilde{}$ conjugation:

$$\bar{A}^+_{\dot\alpha} = -\widetilde{A^+_\alpha}\,. \tag{7.64}$$

The bridge undergoes the following gauge transformations:

$$e^{ib'} = e^{i\lambda} e^{ib} e^{-i\tau}\,,$$
$$D^+_\alpha \lambda = \bar{D}^+_{\dot\alpha} \lambda = 0\,, \qquad \tilde{\lambda} = \lambda\,. \tag{7.65}$$

The pregauge parameter is now a Grassmann analytic real superfield. As in the case $N = 1$, with the help of the bridge one can define a λ gauge frame in which Grassmann analyticity becomes manifest:

$$\phi(X,u) = e^{ib}\Phi(X,u) \quad \Rightarrow \quad D^+_\alpha \phi = \bar{D}^+_{\dot\alpha} \phi = 0$$
$$\Rightarrow \quad \phi = \phi(\zeta, u)\,. \tag{7.66}$$

In other words, in the λ frame the covariant derivatives $\mathcal{D}^+_{\alpha,\dot\alpha}$ coincide with the rigid ones:

$$(\mathcal{D}^+_\alpha)_\lambda = D^+_\alpha\,, \qquad (\bar{\mathcal{D}}^+_{\dot\alpha})_\lambda = D^+_\alpha\,. \tag{7.67}$$

A major difference between the cases $N = 1$ and $N = 2$ is the fact that the $N = 2$ bridge $b(X,u)$ is not an unconstrained prepotential of the theory, it is subject to further restrictions. They originate from the following. The connections $A^i_\alpha(X)$, $\bar{A}_{\dot\alpha i}(X)$ do not depend on the harmonic variables, so their $+$ projections

$$A^+_{\alpha,\dot\alpha} = u^+_i A^i_{\alpha,\dot\alpha} \tag{7.68}$$

depend only linearly on u^+_i. This property can be formulated as an additional differential geometry constraint:†

$$[\mathcal{D}^{++}, \mathcal{D}^+_\alpha] = [\mathcal{D}^{++}, \bar{\mathcal{D}}^+_{\dot\alpha}] = 0\,. \tag{7.69}$$

---

* A similar interpretation of the constraints was proposed by Rosly [R4].

† The derivatives $\mathcal{D}^+_\alpha, \mathcal{D}^+_{\dot\alpha}$ (subject to the constraints (7.60)) and $\mathcal{D}^{++}$ constitute the maximal set of (anti)commuting covariant derivatives called a 'CR structure' in refs. [R4, R5, R6, S4].

Indeed, in the τ frame,

$$(\mathcal{D}^{++})_\tau = D^{++} \tag{7.70}$$

($D^{++}\tau(X) = 0$, so there is no need for a harmonic connection), so from (7.69) one easily derives (7.68). However, the general solution (7.62) of the constraints (7.60) does not automatically solve (7.69). Substituting (7.62) and (7.70) into (7.69) one finds

$$\begin{aligned} 0 &= D^{++}A^+_{\alpha,\dot\alpha} = -iD^{++}\left(e^{-ib}D^+_{\alpha,\dot\alpha}e^{ib}\right) \\ &= ie^{-ib}\left[D^+_{\alpha,\dot\alpha}\left(e^{ib}D^{++}e^{-ib}\right)\right]e^{ib} \end{aligned} \tag{7.71}$$

or, introducing the notation

$$V^{++} = -ie^{ib}D^{++}e^{-ib} = \widetilde{V^{++}}, \tag{7.72}$$

one can rewrite (7.71) as an analyticity condition on V^{++}:

$$D^+_\alpha V^{++} = \bar{D}^+_{\dot\alpha} V^{++} = 0. \tag{7.73}$$

Once again, we have arrived at the already familiar $N = 2$ SYM Grassmann analytic prepotential V^{++} of Section 7.1.3. It transform under the λ gauge group only:

$$V^{++\prime} = -ie^{i\lambda}D^{++}e^{-i\lambda} + e^{i\lambda}V^{++}e^{-i\lambda}, \tag{7.74}$$

which immediately suggests its geometric meaning. Indeed, passing to the λ gauge frame according to (7.66) we have made the spinor covariant derivatives $\mathcal{D}^+_{\alpha,\dot\alpha}$ short (see (7.67)), but the harmonic one has acquired a connection:

$$(\mathcal{D}^{++})_\lambda = e^{ib}D^{++}e^{-ib} = D^{++} + iV^{++}, \tag{7.75}$$

with V^{++} given in (7.72). So, V^{++} is the analytic harmonic connection in the λ frame.

So far we have only dealt with the constraints (7.60) which are projections of the original ones (7.57)–(7.59). In fact, it turns out that these projections together with eq. (7.69) are equivalent to the original constraints. The argument goes as follows. In the τ frame eq. (7.69) simply means (7.68), i.e., $\mathcal{D}^+_{\alpha,\dot\alpha} = u^+_i\mathcal{D}^i_{\alpha,\dot\alpha}$. Substituting this into (7.60) and using the fact that u^+_i are arbitrary commuting variables one finds

$$\left[\mathcal{D}^{(i}_\alpha, \mathcal{D}^{j)}_\beta\right] = \left[\bar{\mathcal{D}}^{(i}_{\dot\alpha}, \bar{\mathcal{D}}^{j)}_{\dot\beta}\right] = \left[\mathcal{D}^{(i}_\alpha, \bar{\mathcal{D}}^{j)}_{\dot\beta}\right] = 0, \tag{7.76}$$

which is the same as (7.57)–(7.59) (assuming the conventional part of (7.59) which serves as a definition of $\mathcal{D}_{\alpha\dot\beta}$).

At this point we can claim that the analytic superfield V^{++} does not satisfy any further constraints, so it can be regarded as the unconstrained prepotential of the theory. In order to express everything else in terms of V^{++} one has to solve eq. (7.72) for the bridge b and then find $A^+_{\alpha,\dot\alpha}$ from (7.62). According to (7.69) the latter are guaranteed to be of the form (7.68), so one finds expressions for $A^i_{\alpha,\dot\alpha}(X)$, and this completes the τ-frame differential geometry formalism. The main question now is: How does one solve eq. (7.72)? It is a non-linear differential equation on the two-sphere (although it can be rewritten in the form $D^{++}e^{-ib} = ie^{-ib}V^{++}$, the unitarity of e^{-ib} still makes this equation non-linear). The only conceivable way of solving such an equation for a general non-Abelian V^{++} is in terms of a perturbation expansion. This would be sufficient for the purposes of quantum field theory, but the task is further complicated by the τ gauge freedom in $b(X,u)$ for a given V^{++} (see (7.65)). Instead of attacking the problem directly, we prefer to recast it in an equivalent form where we can find an explicit perturbative solution [Z3, Z4]. The idea is to work entirely in the λ gauge frame, thus avoiding the use of bridges which are needed to go back to the τ frame. In this case the differential geometry formalism involves the expressions (7.67) for $(\mathcal{D}^+_{\alpha,\dot\alpha})_\lambda$ and (7.75) for $(\mathcal{D}^{++})_\lambda$, and has to be completed with expressions for the remaining covariant derivatives $(\mathcal{D}^{--})_\lambda$, $(\mathcal{D}^-_{\alpha,\dot\alpha})_\lambda$. The second harmonic derivative $(\mathcal{D}^{--})_\lambda$ can be found from the conventional constraint:

$$\left[(\mathcal{D}^{++})_\lambda, (\mathcal{D}^{--})_\lambda\right] = D^0 \,. \tag{7.77}$$

It obviously holds in the τ frame, so it must hold in the λ frame as well. Note that the charge-counting derivative D^0 is not covariantized, since we only deal with harmonic functions with fixed $U(1)$ charge. Writing down $(\mathcal{D}^{--})_\lambda$ in the form

$$(\mathcal{D}^{--})_\lambda = e^{ib}D^{--}e^{-ib} = D^{--} + iV^{--}\,, \qquad V^{--} = -ie^{ib}(D^{--}e^{-ib})\,, \tag{7.78}$$

we can rewrite (7.77) as an equation for V^{--} in terms of V^{++}:

$$D^{++}V^{--} - D^{--}V^{++} + i\left[V^{++}, V^{--}\right] = 0\,. \tag{7.79}$$

This is not only a linear differential equation on S^2, but it has a unique solution* because of the negative charge of V^{--} (see (4.26)). It is not hard to check that the solution is given by the following power series:

$$V^{--}(X,u) = \sum_{n=1}^{\infty} \int du_1 \dots du_n \, \frac{(-i)^{n+1} V^{++}(X,u_1)\dots V^{++}(X,u_n)}{(u^+u_1^+)(u_1^+u_2^+)\dots(u_n^+u^+)}\,. \tag{7.80}$$

* A word of caution is due here: For some special choices of V^{++} the solution may not exist at all. Take, for instance, an $SU(2)$ gauge group and a V^{++} of the form $(V^{++})_i^j = igu_i^+u^{+j}$. Then the solution of (7.79) is $(V^{--})_i^j = ig(1-g)^{-1}u_i^-u^{-j}$, which is singular for $g=1$, i.e., there exists no solution for that value of g [I5]. However, it can be shown that for sufficiently small V^{++} the solution (7.80) always exists.

Indeed, for $n \geq 2$ the derivative D^{++} acting on the n-th term in (7.80) gives (see (4.47))

$$\begin{aligned} D^{++}V^{--}_{(n)} &\equiv D^{++}\int du_1 \dots du_n \, \frac{(-i)^{n+1}V^{++}(1)\dots V^{++}(n)}{(u^+u_1^+)\dots(u_n^+u^+)} \\ &= \int du_1 \dots du_n \, (-i)^{n+1}V^{++}(1)\dots V^{++}(n) \\ &\quad \times \left[\frac{\delta(u,u_1)}{(u_1^+u_2^+)\dots(u_n^+u^+)} - \frac{\delta(u,u_n)}{(u^+u_1^+)\dots(u_{n-1}^+u_n^+)} \right] \\ &= -iV^{++}V^{--}_{(n-1)} + iV^{--}_{(n-1)}V^{++}, \qquad n \geq 2. \end{aligned} \tag{7.81}$$

For $n = 1$ we have

$$\begin{aligned} D^{++}V^{--}_{(1)} &\equiv D^{++}\int du_1 \, \frac{V^{++}(u_1)}{(u^+u_1^+)^2} = \int du_1 \, D^{--}\delta^{(2,-2)}(u,u_1)V^{++}(u_1) \\ &= D^{--}V^{++}. \end{aligned} \tag{7.82}$$

Putting all of this together we see that (7.79) is satisfied.

So, expressing V^{--} in terms of V^{++} turns out much easier than finding the bridge $b(X,u)$. Note that V^{--} is not an analytic superfield. It is also a non-local functional of V^{++} in the harmonic sector (although local in the (x,θ) sector, since all V^{++}'s are taken at the same (x,θ) point).

The remaining covariant derivatives $(\mathcal{D}^-_{\alpha,\dot\alpha})_\lambda$ are easily found from the conventional constraint:

$$\left[(\mathcal{D}^{--})_\lambda, (\mathcal{D}^+_{\alpha,\dot\alpha})_\lambda\right] = (\mathcal{D}^-_{\alpha,\dot\alpha})_\lambda\,, \tag{7.83}$$

which follows from the obvious τ frame relation. This is all that one needs for the differential geometry formalism in the λ frame. In particular, here is the expression for the field strength (curvature) tensor in terms of V^{--} (it is obtained from (7.57) by multiplication with $u_i^+u_j^-$ and then passing to the λ frame):

$$W_\lambda = -\frac{1}{4}(\bar{D}^+)^2V^{--}\,, \qquad \overline{W}_\lambda = -\frac{1}{4}(D^+)^2V^{--} = \widetilde{W}_\lambda\,. \tag{7.84}$$

The manifest u independence of $\overline{W}$, W in the τ frame becomes a covariant property in the λ frame:

$$(\mathcal{D}^{++})_\lambda\overline{W}_\lambda = (\mathcal{D}^{--})_\lambda\overline{W}_\lambda = 0\,, \qquad (\mathcal{D}^{++})_\lambda W_\lambda = (\mathcal{D}^{--})_\lambda W_\lambda = 0\,, \tag{7.85}$$

The first of these constraints (with $\mathcal{D}^{++}$), e.g., for $\overline{W}_\lambda$, can be checked directly using (7.69), (7.79) and (7.73). The second one (with $\mathcal{D}^{--}$) is derived by going back to the τ frame where $D^{++}\overline{W}_\tau = 0$ implies $D^{--}\overline{W}_\tau = 0$. The remaining properties of $\overline{W}$, namely its chirality:

$$\mathcal{D}_{\alpha i}\overline{W} = 0 \tag{7.86}$$

and the Bianchi identity:

$$\mathcal{D}^{\alpha i}\mathcal{D}^{j}_{\alpha}W = \bar{\mathcal{D}}^{i}_{\dot{\alpha}}\bar{\mathcal{D}}^{\dot{\alpha} j}\overline{W} \tag{7.87}$$

can also be easily proved.

In conclusion, we may say that the entire $N = 2$ SYM differential geometry formalism can indeed be expressed directly in terms of the λ-frame analytic harmonic connection V^{++}, which is thus the single unconstrained prepotential of the theory.

7.2.4 V^{++} versus Mezincescu's prepotential

Like any other analytic superfield, the $N = 2$ SYM prepotential $V^{++}(\zeta, u)$ and its gauge parameter $\lambda(\zeta, u)$ can be expressed in terms of general harmonic superfields in $\mathbb{HR}^{4+2|8}$:

$$V^{++}(\zeta, u) = (D^+)^4 M^{--}(X, u)\,, \tag{7.88}$$
$$\lambda(\zeta, u) = (D^+)^4 \rho^{(-4)}(X, u) \tag{7.89}$$

(compare with the similar solution of the $N = 1$ chirality condition $\bar{D}_{\dot{\alpha}}\phi = 0 \rightarrow \phi(\zeta_L) = \bar{D}^2\psi(X)$). The superfields M^{--} and $\rho^{(-4)}$ are unconstrained.

The gauge transformation of the 'preprepotential' M^{--} depends on two parameters:

$$\delta M^{--} = -D^{++}\rho^{(-4)} - i\left[M^{--}, (D^+)^4\rho^{(-4)}\right] + D^{+\alpha}\xi^{(-3)}_{\alpha} + \bar{D}^{+}_{\dot{\alpha}}\bar{\xi}^{(-3)\dot{\alpha}}\,, \tag{7.90}$$

where the new (pregauge) parameter $\xi^{(-3)}_{\alpha}(X, u)$ appears as the arbitrariness in M^{--} in eq. (7.88). We stress that it has an unusually high dimension $[\xi] = -5/2$ and a non-trivial $SL(2, \mathbb{C}) \times SU(2)$ index structure.

Let us now compare the harmonic expansions of the gauge superfield M^{--} and the gauge parameters $\rho^{(-4)}$, $\xi^{(-3)}_{\alpha}$ in the central basis of $\mathbb{HR}^{4+2|8}$:

$$\begin{aligned} M^{--}(X, u) &= M^{(ij)}(X)u^-_i u^-_j + M^{(ijkl)}(X)u^+_i u^-_j u^-_k u^-_l + \cdots\,, \\ \rho^{(-4)}(X, u) &= \rho^{(ijkl)}(X)u^-_i u^-_j u^-_k u^-_l + \cdots\,, \\ \xi^{(-3)}_{\alpha}(X, u) &= \xi^{(ijk)}_{\alpha}(X)u^-_i u^-_j u^-_k + \cdots\,. \end{aligned} \tag{7.91}$$

From (7.90) we see that there exists a supersymmetric gauge with a finite set of fields:

$$M^{--}(X, u) = M^{(ij)}(X)u^-_i u^-_j \quad \Leftrightarrow \quad (D^{++})^3 V^{++} = 0\,. \tag{7.92}$$

The linearized form of the gauge transformation of the ordinary superfield $M^{(ij)}(X)$ is

$$\delta M^{(ij)} = D^{\alpha}_k \xi^{(ijk)}_{\alpha} + \bar{D}_{\dot{\alpha}k}\bar{\xi}^{(ijk)\dot{\alpha}}\,. \tag{7.93}$$

It coincides with the pioneering result of L. Mezincescu [M3], obtained by solving the constraints of the $N = 2$ Maxwell theory (see also [K10]).

The price for using the ordinary superfield preprepotential $M^{(ij)}(X)$ is the loss of the transparent geometric and group structure. Neither $M^{(ij)}(X)$ ($[M] = -2$), nor the gauge parameter $\xi^{(ijk)}_\alpha$ seem to have any reasonable geometric meaning (in contrast to V^{++} and $N = 1$ SYM prepotential [I3, I4]). In particular, it is far from obvious how to generalize the linearized transformation (7.93) to the non-Abelian case.

It should be pointed out that this approach to $N = 2$ SYM resulted in one of the first proofs of the ultraviolet finiteness of $N = 4$ SYM and a number of $N = 2$ SYM–matter systems [H16].

7.3 $N = 2$ SYM action

The action for $N = 2$ SYM can be written down in a manifestly gauge invariant form in terms of $\overline{W}$ (using its chirality (7.86)):

$$S^{N=2}_{\rm SYM} = \frac{1}{4}\int d\zeta_R \,{\rm Tr}(\overline{W}\,\overline{W})\,. \tag{7.94}$$

The chiral superspace integral is defined as follows:

$$\int d\zeta_R = \int d^4x\, (\bar{D})^4\,, \qquad \int d^{12}X = \int d\zeta_R\, (D)^4 \tag{7.95}$$

and

$$(D)^4 \equiv \frac{1}{48} D^{\alpha i} D^j_\alpha D^\beta_i D_{\beta j} = \frac{1}{16}(D^{+\alpha} D^+_\alpha)(D^{-\beta} D^-_\beta)$$

and similarly for $(\bar{D})^4$.

Note that this chiral superspace integral is real owing to the reality property (7.87) of W, $\overline{W}$. As in the $N = 1$ case (7.56), it does not matter in what kind of gauge frame $\overline{W}$ is written down, since the trace is always manifestly chiral.

Our task now is to rewrite this action in terms of the prepotential V^{++}. For simplicity we do this only in the Abelian case (the non-Abelian form will be derived independently). Since $\overline{W}$ does not depend on $u^\pm$, one can insert a harmonic integral in (7.94). Then, using the expression (7.84) for $\overline{W}$, its $u^\pm$ independence (7.85) and chirality (7.86), the identity $D^{+\alpha} D^+_\alpha = D^{++}(D^{+\alpha} D^-_\alpha)$, the relations (7.79) and (7.73), one obtains

$$\begin{aligned} S^{N=2}_{\rm Maxwell} &= \frac{1}{4}\int d\zeta_R\, du\, \overline{W}^2 = -\frac{1}{16}\int d\zeta_R\, du\, D^{+\alpha} D^+_\alpha (V^{--}\overline{W}) \\ &= \frac{1}{16}\int d\zeta_R\, du D^{+\alpha} D^-_\alpha (D^{--} V^{++}\overline{W}) \\ &= -\frac{1}{16}\int d\zeta_R\, du\, D^{-\alpha} D^-_\alpha (V^{++}\overline{W}) \end{aligned}$$

$$
\begin{aligned}
&= \frac{1}{64}\int d\zeta_R\, du\, D^{-\alpha}D^{-}_{\alpha}(V^{++}D^{+\beta}D^{+}_{\beta}V^{--}) \\
&= \frac{1}{4}\int d\zeta_R\, du\, D^4(V^{++}V^{--}) = \frac{1}{4}\int d^{12}X\, du\, V^{++}V^{--}\,. \qquad (7.96)
\end{aligned}
$$

Further, the linearized part of (7.80) is just

$$
V^{--}(X,u) = \int du_1\, \frac{V^{++}(X,u_1)}{(u^+u_1^+)^2}\,, \tag{7.97}
$$

so one finds an expression in terms of V^{++} which coincides with the linearized (Abelian) action (7.26).

The full non-Abelian form of the action [Z4] can most easily be found by generalizing the Abelian term (7.26) in close analogy with the perturbative solution (7.80) for the connection V^{--}:

$$
S^{N=2}_{\text{SYM}} = \frac{1}{2}\sum_{n=2}^{\infty}\frac{(-i)^n}{n}\text{Tr}\int d^{12}X\, du_1\ldots du_n\, \frac{V^{++}(X,u_1)\ldots V^{++}(X,u_n)}{(u_1^+u_2^+)\ldots(u_n^+u_1^+)}\,. \tag{7.98}
$$

Since we already know that the linearized term in (7.98) is the right one, it will be sufficient to show that the full action is invariant under the non-Abelian gauge group. To this end let us first compute the variation of the action with respect to the prepotential V^{++}. Using (7.80), we find

$$
\begin{aligned}
\delta S &= \frac{1}{2}\text{Tr}\sum_{n=2}^{\infty}(-i)^n \\
&\quad \times \int d^{12}X\, du_1\ldots du_n\, \frac{\delta V^{++}(X,u_1)V^{++}(X,u_2)\ldots V^{++}(X,u_n)}{(u_1^+u_2^+)\ldots(u_n^+u_1^+)} \\
&= \frac{1}{2}\text{Tr}\int d^{12}X\, du_1\, \delta V^{++}(1)\sum_{n=2}^{\infty}(-i)^n \\
&\quad \times \int du_2\ldots du_n\, \frac{V^{++}(2)\ldots V^{++}(n)}{(u_1^+u_2^+)\ldots(u_n^+u_1^+)} \\
&= \frac{1}{2}\text{Tr}\int d^{12}X\, du\, \delta V^{++}(X,u)V^{--}(X,u)\,. \qquad (7.99)
\end{aligned}
$$

Now, the infinitesimal form of the gauge transformation (7.74) is

$$
\delta V^{++} = -D^{++}\lambda + i[\lambda, V^{++}] = -\mathcal{D}^{++}\lambda\,. \tag{7.100}
$$

Inserting this in (7.99), integrating D^{++} by parts, using the cyclic property of the trace and the relation between V^{++} and V^{--} (7.79), we obtain

$$
\delta S = \frac{1}{2}\text{Tr}\int d^{12}X\, du\, \lambda D^{--}V^{++} = \frac{1}{2}\text{Tr}\int du\, d\zeta^{(-4)}\, (D^+)^4[\lambda D^{--}V^{++}] = 0 \tag{7.101}
$$

since both λ and V^{++} are analytic.

The expression (7.99) for the variation of the action allows us to find a very simple form of the field equation for V^{++}:

$$\frac{\delta S}{\delta V^{++}} = 0 \quad \Rightarrow \quad (D^+)^4 V^{--} = 0 \qquad \text{or} \qquad (\bar{D}^+)^2 \overline{W}_\lambda = 0 \,. \tag{7.102}$$

For completeness we present the component form of the non-Abelian action above:

$$\begin{aligned} S^{N=2}_{\text{SYM}} &= \text{Tr} \int d^4x \left(-\frac{1}{4} F_{mn} F^{mn} + \mathcal{D}_m \phi \mathcal{D}^m \bar{\phi} - \frac{1}{2} [\phi, \bar{\phi}]^2 - i \psi^i \sigma^m \mathcal{D}_m \bar{\psi}_i \right. \\ &\left. - \frac{1}{\sqrt{2}} \psi^i [\bar{\phi}, \psi_i] - \frac{1}{\sqrt{2}} \bar{\psi}_i [\phi, \bar{\psi}^i] + \frac{1}{4} D^{ij} D_{ij} \right), \end{aligned} \tag{7.103}$$

where the fields are defined in the Wess–Zumino gauge (7.25).

Various possibilities of generating a mass for V^{++} coupled to hypermultiplets, via the super-Higgs effect or due to the appearance of central charges in $N = 2$ superalgebra (e.g., as a result of dimensional reduction from six-dimensional theory), were discussed in [D1, D2, D3, D25, E9, Z2]. We shall not dwell on these issues here.

In conclusion, we point out that the $N = 4$ SYM action can be written down in terms of $N = 2$ superfields as a sum of the $N = 2$ SYM action and the action of a q^+ hypermultiplet in the adjoint representation of the gauge group:

$$\begin{aligned} S^{N=4}_{\text{SYM}} &= S^{N=2}_{\text{SYM}} + \frac{1}{2} \text{Tr} \int du \, d\zeta^{(-4)} \, q^+_a \mathcal{D}^{++} q^{+a} \,, \\ \mathcal{D}^{++} q^{+a} &= D^{++} q^{+a} + i [V^{++}, q^{+a}] \,. \end{aligned} \tag{7.104}$$

Here $a = 1, 2$ is the index of the Pauli–Gürsey rigid $SU(2)$ symmetry (5.50), (5.51). The two extra supersymmetries are realized in terms of the gauge and matter superfields as follows:

$$\delta V^{++} = (\epsilon^{\alpha a} \theta^+_\alpha + \bar{\epsilon}^a_{\dot{\alpha}} \bar{\theta}^{+\dot{\alpha}}) \, q^+_a \,, \qquad \delta q^+_a = -\frac{1}{2} (D^+)^4 [(\epsilon^\alpha_a \theta^-_\alpha + \bar{\epsilon}_{\dot{\alpha} a} \bar{\theta}^{-\dot{\alpha}}) V^{--}] \,, \tag{7.105}$$

where ϵ^α_a and $\bar{\epsilon}^{\dot{\alpha} a} \equiv \overline{(\epsilon^\alpha_a)}$ are the relevant Grassmann parameters. Note that the variation of q^{+a} can be rewritten in the following manifestly gauge-invariant form:

$$\begin{aligned} \delta q^+_a &= \frac{1}{8} \{ (D^+)^2 [(\epsilon^\alpha_a \theta^-_\alpha + \bar{\epsilon}_{\dot{\alpha} a} \bar{\theta}^{-\dot{\alpha}}) W_\lambda] + (\bar{D}^+)^2 [(\epsilon^\alpha_a \theta^-_\alpha + \bar{\epsilon}_{\dot{\alpha} a} \bar{\theta}^{-\dot{\alpha}}) \overline{W}_\lambda] \\ &- (\epsilon^\alpha_a \theta^-_\alpha + \bar{\epsilon}_{\dot{\alpha} a} \bar{\theta}^{-\dot{\alpha}}) (D^+)^2 W_\lambda \} \,, \end{aligned} \tag{7.106}$$

where $\overline{W}_\lambda$, W_λ were defined in (7.84).

Alternatively, the $N = 4$ SYM action is equivalent to the coupling of $N = 2$ SYM to one ω hypermultiplet in the adjoint representation:

$$S^{N=4}_{\text{SYM}} = S^{N=2}_{\text{SYM}} - \frac{1}{2}\text{Tr}\int du\, d\zeta^{(-4)}\, (\mathcal{D}^{++}\omega)^2 \,. \tag{7.107}$$

The relation between the two forms is given in (5.55). In this case the extra supersymmetries are $\delta V^{++} = \epsilon^{\alpha i}u^+_i\theta^+_\alpha\omega + (\tilde{}\ \text{conj})$, $\delta\omega = -1/2(D^+)^4(\epsilon^{\alpha i}u^-_i\theta^-_\alpha V^{--}) + (\tilde{}\,\text{conj})$.

8
Harmonic supergraphs

In this chapter, which is based on refs. [G14, G15], we develop a manifestly supersymmetric quantization scheme for the $N = 2$ matter and SYM theories. The propagators of the harmonic superfields involve singular harmonic distributions. One might suspect, in principle, that this could lead to unwanted harmonic divergences, but we show that this is not the case. We give several examples of supergraph calculations, in particular, we show how the one-loop ultraviolet divergences cancel in the $N = 4$ SYM theory written down in terms of $N = 2$ superfields. We also present a two-loop calculation which produces a non-trivial finite result. As another application of the supergraph technique we prove in a direct and very simple way the ultraviolet finiteness of two-dimensional $N = 4$ sigma models.

8.1 Analytic delta functions

In order to find the propagators (Green's functions) for the various harmonic superfields, we need a delta function for the analytic superspace. To begin with, we recall the definition of the delta function for the ordinary $N = 2$ superspace:

$$\int d^{12}X_2\, \delta^{12}(X_1 - X_2)\Phi(X_2) = \Phi(X_1)\,, \tag{8.1}$$

where

$$\delta^{12}(X_1 - X_2) \equiv \delta^4(x_1 - x_2)\delta^8(\theta_1 - \theta_2)$$

and the Grassmann delta function is normalized so that

$$\int d^8\theta\, \delta^8(\theta) = 1\,.$$

It is straightforward to extend this definition to the $N = 2$ harmonic superspace:

$$\int du_2\, d^{12}X_2\, \delta^{12}(X_1 - X_2)\, \delta^{(q,-q)}(u_1, u_2)\Phi^{(q)}(X_2, u_2) = \Phi^{(q)}(X_1, u_1) \tag{8.2}$$

(note the matching charges with respect to the first and second harmonic variables).

Now, consider an analytic superfield treated as a function of the coordinates of the full superspace, $\Phi^{(q)}(\zeta(X,u),u)$, and write down

$$\int du_2\, d^{12}X_2\, \delta^{12}(X_1 - X_2)\, \delta^{(q,-q)}(u_1,u_2)\Phi^{(q)}(\zeta(X_2,u_2),u_2) \\ = \Phi^{(q)}(\zeta(X_1,u_1),u_1)\,. \tag{8.3}$$

Using the identity (5.42) and regarding ζ as independent variables one finds

$$\int du_2\, d^{(-4)}\zeta_2\, [(D_2^+)^4\delta^{12}(X_1 - X_2)]\, \delta^{(q,-q)}(u_1,u_2)\Phi^{(q)}(\zeta_2,u_2) = \Phi^{(q)}(\zeta_1,u_1)\,. \tag{8.4}$$

Hence one reads off the definition of the analytic delta function:

$$\begin{aligned} \delta_A^{(q,4-q)}(\zeta_1,u_1|\zeta_2,u_2) &= (D_2^+)^4\delta^{12}(X_1 - X_2)\, \delta^{(q,-q)}(u_1,u_2) \\ &= (D_1^+)^4\delta^{12}(X_1 - X_2)\, \delta^{(q-4,4-q)}(u_1,u_2)\,. \end{aligned} \tag{8.5}$$

The second form is derived with the help of the property

$$D_{1i}\delta^{12}(X_1 - X_2) = -D_{2i}\delta^{12}(X_1 - X_2)$$

and the identity (4.36).

The analytic delta function (8.5) can also be written down in an alternative form explicitly involving the analytic superspace coordinates:

$$\begin{aligned} \delta_A^{(q,4-q)}(\zeta_1,u_1|\zeta_2,u_2) &= \delta^4(x_{A1} - x_{A2})\, [\theta_1^+ - (u_1^+u_2^-)\theta_2^+]^4\, \delta^{(q-4,4-q)}(u_1,u_2) \\ &= \delta^4(x_{A1} - x_{A2})\, [(u_2^+u_1^-)\theta_1^+ - \theta_2^+]^4\, \delta^{(q,-q)}(u_1,u_2)\,. \end{aligned} \tag{8.6}$$

Note that the delta function in the form (8.5) is not manifestly analytic at both points; similarly, in the form (8.6) it is not manifestly supersymmetric. Both properties are only achieved because of the presence of the harmonic delta function. As we shall see later on, in some supergraph calculations this harmonic delta function, which will appear as a part of a propagator, may be multiplied by other singular harmonic distributions. In such a case the harmonic delta function must be regularized and neither of the two expressions in (8.5) will be analytic anymore. Then we will have to use an alternative, manifestly analytic form of the delta function:

$$\delta_A^{(q,4-q)}(1|2) = -\frac{1}{2\Box}(D_1^+)^4(D_2^+)^4\delta^{12}(X_1 - X_2)\,(D_2^{--})^2\delta^{(q-4,4-q)}_{\text{regularized}}(u_1,u_2)\,. \tag{8.7}$$

When the regularization can be removed, the new form (8.7) becomes equivalent to the short form (8.5) owing to the identity

$$-\frac{1}{2}(D^+)^4(D^{--})^2\Phi(\zeta,u) = \Box\Phi(\zeta,u) \tag{8.8}$$

valid for any analytic superfield Φ. In (8.7) such a superfield is $(D_1^+)^4\delta^{12}(X_1 - X_2)\delta^{(q-4,4-q)}(u_1,u_2) \equiv \delta_A^{(q,4-q)}(1|2)$, so one arrives at (8.5).

In conclusion, we mention that the analytic delta function above is the analog of the chiral ones in $N = 1$ superspace. For the left-handed chiral superspace one has

$$\delta_L^6(\zeta_{L1} - \zeta_{L2}) = (\bar{D}_2)^2\delta^8(X_1 - X_2) = (\bar{D}_1)^2\delta^8(X_1 - X_2) \tag{8.9}$$

or, in terms of the chiral superspace coordinates alone,

$$\delta_L^6(\zeta_{L1} - \zeta_{L2}) = \delta^4(x_{L1} - x_{L2})\,(\theta_1 - \theta_2)^2 .$$

8.2 Green's functions for hypermultiplets

The starting point for deriving Green's functions is the free action with sources. In the case of the q^+ hypermultiplet it is

$$S = \int du\, d\zeta^{(-4)}\, (-\tilde{q}^+ D^{++} q^+ + \tilde{q}^+ J^{(+3)} + q^+ \tilde{J}^{(+3)}) . \tag{8.10}$$

The variation with respect to $\tilde{q}^+$ gives the equation of motion

$$D^{++}q^+(\zeta,u) = J^{(+3)}(\zeta,u) . \tag{8.11}$$

Its solution is given by a Green's function,

$$q^+(\zeta_1,u_1) = \int du_2\, d\zeta_2^{(-4)} G^{(1,1)}(\zeta_1,u_1|\zeta_2,u_2) J^{(+3)}(\zeta_2,u_2) , \tag{8.12}$$

which must be analytic with respect to both arguments and must satisfy the equation

$$D_1^{++}G^{(1,1)}(1|2) = \delta_A^{(3,1)}(1|2) . \tag{8.13}$$

It is not hard to see that the following manifestly analytic expression:

$$G^{(1,1)}(1|2) = -\frac{1}{\Box}(D_1^+)^4(D_2^+)^4\delta^{12}(X_1 - X_2)\frac{1}{(u_1^+u_2^+)^3} \tag{8.14}$$

is the solution of eq. (8.13). Indeed, using the property (4.47) of the harmonic distribution $(u_1^+u_2^+)^{-3}$ and the identity (8.8) one finds

$$\begin{aligned} D_1^{++}G^{(1,1)}(1|2) &= -\frac{1}{\Box}(D_1^+)^4(D_2^+)^4\delta^{12}(X_1 - X_2)\,\frac{1}{2}(D_1^{--})^2\delta^{(3,-3)}(1|2) \\ &= \delta_A^{(3,1)}(1|2) . \end{aligned} \tag{8.15}$$

Note that the Green's function (8.14) is antisymmetric:

$$G^{(1,1)}(1|2) = -G^{(1,1)}(2|1) \tag{8.16}$$

and real in the sense of $\widetilde{\ }$ conjugation.

There exists an alternative form of the Green's function (8.14) involving analytic superspace coordinates only:

$$G^{(1,1)}(1|2) = \frac{1}{4i\pi^2} \frac{(u_1^+ u_2^+)}{\hat{x}_{12}^2} \,. \tag{8.17}$$

Here

$$\begin{aligned} \hat{x}_{12}^m &= x_{A1}^m - x_{A2}^m + \frac{2i}{(u_1^+ u_2^+)} [(u_1^- u_2^+)\theta_1^+ \sigma^m \bar{\theta}_1^+ - (u_1^+ u_2^-)\theta_2^+ \sigma^m \bar{\theta}_2^+ \\ &\quad + \theta_1^+ \sigma^m \bar{\theta}_2^+ + \theta_2^+ \sigma^m \bar{\theta}_1^+] \end{aligned} \tag{8.18}$$

is a coordinate difference invariant under supersymmetry. The easiest way to prove that (8.17) solves eq. (8.13) is to make a finite supersymmetry transformation after which $\theta_1 = \bar{\theta}_1 = 0$. In this frame the operator D_1^{++} is reduced to just a partial harmonic derivative. Then one expands the expression in (8.17) in θ_2^+, $\bar{\theta}_2^+$ and sees that it contains only one term with a harmonic singularity:

$$\frac{(u_1^+ u_2^-)^2}{(u_1^+ u_2^+)} (\theta_2^+)^4 4i\pi^2 \delta^4(x_{12}) \,,$$

where the identity

$$\Box_1 \frac{1}{x_{12}^2} = 4i\pi^2 \delta^4(x_{12}) \tag{8.19}$$

has been used. This expression clearly satisfies (8.13) in the frame $\theta_1 = \bar{\theta}_1 = 0$.

The Green's function for the ω hypermultiplet is equally easy to find. There the analogs of (8.10) and (8.11) are

$$S = \int du\, d\zeta^{(-4)} \left[-\frac{1}{2}(D^{++}\omega)^2 + \omega J^{(+4)} \right], \tag{8.20}$$

$$-(D^{++})^2 \omega(\zeta, u) = J^{(+4)}(\zeta, u) \,. \tag{8.21}$$

The solution to the equation of motion is given by the expression

$$\omega(\zeta_1, u_1) = \int du_2\, d\zeta_2^{(-4)}\, G^{(0,0)}(\zeta_1, u_1|\zeta_2, u_2) J^{(+4)}(\zeta_2, u_2) \,, \tag{8.22}$$

provided the Green's function $G^{(0,0)}(1|2)$ satisfies the equation

$$(D_1^{++})^2 G^{(0,0)}(1|2) = \delta_{\mathrm{A}}^{(4,0)}(1|2) \,. \tag{8.23}$$

The solution of (8.23) differs from (8.14) by a harmonic factor:

$$G^{(0,0)}(1|2) = \frac{1}{\Box}(D_1^+)^4(D_2^+)^4\delta^{12}(X_1 - X_2)\frac{(u_1^- u_2^-)}{(u_1^+ u_2^+)^3}\,. \tag{8.24}$$

Indeed,

$$\begin{aligned}
(D_1^{++})^2\frac{(u_1^- u_2^-)}{(u_1^+ u_2^+)^3} &= D_1^{++}\left[\frac{(u_1^+ u_2^-)}{(u_1^+ u_2^+)^3} + \frac{1}{2}(u_1^- u_2^-)(D_1^{--})^2\delta^{(3,-3)}(1|2)\right] \\
&= \frac{1}{2}(u_1^+ u_2^-)(D_1^{--})^2\delta^{(3,-3)}(1|2) \\
&= \frac{1}{2}(D_1^{--})^2\left[(u_1^+ u_2^-)\delta^{(3,-3)}(1|2)\right] \\
&= \frac{1}{2}(D_1^{--})^2\delta^{(4,-4)}(1|2)\,.
\end{aligned} \tag{8.25}$$

Here we have used (4.34) and (4.47). Once again, (8.8) leads to (8.23). It is clear that $G^{(0,0)}(1|2)$ is real and symmetric.

The alternative form of this Green's function is

$$G^{(0,0)}(1|2) = -\frac{1}{4i\pi^2}\frac{(u_1^+ u_2^+)(u_1^- u_2^-)}{\hat{x}_{12}^2}\,. \tag{8.26}$$

8.3 $N = 2$ SYM: Gauge fixing, Green's functions and ghosts

The starting point for the quantization of the $N = 2$ SYM theory is the linearized form of the action (see (7.26)):

$$S_{\text{lin}}^{N=2} = \frac{1}{4}\text{Tr}\int du_1\, du_2\, d^{12}X\, \frac{V^{++}(1)V^{++}(2)}{(u_1^+ u_2^+)^2}\,. \tag{8.27}$$

Inserting a second superspace integral with a delta function and splitting the full superspace integral

$$\int d^{12}X = \int d\zeta^{(-4)}(D^+)^4\,, \tag{8.28}$$

we can rewrite (8.27) in the following form

$$S_{\text{lin}}^{N=2} = -\frac{1}{4}\text{Tr}\int du_1\, du_2\, d\zeta_1^{(-4)}\, d\zeta_2^{(-4)}\; V^{++}(1)\Box\Pi^{(2,2)}(1|2)V^{++}(2)\,, \tag{8.29}$$

where

$$\Pi^{(2,2)}(1|2) = -\frac{1}{\Box}(D_1^+)^4(D_2^+)^4\delta^{12}(X_1 - X_2)\frac{1}{(u_1^+ u_2^+)^2}\,. \tag{8.30}$$

The analytic distribution (8.30) is a projection operator, i.e., it has the property

$$\int du_2\, d\zeta_2^{(-4)}\ \Pi^{(2,2)}(1|2)\Pi^{(2,2)}(2|3) = \Pi^{(2,2)}(1|3)\,. \tag{8.31}$$

To see this one uses $(D_2^+)^4$ from, e.g., $\Pi^{(2,2)}(2|3)$ to restore the full measure $d^{12}X_2$ (see (8.28)) and then integrates out X_2 with the help of $\delta^{12}(X_2 - X_3)$. The result is

$$\int du_2\, d\zeta_2^{(-4)}\ \Pi^{(2,2)}(1|2)\Pi^{(2,2)}(2|3) = \int du_2\, \frac{(D_1^+)^4(D_2^+)^4(D_3^+)^4}{\Box^2(u_1^+u_2^+)^2(u_2^+u_3^+)^2}\delta^{12}(X_1-X_2)\,. \tag{8.32}$$

Further, with the help of the identity

$$(D_2^+)^4\delta^{12}(X_1-X_2) = \frac{1}{4}D_2^0(D_2^+)^4\delta^{12}(X_1-X_2) = \frac{1}{4}D_2^{++}D_2^{--}(D_2^+)^4\delta^{12}(X_1-X_2) \tag{8.33}$$

one pulls out the harmonic derivative D_2^{++}, integrates by parts and obtains (using (4.47))

$$\begin{aligned}\int du_2\, d\zeta_2^{(-4)}\ &\Pi^{(2,2)}(1|2)\Pi^{(2,2)}(2|3) = \\ &-\frac{(D_1^+)^4(D_3^+)^4}{4\Box^2}\int du_2 \left[\frac{D_2^{--}\delta^{(2,-2)}(u_2,u_1)}{(u_2^+u_3^+)^2}\right. \\ &+\left.\frac{D_2^{--}\delta^{(2,-2)}(u_2,u_3)}{(u_1^+u_2^+)^2}\right] D_2^{--}(D_2^+)^4\delta^{12}(X_1-X_3)\,.\end{aligned} \tag{8.34}$$

The next step is to integrate D_2^{--} by parts and to use the harmonic delta functions to do the u_2 integral, thus identifying $(D_2^+)^4$ with $(D_1^+)^4$ or $(D_3^+)^4$. The only non-vanishing contribution is obtained when D_2^{--} hits $D_2^{--}(D_2^+)^4\delta^{12}(X_1 - X_3)$ and the identity (8.8) can be applied. The result is just $\Pi^{(2,2)}(1|3)$, so (8.31) does indeed take place.

The operator (8.30) projects out superisospin 0 from the analytic superfield $V^{++}(\zeta, u)$. In general, V^{++} contains superspin 0 multiplets with superisospins $0, 1, 2, \ldots$ (see Chapter 7). The condition which annihilates all but the superisospin 0 multiplets in V^{++} is

$$D^{++}V^{++} = 0 \tag{8.35}$$

(see the discussion of the tensor multiplet in Section 6.2). This is precisely the condition satisfied by the operator (8.30):

$$D_1^{++}\Pi^{(2,2)}(1|2) = D_2^{++}\Pi^{(2,2)}(1|2) = 0\,. \tag{8.36}$$

Indeed,

$$D_1^{++}\Pi^{(2,2)}(1|2) = -\frac{1}{\Box}(D_1^+)^4(D_2^+)^4D_1^{--}\delta^{(2,-2)}(u_1,u_2)\delta^{12}(X_1-X_2) = 0\,, \tag{8.37}$$

since one D^{--} is not sufficient to prevent $(D_1^+)^4(D_2^+)^4$ from vanishing on the harmonic δ function. Note that it is precisely the property (8.36) that is responsible for the invariance of the linearized action (8.29) under the gauge transformations $\delta V^{++} = -D^{++}\lambda$.

One can say that there is a close analogy between (8.29) and the linearized action for $N = 0$ SYM:

$$S^{N=0}_{\text{lin}} = \frac{1}{2}\text{Tr}\int d^4x\ A^m \Box \Pi_{mn} A^n\,, \tag{8.38}$$

where

$$\Pi_{mn} = \eta_{mn} - \frac{\partial_m \partial_n}{\Box} \tag{8.39}$$

is the projection operator for spin 1 in the field $A_m(x)$, satisfying the condition

$$\partial^m \Pi_{mn} = 0\,. \tag{8.40}$$

Condition (8.35) is a gauge-fixing condition. In particular, among the components of this equation one finds the usual Lorentz gauge condition on the gauge field,

$$\partial^m A_m(x) = 0\,. \tag{8.41}$$

The way to show that (8.41) is a proper gauge-fixing condition is to perform a gauge transformation $\delta A_m = -\partial_m \lambda$. The resulting condition on the parameter $\Box\lambda = \partial^m A_m(x)$ can be solved since the operator $\Box$ is invertible. In the case of eq. (8.35) the argument is the same. The condition on the gauge parameter $\lambda(\zeta, u)$ following from the linearized variation of (8.35),

$$(D^{++})^2\lambda = D^{++}V^{++}\,, \tag{8.42}$$

has a unique solution, since the operator $(D^{++})^2$ in (8.42) is just the kinetic operator for the ω hypermultiplet which is invertible (see Section 8.2).

Projection operators like (8.30) and (8.39) are non-invertible, so one cannot directly find Green's functions for gauge fields. The remedy is well known in the case $N = 0$: It consists in adding a gauge-fixing term to the action (8.38):

$$S^{N=0}_{\text{gf}} = \frac{1}{2\alpha}\text{Tr}\int d^4x\ A^m \Box\,(\eta_{mn} - \Pi_{mn})\,A^n\,. \tag{8.43}$$

Here

$$\eta_{mn} - \Pi_{mn} = \frac{\partial_m \partial_n}{\Box} \tag{8.44}$$

is the projection operator for the remaining (gauge) degree of freedom in A_m (in this case spin 0). Putting the two terms together one obtains the gauge-fixed action

$$S^{N=0} = \frac{1}{2\alpha}\text{Tr}\int d^4x\ A^m \Box A_m + \frac{1}{2}\left(1 - \frac{1}{\alpha}\right)\text{Tr}\int d^4x\ A^m \Box \Pi_{mn} A^n\,. \tag{8.45}$$

A particularly simple and convenient choice is the Fermi–Feynman gauge $\alpha = 1$:

$$S_{\rm FF}^{N=0} = \frac{1}{2}{\rm Tr}\int d^4x\ A^m \Box A_m\,. \tag{8.46}$$

In the case $N = 2$ we can do exactly the same. The gauge-fixing term can be written down as follows:

$$\begin{aligned} S_{\rm gf}^{N=2} &= -\frac{1}{4\alpha}{\rm Tr}\int du_1\, du_2\, d\zeta_1^{(-4)} d\zeta_2^{(-4)} \\ &\quad \times V^{++}(1)\Box\left(\delta_{\rm A}^{(2,2)}(1|2) - \Pi^{(2,2)}(1|2)\right) V^{++}(2)\,. \end{aligned} \tag{8.47}$$

Putting together (8.29) and (8.47) and doing the $du_2\, d\zeta_2$ integral in the term with the delta function, we obtain the following gauge-fixed linearized $N = 2$ SYM action:

$$S_{\rm SYM}^{N=2} = -\frac{1}{4\alpha}{\rm Tr}\int du\, d\zeta^{(-4)}\ V^{++}\Box V^{++} - \frac{1}{4}\left(1 - \frac{1}{\alpha}\right){\rm Tr}\int V\Box\Pi V\,. \tag{8.48}$$

Once again, the simplest choice is the Fermi–Feynman gauge $\alpha = 1$:

$$S_{\rm FF}^{N=2} = -\frac{1}{4}{\rm Tr}\int du\, d\zeta^{(-4)}\ V^{++}\Box V^{++}\,. \tag{8.49}$$

It is now obvious that the kinetic operator in (8.49) is no longer degenerate, so one can immediately find the $N = 2$ SYM Green's function in the Fermi–Feynman gauge:

$$G_{\rm FF}^{(2,2)}(1|2) = \frac{2}{\Box}\delta_{\rm A}^{(2,2)}(1|2) \tag{8.50}$$

or, more explicitly,

$$G_{\rm FF}^{(2,2)}(1|2) = \frac{1}{2i\pi^2\,(x_{A1} - x_{A2})^2}\left[\theta_1^+ - (u_1^+ u_2^-)\theta_2^+\right]^4 \delta^{(-2,2)}(u_1, u_2)\,. \tag{8.51}$$

Note that due to the presence of the Grassmann and harmonic delta functions in (8.51) one can replace the analytic x_A by the central basis ones.

The final step in the quantization of the SYM theory is the introduction of Faddeev–Popov ghosts. As usual, they are similar to matter fields of the same type as the gauge parameters, except for their statistics. In complete analogy with the $N = 0$ Faddeev–Popov term

$$S_{\rm FP}^{N=0} = {\rm Tr}\int d^4x\ \bar c\,[-\partial^m\,(\partial_m + iA_m)]c\,, \tag{8.52}$$

one can write down in the case $N = 2$:

$$S_{\rm FP}^{N=2} = {\rm Tr}\int du\, d\zeta^{(-4)}\ \tilde c\, D^{++}(D^{++} + iV^{++})\, c\,. \tag{8.53}$$

Here $c, \tilde{c}$ are anticommuting ghost analytic superfields similar to the ω hypermultiplet. Indeed, the gauge-fixing condition (8.42) on the parameter resembles the equation of motion for ω. Therefore, the kinetic term in (8.53) reproduces that of ω in (5.58), just like the kinetic term in (8.52) is related to that of a scalar field.

In conclusion, we point out that the gauge-fixed action (8.49) together with the Faddeev–Popov term (8.53) is invariant under the following BRST transformations:

$$\begin{aligned} \delta V^{++} &= \xi\,(D^{++} + iV^{++})\,c\,, \\ \delta\tilde{c} &= \frac{1}{2}\xi\,(D_1^+)^4 \int du_2\, \frac{(u_1^- u_2^-)}{(u_1^+ u_2^+)^3} D_2^{++} V^{++}(2)\,, \\ \delta c &= i\xi\, c^2\,, \end{aligned} \tag{8.54}$$

where ξ is an anticommuting constant parameter.

8.4 Feynman rules

The Feynman rules that we present in this section are based on the Green's functions found in Sections 8.1 and 8.2, as well as on the various interactions of matter and SYM superfields studied in Chapters 5 and 7. As usual (see, for instance, [I1]), these rules are derived from a generating functional of the form

$$Z = \int \mathcal{D}\tilde{q}^+\, \mathcal{D}q^+\, \mathcal{D}\omega\, \mathcal{D}V^{++}\, \mathcal{D}\tilde{c}\, \mathcal{D}c\; e^{iS}\,, \tag{8.55}$$

where the action of hypermultiplets of the q^+ and ω types, of the SYM multiplet V^{++} (gauge fixed) and the ghosts $\tilde{c}, c$ also involves their sources:

$$\begin{aligned} S &= S[\tilde{q}^+, q^+, \omega, V^{++}, \tilde{c}, c] \\ &\quad + \int du\, d\zeta^{(+4)} \big(\tilde{q}^+ J^{(+3)} + q^+ \tilde{J}^{(+3)} + \omega J^{(+4)} \\ &\quad + V^{++} J^{(+2)} + \tilde{c}\eta^{(+4)} + \tilde{\eta}^{(+4)} c\big). \end{aligned} \tag{8.56}$$

Here we give the Feynman rules in momentum space; their configuration space analogs are easily obtained. We begin with the propagator for a hypermultiplet superfield q^+ in a certain representation (real or complex) of the gauge group (see (8.14)):

$$1\,r \xrightarrow{\quad p \quad} 2\,s$$

$$\langle \tilde{q}^{+r}(p, \theta_1, u_1) q_s^+(p, \theta_2, u_2)\rangle = \frac{i}{p^2} \frac{(D_1^+)^4 (D_2^+)^4}{(u_1^+ u_2^+)^3} \delta^8(\theta_1 - \theta_2)\delta_s^r\,. \tag{8.57}$$

The propagator for a real hypermultiplet ω_a is given by (8.24):

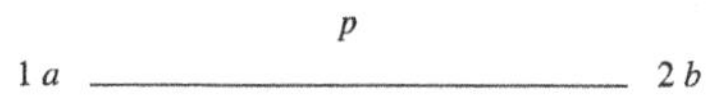

1 a —— p —— 2 b

$$\langle \omega_a(1)\omega_b(2)\rangle = \frac{i}{p^2}(D_1^+)^4(D_2^+)^4\delta^8(\theta_1-\theta_2)\frac{(u_1^-u_2^-)}{(u_1^+u_2^+)^3}\delta_{ab}\,. \tag{8.58}$$

The ghost propagator is just the same:

1 a — — p — — 2 b

$$\langle \tilde{c}_a(1)c_b(2)\rangle = \frac{i}{p^2}(D_1^+)^4(D_2^+)^4\delta^8(\theta_1-\theta_2)\frac{(u_1^-u_2^-)}{(u_1^+u_2^+)^3}\delta_{ab}\,. \tag{8.59}$$

The SYM propagator in the Fermi–Feynman gauge is (see (8.50)):

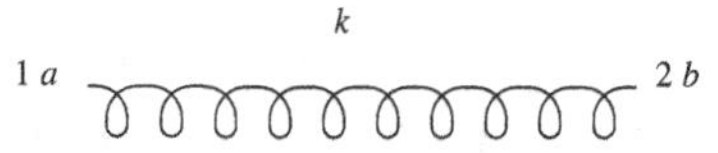

1 a ~~~ k ~~~ 2 b

$$\langle V^{++}(1)V^{++}(2)\rangle = \frac{2i}{k^2}(D_1^+)^4\delta^8(\theta_1-\theta_2)\delta^{(-2,2)}(u_1,u_2)\delta_{ab}\,. \tag{8.60}$$

Note the important difference between (8.60) and all the matter propagators (8.57), (8.58), (8.59). The latter contain non-trivial harmonic distributions, hence the necessity to have eight spinor derivatives $(D_1^+)^4(D_2^+)^4$, which make the matter propagators explicitly analytic. On the other hand, the SYM propagator contains a harmonic δ function, so only four spinor derivatives $(D_1^+)^4$ are sufficient to make it analytic with respect to the first as well as to the second argument. Nevertheless, there exists an alternative 'longer' form of the same propagator where analyticity is manifest (see (8.7)):

$$\langle V^{++}(1)V^{++}(2)\rangle = \frac{4i}{k^4}(D_1^+)^4(D_2^+)^4\delta^8(\theta_1-\theta_2)(D_2^{--})^2\delta^{(2,-2)}(u_1,u_2)\delta_{ab}\,. \tag{8.61}$$

This alternative form is to be used when the harmonic δ function in (8.61) needs to be regularized.

Now we start considering the matter and SYM couplings.* Firstly, a q^+ hypermultiplet can interact with SYM in a minimal way (see (7.17)):

$$S_{q^+|\mathrm{SYM}} = -ig\int du\, d\zeta^{(-4)}\, \tilde{q}^{+r}V_a^{++}(T_a)_r^s q_s^+\,, \tag{8.62}$$

* We use the following conventions. The generators T_a of the Yang–Mills group form the algebra $[T_a, T_b] = if_{abc}T_c$; the trace of two generators is $\mathrm{Tr}(T_aT_b) = T(R)\delta_{ab}$; in the adjoint representation $(t_a)_{bc} = -if_{abc}$ and $\mathrm{Tr}(t_at_b) = f_{acd}f_{bcd} = \delta_{ab}$. The Yang–Mills coupling constant is introduced by the replacement $V^{++} \to gV^{++}$.

where T_a are the generators of the matter representation of the Yang–Mills group. The corresponding vertex

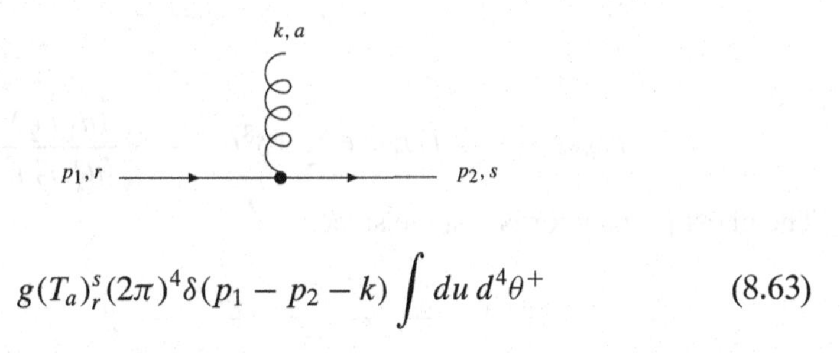

$$g(T_a)_r^s(2\pi)^4\delta(p_1 - p_2 - k)\int du\, d^4\theta^+ \tag{8.63}$$

contains the θ^+ and harmonic integrals remaining from the analytic measure $du\, d\zeta^{(-4)}$ after converting the space-time integral into a momentum one with measure $(2\pi)^{-4}d^4p$. Further, q^+ hypermultiplets can self-interact in many ways (see Chapter 5). As an example we take the Taub–NUT self-interaction (5.85):

$$S_{q^+|q^+} = -\frac{\lambda}{2}\int du\, d\zeta^{(-4)}\, \tilde{q}^{+r}\tilde{q}^{+s}q_r^+q_s^+ \tag{8.64}$$

giving rise to the vertex

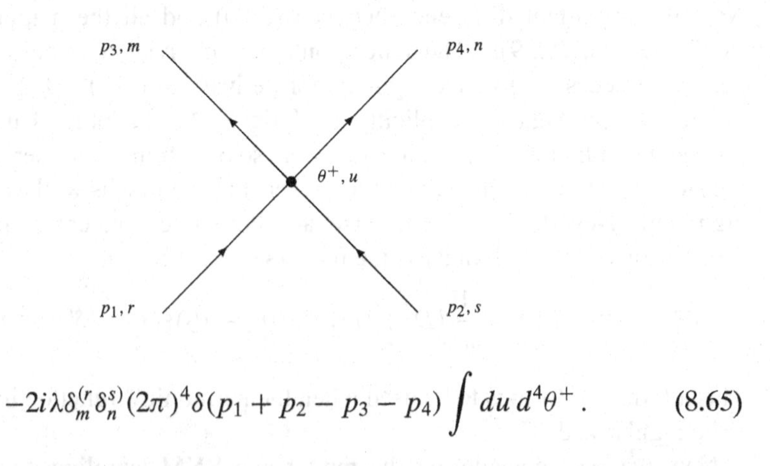

$$-2i\lambda\delta_m^{(r}\delta_n^{s)}(2\pi)^4\delta(p_1 + p_2 - p_3 - p_4)\int du\, d^4\theta^+ . \tag{8.65}$$

The minimal coupling of ω hypermultiplets in the adjoint representation to SYM is (see (7.107))

$$S_{\omega|\mathrm{SYM}} = -\frac{1}{2}\int du\, d\zeta^{(-4)}\, (\mathcal{D}^{++}\omega)_a(\mathcal{D}^{++}\omega)_a\,, \tag{8.66}$$

where $(\mathcal{D}^{++}\omega)_a = [(D^{++} + igV^{++})\omega]_a = D^{++}\omega_a - gf_{abc}V_b^{++}\omega_c$. There are two vertices corresponding to this coupling:

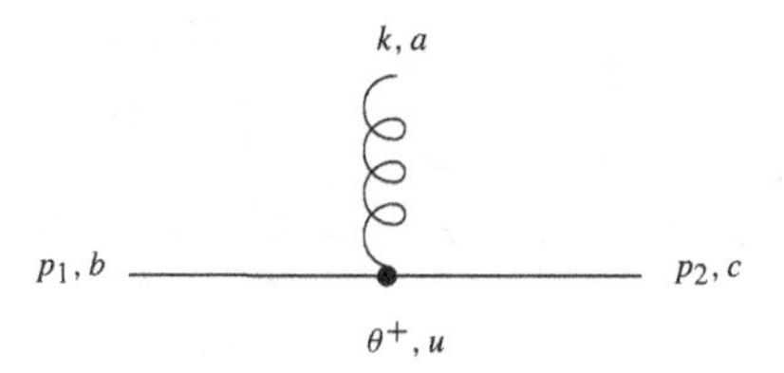

$$-igf_{abc}(2\pi)^4\delta(p_1 - p_2 - k) \int du\, d^4\theta^+ \, (D^{++}_{(b)} - D^{++}_{(c)}) \,, \tag{8.67}$$

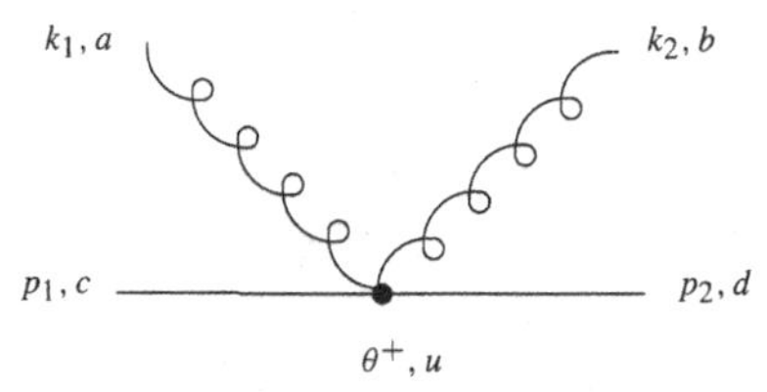

$$-ig^2(f_{ade}f_{ebc} + f_{ace}f_{ebd})(2\pi)^4\delta(p_1 - p_2 - k_1 - k_2) \int du\, d^4\theta^+ \,. \tag{8.68}$$

The ghost-SYM vertex can be read off from (8.53) and is similar to the $\omega V \omega$ one (8.67):

k, a

p_1, b p_2, c

θ^+, u

$$-igf_{abc}(2\pi)^4\delta(p_1 - p_2 - k) \int du\, d^4\theta^+ \, D^{++}_{(b)} \,. \tag{8.69}$$

At each vertex considered so far one integrates over all the internal momenta (with measure $(2\pi)^{-4}d^4p$). An integral $\int du\, d^4\theta^+$ (coming from $\int du\, d\zeta^{(-4)}$ in the coordinate representation) is also present. Inspecting the propagators one can see that at each analytic vertex there are factors of $(D^+)^4$ coming from the propagators which can be used to restore the full Grassmann measure $d^8\theta$ at the vertex. This will often be the first step in the supergraph calculations shown below.

Finally, there is an infinite number of $N = 2$ SYM self-couplings which are given in (7.98). For example, the cubic interaction term is

$$\begin{aligned} S^3_{\rm SYM} &= \frac{1}{2g^2}\frac{(-ig)^3}{3}\mathrm{Tr}\int d^{12}X\, du_1\, du_2\, du_3 \\ &\quad \times \frac{V^{++}(X,u_1)V^{++}(X,u_2)V^{++}(X,u_3)}{(u_1^+u_2^+)(u_2^+u_3^+)(u_3^+u_1^+)} \end{aligned}$$

$$= \frac{ig}{12}\mathrm{Tr}\int d^{12}X\, du_1\, du_2\, du_3\, \frac{[V^{++}(1), V^{++}(2)]V^{++}(3)}{(u_1^+u_2^+)(u_2^+u_3^+)(u_3^+u_1^+)}$$

$$= -\frac{g}{12}\int d^{12}X\, du_1\, du_2\, du_3\, \frac{f_{abc}V_a^{++}(1)V_b^{++}(2)V_c^{++}(3)}{(u_1^+u_2^+)(u_2^+u_3^+)(u_3^+u_1^+)}, \quad (8.70)$$

where we have used the cyclic property of the trace and the antisymmetry of the harmonic integrals. The corresponding three-particle vertex is

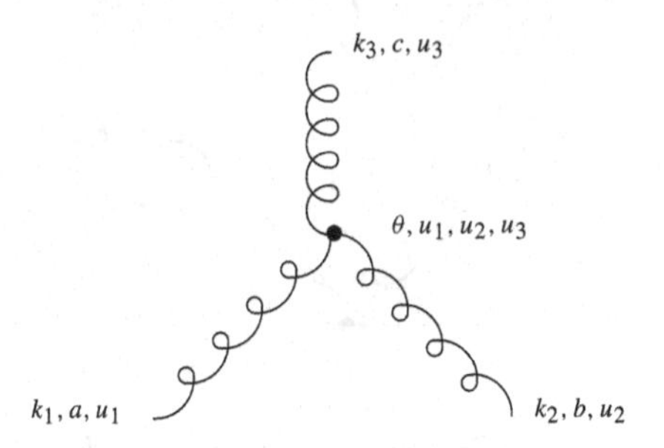

$$-\frac{ig}{2}f_{abc}(2\pi)^4\delta(k_1+k_2+k_3)\int d^8\theta\, du_1\, du_2\, du_3\, \frac{1}{(u_1^+u_2^+)(u_2^+u_3^+)(u_3^+u_1^+)}. \quad (8.71)$$

Note two important new features. Unlike all the other vertices above, this one is non-local in harmonic space and involves an integral with the full Grassmann measure $d^8\theta$ (not just the analytic one). The same applies to all the n-particle SYM vertices.

8.5 Examples of supergraph calculations. Absence of harmonic divergences

Our first example is the one-loop correction to the four-point function for a self-interacting q^+ hypermultiplet (Figure 8.1). It will help us illustrate some of the main rules for handling harmonic supergraphs.

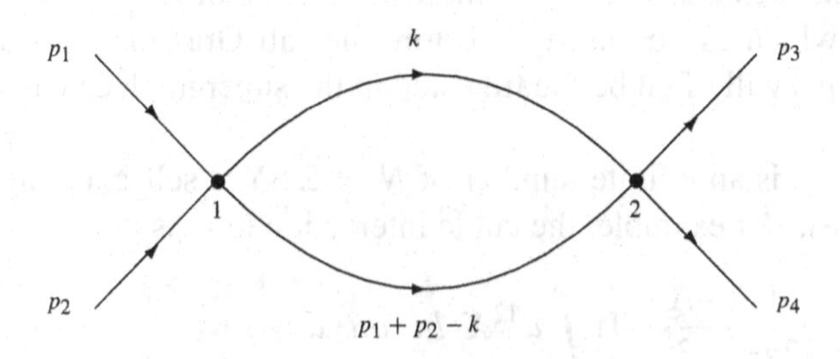

Fig. 8.1. One-loop correction to the four-point function of a self-interacting q^+ hypermultiplet.

The corresponding analytic expression is

$$\Gamma = -\frac{\lambda^2}{4}\int \frac{d^4p_1\dots d^4p_4\,d^4k}{(2\pi)^{16}k^2(p_1+p_2-k)^2}\,du_1\,du_2\,d^4\theta_1^+\,d^4\theta_2^+\;\delta(p_1+p_2-p_3-p_4)$$
$$\times\,\tilde{q}^+(p_1,\theta_1,u_1)\tilde{q}^+(p_2,\theta_1,u_1)q^+(p_3,\theta_2,u_2)q^+(p_4,\theta_2,u_2)$$
$$\times\frac{(D_1^+)^4(D_2^+)^4}{(u_1^+u_2^+)^3}\delta^8(\theta_1-\theta_2)\frac{(D_1^+)^4(D_2^+)^4}{(u_1^+u_2^+)^3}\delta^8(\theta_1-\theta_2)\,. \tag{8.72}$$

Note the presence of the two harmonic distributions $(u_1^+u_2^+)^{-3}$ with coincident singularities. For the time being we assume that they are regularized. Later on we shall see that in the process of doing one of the θ integrals there will appear positive powers of $(u_1^+u_2^+)$ which will cancel out one of the singular factors.

The general rule for handling expressions like (8.72) is to first do all but one of the θ integrals using the Grassmann δ functions from the propagators. For this purpose one has to restore the full measures $d^8\theta_1\,d^8\theta_2$. This can be achieved by removing $(D_1^+)^4(D_2^+)^4$ from one of the propagators and using (8.28) (note that the other propagator and the external line superfields are analytic, so D_1^+, D_2^+ do not act on them). Then one applies the identity

$$\delta^8(\theta_1-\theta_2)(D_1^+)^4(D_2^+)^4\delta^8(\theta_1-\theta_2) = (u_1^+u_2^+)^4\delta^8(\theta_1-\theta_2) \tag{8.73}$$

and does the θ_2 integral. Note that the factor $(u_1^+u_2^+)^4$ from (8.73) cancels out one of the singular harmonic distributions in (8.72), so one can safely remove the regularization. Thus one obtains

$$\Gamma = -\frac{\lambda^2}{4}\int \frac{d^4p_1\dots d^4p_4\,d^4k}{(2\pi)^{16}k^2(p_1+p_2-k)^2}\frac{du_1\,du_2\,d^8\theta}{(u_1^+u_2^+)^2}\delta(p_1+p_2-p_3-p_4)$$
$$\times\,\tilde{q}^+(p_1,\theta,u_1)\tilde{q}^+(p_2,\theta,u_1)q^+(p_3,\theta,u_2)q^+(p_4,\theta,u_2)\,. \tag{8.74}$$

This expression is still non-local in harmonic space, like the effective action in x space. The non-locality can be removed if one puts the external lines on shell, i.e., $D^{++}\tilde{q}^+ = 0$ which implies $\tilde{q}^+ = \bar{q}^i u_i^+$. Then

$$\tilde{q}^+(u_1)\tilde{q}^+(u_1) = \frac{1}{2}D_1^{++}D_1^{--}\left[\tilde{q}^+(u_1)\tilde{q}^+(u_1)\right], \tag{8.75}$$

and the u_2 integral can be computed:

$$\begin{aligned}
&\frac{1}{2}\int du_1\,du_2\;D_1^{++}D_1^{--}(\tilde{q}_1^+\tilde{q}_1^+)\frac{q_2^+q_2^+}{(u_1^+u_2^+)^2}\\
&= -\frac{1}{2}\int du_1\,du_2\;D_1^{--}(\tilde{q}_1^+\tilde{q}_1^+)q_2^+q_2^+D_1^{--}\delta^{(2,-2)}(u_1,u_2)\\
&= \frac{1}{2}\int du\;\tilde{q}^+\tilde{q}^+(D^{--})^2(q^+q^+)\\
&= \bar{q}^i\bar{q}^j q^k q^l\int du\;u_i^+u_j^+u_k^-u_l^- = \frac{1}{3}\bar{q}^i\bar{q}^j q_i q_j\,.
\end{aligned} \tag{8.76}$$

Remaining off shell one can remove the harmonic non-locality by fixing the superisospins of the external fields (just as one fixes the external leg momenta), e.g.,

$$q^+_{(k)}(p,\theta,u) = (u^+)^{i_1\dots i_{k+1}}(u^-)^{i_{k+2}\dots i_{2k+1}} q_{(i_1\dots i_{2k+1})}(p,\theta)\,. \tag{8.77}$$

One then multiplies the four such fields in the numerator and decomposes the product of harmonic u_1 and u_2 monomials into $SU(2)$ irreps. Among them only the singlet contributes to the harmonic integral $\int du_1\, du_2$. This singlet has charges (+2, +2), so it must be of the form of the denominator. Thus one avoids any potential harmonic divergences. The remaining u integral is computed easily.

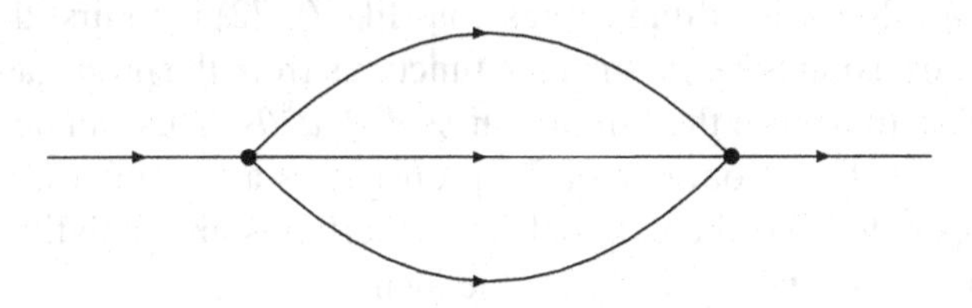

Fig. 8.2. An example of a two-loop supergraph.

The treatment of the graph in Figure 8.2 is similar. One removes the spinor derivatives and restores the Grassmann measures at the vertices. The application of (8.73) then produces enough factors of $(u_1^+u_2^+)$ to cancel all the harmonic denominators but one, thus eliminating the risk of coincident harmonic singularities. The harmonic integrals can be computed on shell and one can see that the quadratically divergent part of the graph vanishes.

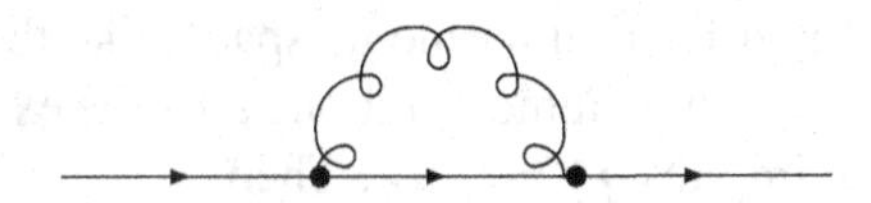

Fig. 8.3. One-loop self-energy for q^+ coupled to SYM.

A second kind of example is the one-loop self-energy correction for q^+ coupled to SYM shown in Figure 8.3. Here the problem is the SYM propagator which contains a harmonic delta function. The latter must be regularized because it multiplies the singular distribution from the q^+ propagator. Then one has to use the longer form (8.61) of the SYM propagator which is manifestly analytic. After restoring the Grassmann measures and doing the θ_2 integral in the same way as in the previous examples one obtains a factor $(u_1^+u_2^+)^4$ which cancels out the denominator $(u_1^+u_2^+)^{-3}$ from the matter propagator. Thus the occurrence of coincident harmonic singularities is avoided and one can remove the regularization of the delta function. At this stage the graph contribution amounts to (suppressing the momentum integral):

$$\begin{aligned}\Gamma &\sim \int d^8\theta\, du_1\, du_2\, (u_1^+ u_2^+)(D_2^{--})^2 \delta^{(-2,2)}(u_1,u_2)\tilde{q}^+(\theta,u_1)q^+(\theta,u_2) \\ &= -2\int d^8\theta\, du\, \tilde{q}^+ D^{--} q^+ = -2\int du\, d^4\theta^+\, \tilde{q}^+ (D^+)^4 D^{--} q^+ = 0\,. \end{aligned} \tag{8.78}$$

Here we have integrated by parts $(D_2^{--})^2$ and have used the delta function and the analyticity of q^+. We see that the graph vanishes, so the (divergent) momentum integral does not need any special treatment.

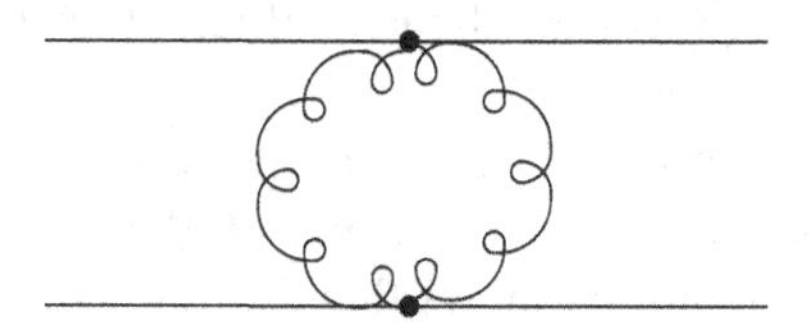

Fig. 8.4. Harmonic singularity of the type $(u_1^+ u_2^+)^2[\delta(u_1,u_2)]^2$.

A further example of a potentially dangerous graph is shown in Figure 8.4. The external lines are of the ω type and the vertices are given in (8.68). This time we deal with two harmonic delta functions with coincident singularities. Once again we have to use (8.61). After the θ integration we get

$$\Gamma \sim \int d^8\theta\, du_1\, du_2\, (u_1^+ u_2^+)^4 \left[(D_2^{--})^2 \delta^{(-2,2)}(u_1,u_2)\right]^2 (\omega(1))^2\, (\omega(2))^2\,. \tag{8.79}$$

This expression is still not well defined. The problem can be illustrated by the following example [G30]. The square of a delta function $[\delta(x)]^2$ is, of course, ill defined. Multiplying it by a single power of x improves the situation a little,

$$x[\delta(x)]^2 = c\delta(x)\,, \tag{8.80}$$

although the constant in (8.80) is not uniquely fixed. However, the product

$$x^2[\delta(x)]^2 = 0 \tag{8.81}$$

is uniquely defined and vanishes. The situation in (8.79) is not yet as good as this, since the derivatives D^{--} can neutralize the factors $(u_1^+ u_2^+)$. Fortunately, we can pull out some more of the latter in the following way:

$$\begin{aligned}\int d^8\theta[\omega(1)]^2[\omega(2)]^2 &= \int d^4\theta_1^+[\omega(1)]^2(D_1^+)^4[\omega(2)]^2 \\ &= \int d^4\theta_1^+[\omega(1)]^2(D_1^+)^4(D_2^+)^4\Omega^{(-4)}(2)\,. \end{aligned} \tag{8.82}$$

Here we have written down the analytic superfield $[\omega(2)]^2$ in a manifestly analytic form with some non-analytic 'prepotential' $\Omega^{(-4)}(2)$. Further,

$$(D_1^+)^4(D_2^+)^4 = \frac{1}{256}(D_1^+)^2(\bar{D}_1^+)^2(\bar{D}_2^+)^2(D_2^+)^2 = \frac{1}{16}(u_1^+u_2^+)^2(D_1^+)^2(\bar{D})^4(D_2^+)^2\,, \tag{8.83}$$

so we find

$$\Gamma \sim \int du_1\, du_2\, d^4\theta_1^+\, (u_1^+u_2^+)^6\left[(D_2^{--})^2\delta^{(-2,2)}(u_1,u_2)\right]^2$$
$$\times\, [\omega(1)]^2(D_1^+)^2(\bar{D})^4(D_2^+)^2\Omega^{(-4)}(2)\,. \tag{8.84}$$

The powers of $(u_1^+u_2^+)$ in (8.84) are already sufficient to allow us to apply (8.81), after which the graph vanishes.

The examples shown above illustrate the possibilities of having coincident harmonic singularities and the mechanisms of their suppression. A careful analysis of all possible situations shows that dealing with harmonic supergraphs is completely safe, no new specific harmonic divergences occur [G24].

In conclusion, we give an example of a supergraph computation with a non-trivial result. It is the one-loop correction to the V^{++} self-energy in $N = 2$ SYM. The relevant graphs are shown in Figures 8.5a,b,c.

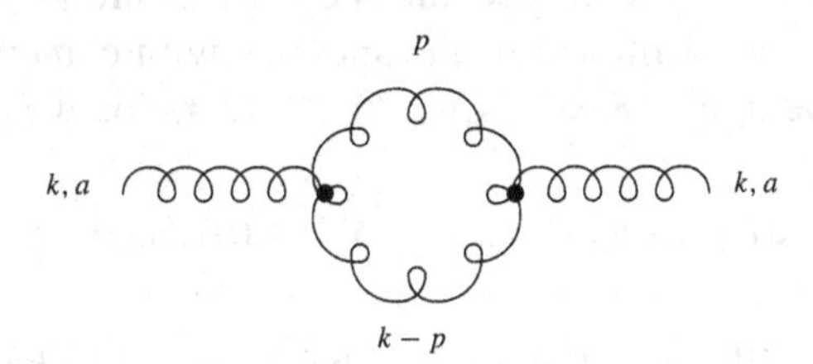

Fig. 8.5. (a) SYM contribution to the one-loop V^{++} self-energy.

The SYM contribution (Figure 8.5a) is

$$\begin{aligned}\Gamma^{\rm SYM} &= \frac{g^2}{2}\int \frac{d^4k\, d^4p}{(2\pi)^8 p^2(k-p)^2}\, d^8\theta_1\, d^8\theta_2\, du_1\, du_2\, du_3\, dw_1\, dw_2\, dw_3 \\ &\times \frac{V_a^{++}(k,\theta_1,u_1)V_a^{++}(k,\theta_2,w_1)}{(u_1^+u_2^+)(u_2^+u_3^+)(u_3^+u_1^+)(w_1^+w_2^+)(w_2^+w_3^+)(w_3^+w_1^+)} \\ &\times (D^+(\theta_1,u_2))^4\delta^8(\theta_1-\theta_2)\delta^{(-2,2)}(u_2,w_2) \\ &\times (D^+(\theta_1,u_3))^4\delta^8(\theta_1-\theta_2)\delta^{(-2,2)}(u_3,w_3)\end{aligned} \tag{8.85}$$

(the overall factor of 1/2 is a symmetry factor). We have used the short form (8.60) of the SYM propagator, since in (8.85) there are no coincident harmonic singularities. Unlike the previous examples, the Grassmann measures are already complete, so we can start pulling the derivatives $D^+(\theta_1,u_2)$ off the first delta function. The result vanishes unless all of them hit the second delta function, since

$$\delta^8(\theta_1 - \theta_2)(D)^k\delta^8(\theta_1 - \theta_2) = 0 \qquad \text{if} \qquad k < 8\,. \tag{8.86}$$

Next we apply (8.73) and do the w_2 and w_3 integrals with the help of the harmonic delta functions. The result is

$$\begin{aligned}\Gamma^{\text{SYM}} &= \frac{g^2}{2}\int \frac{d^4k\, d^4p}{(2\pi)^8 p^2(k-p)^2}\, d^8\theta\, du_1\, du_2\, du_3\, dw_1 \\ &\quad \times \frac{(u_2^+u_3^+)^2 V_a^{++}(k,\theta,u_1)V_a^{++}(k,\theta,w_1)}{(u_1^+u_2^+)(u_1^+u_3^+)(w_1^+u_2^+)(w_1^+u_3^+)}\,.\end{aligned} \tag{8.87}$$

To do the u_3 and w_1 integrals we pull out D_2^{++} and D_3^{++} from $(u_2^+u_3^+)^2$ and integrate by parts. The final result is

$$\Gamma^{\text{SYM}} = -g^2\int \frac{d^4k\, d^4p\; d^8\theta\, du_1\, du_2}{(2\pi)^8 p^2(k-p)^2}\, \frac{(u_1^-u_2^-)}{(u_1^+u_2^+)}\, V_a^{++}(1)V_a^{++}(2)\,. \tag{8.88}$$

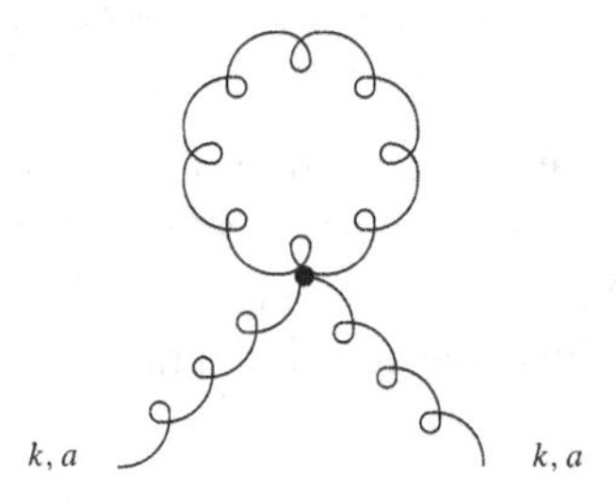

Fig. 8.5. (b) Tadpole contribution to the one-loop V^{++} self-energy.

There also exists a tadpole pure SYM diagram of the same order (Figure 8.5b). In the case $N = 0$ it is known to vanish in dimensional regularization. Here we find a simpler reason for this. Indeed, the graph contribution is given by the harmonic and Grassmann integral

$$\begin{aligned}&\int d^8\theta\, du_1\, du_2\, du_3\, du_4\; V_a^{++}(1)V_a^{++}(2) \\ &\quad\times \frac{[(D_3^+)^4(D_4^+)^4\delta^8(\theta_3-\theta_4)]_{\theta_3=\theta_4}(D_3^{--})^2\delta^{(2,-2)}_{\text{reg}}(u_3,u_4)}{(u_1^+u_2^+)(u_2^+u_3^+)(u_3^+u_4^+)(u_4^+u_1^+)}\end{aligned} \tag{8.89}$$

(the four-point vertex can be read off from the action (7.98); we drop the momentum integral). Here we had to use the longer (regularized) form (8.61) of the SYM propagator, because of the harmonic singularity at the vertex. Now, $[(D_3^+)^4(D_4^+)^4\delta^8(\theta_3-\theta_4)]_{\theta_3=\theta_4} = (u_3^+u_4^+)^4$, so the singular denominator is canceled and the harmonic regularization can be removed. Then the remaining factor $(u_3^+ u_4^+)^3$ goes through the derivatives $(D_3^{--})^2$ and is annihilated against the harmonic delta function.

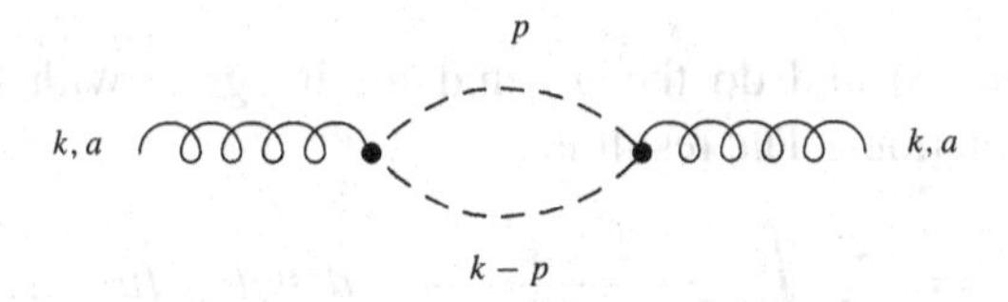

Fig. 8.5. (c) Ghost contribution to the one-loop V^{++} self-energy.

Finally, the ghost contribution in Figure 8.5c is

$$\begin{aligned}\Gamma^{\text{ghost}} &= g^2 \int \frac{d^4k\, d^4p\, du_1\, du_2\, d^4\theta_1^+\, d^4\theta_2^+}{(2\pi)^8 p^2 (k-p)^2} V_a^{++}(1) V_a^{++}(2) \\ &\times (D_1^+)^4 (D_2^+)^4 \delta^8(\theta_1 - \theta_2) D_1^{++} \left(\frac{(u_1^- u_2^-)}{(u_1^+ u_2^+)^3} \right) \\ &\times (D_1^+)^4 (D_2^+)^4 \delta^8(\theta_1 - \theta_2) D_2^{++} \left(\frac{(u_1^- u_2^-)}{(u_1^+ u_2^+)^3} \right) \end{aligned} \tag{8.90}$$

(an overall minus sign is due to the fermion loop). As in our first example, one restores the measures $d^8\theta_1\, d^8\theta_2$, then applies (8.73) and does the θ_2 integral. The factor of $(u_1^+ u_2^+)^4$ from (8.73) cancels out one of the denominators completely and the other partially. Thus, the multiplication of coincident singularities is avoided. The final result is

$$\Gamma^{\text{ghost}} = g^2 \int \frac{d^4k\, d^4p\, d^8\theta\, du_1\, du_2}{(2\pi)^8 p^2 (k-p)^2} \frac{(u_1^+ u_2^-)(u_1^- u_2^+)}{(u_1^+ u_2^+)^2} V_a^{++}(1) V_a^{++}(2)\,. \tag{8.91}$$

Putting together (8.88) and (8.91) we obtain the total one-loop self-energy contribution:

$$\Gamma = -g^2 \int \frac{d^4k\, d^4p\, d^8\theta\, du_1\, du_2}{(2\pi)^8 p^2 (k-p)^2} \frac{V_a^{++}(1) V_a^{++}(2)}{(u_1^+ u_2^+)^2}\,. \tag{8.92}$$

The logarithmically divergent part of (8.92) has the form of the linearized SYM action (8.27), as expected from a renormalizable theory.

One more example is the computation of the one-loop contribution of q^+ hypermultiplet matter to the SYM self-energy (Figure 8.6). The calculation is similar to that for the graph in Figure 8.1 and the result is

$$\Gamma^{q^+} = g^2 T(R) \int \frac{d^4k\, d^4p\, d^8\theta\, du_1\, du_2}{(2\pi)^8 p^2 (k-p)^2} \frac{V_a^{++}(1) V_a^{++}(2)}{(u_1^+ u_2^+)^2}\,, \tag{8.93}$$

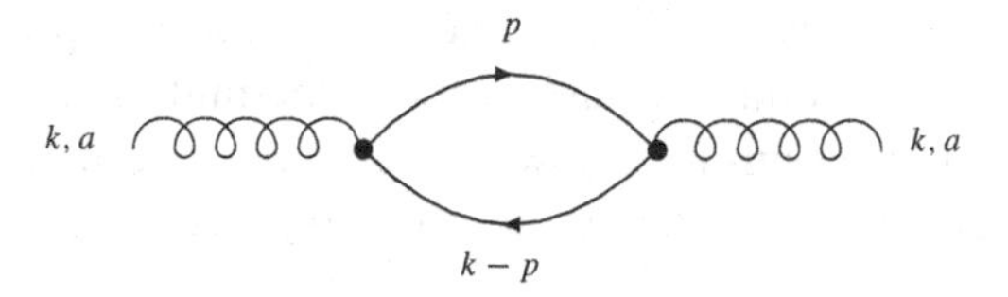

Fig. 8.6. q^+ matter contribution to the one-loop V^{++} self-energy.

where $T(R)$ is the trace of the matter representation. We see that this contribution exactly cancels Γ (8.92), provided $T(R) = 1$ (as for the adjoint representation), or if there are n hypermultiplets q_s in representations R_s such that

$$\sum_{s=1}^{n} T(R_s) = 1 . \tag{8.94}$$

As pointed out in (7.104), $N = 2$ SYM coupled to hypermultiplet matter in the adjoint representation is equivalent to the $N = 4$ SYM theory. The cancellation of the one-loop contributions to the gluon self-energy shown above is a well-known feature of this theory [W12], as well as of a class of $N = 2$ theories satisfying the condition (8.94) [D10, H17].

8.6 A finite four-point function at two loops

In this section we present a two-loop calculation which produces a finite non-trivial result. In the framework of the $N = 4$ SYM theory mentioned above one can define gauge invariant composite operators. The simplest example is the $N = 4$ stress tensor (or supercurrent). When decomposed into $N = 2$ superfields it gives rise to a number of current-like objects. For instance,

$$J^{++} = \mathrm{Tr}(\widetilde{q}^+ q^+) \tag{8.95}$$

is (classically) conserved due to the hypermultiplet field equation:

$$D^{++} J^{++} = \mathrm{Tr}(\mathcal{D}^{++}\widetilde{q}^+ \, q^+ + \widetilde{q}^+ \, \mathcal{D}^{++} q^+) = 0 . \tag{8.96}$$

In fact, its lowest component $J^{++}(\theta = 0) = u^{+i} u^+_j \mathrm{Tr}(\bar{\phi}_i \phi^j)$ is an $SU(2)$ part of the $SU(4)$ R symmetry current of the $N = 4$ theory. The remaining $SU(4)$ generators are obtained by taking different bilinear combinations of $\widetilde{q}^+$, q^+ and the $N = 2$ SYM field strength W.

Let us now consider the correlation function of four such currents:

$$\langle J^{++}(1) J^{++}(2) J^{++}(3) J^{++}(4) \rangle . \tag{8.97}$$

Such correlators are of interest because they provide the simplest test cases for the so-called AdS/CFT correspondence conjecture[M1, G42, W16]. As we know, $N = 4$ SYM is a finite, thus conformally invariant theory. Therefore, the correlator (8.97) should have well-defined (super)conformal properties. Another interesting property is found by using the hypermultiplet Green's function equation (8.13) and by assuming that we keep away from the coincident points. Then we can expect (8.97) to satisfy the following Schwinger–Dyson equation at point 1:

$$D_1^{++}\langle J^{++}(1)J^{++}(2)J^{++}(3)J^{++}(4)\rangle = 0 \qquad \text{if points } x_1 \neq x_2 \neq x_3 \neq x_4 \tag{8.98}$$

and similar equations at the other points. This property generalizes to any correlation function of gauge invariant hypermultiplet composite operators and can be called 'H-analyticity' [H20, H21].

Our task now is to compute the first non-trivial radiative correction to the lowest component (i.e., evaluated at $\theta^+_{1,2,3,4} = 0$) of this correlator. At order g^2 we find three types of two-loop graphs:

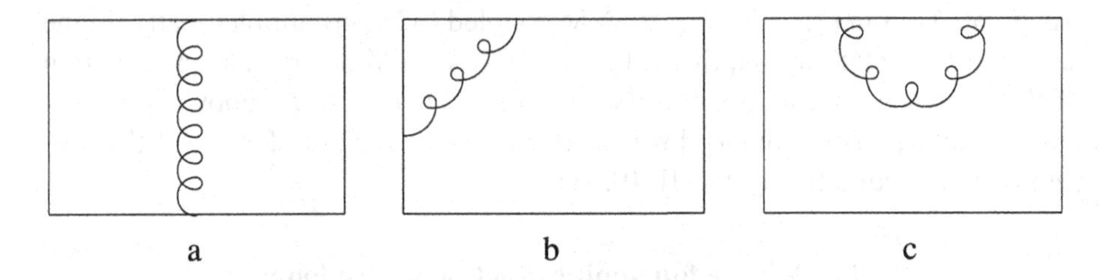

Fig. 8.7. Two-loop contributions to $\langle J^{++}(1)J^{++}(2)J^{++}(3)J^{++}(4)\rangle$.

and those obtained by permuting the points.

The one-loop subgraph appearing in Figure 8.7c is the same as that in Figure 8.3, and therefore vanishes. The graphs in Figures 8.7a,b can be obtained from the building block depicted in Figure 8.8 by attaching hypermultiplet propagators and, for the graphs in Figure 8.7b, by identifying two of the external points.

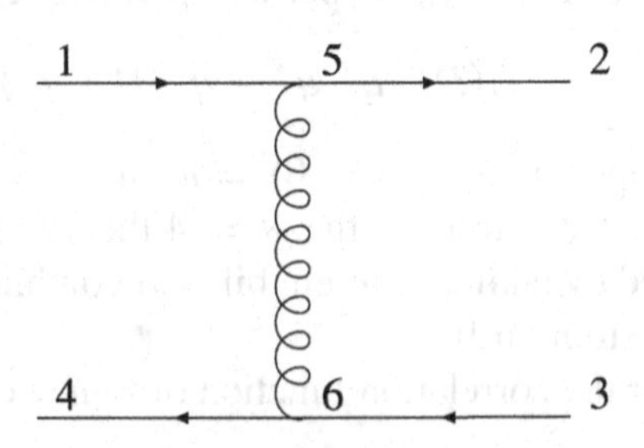

Fig. 8.8. Building block I.

The calculation of this building block is most easily done using the coordinate space form of the Feynman rules, in particular, the hypermultiplet (8.17) and gluon (8.51) propagators. Denoting the corresponding analytic expression by I, we find (up to a numerical factor containing the dependence on the gauge group):

$$I = \int du_{5,6}\, d^4\theta^+_{5,6}\, d^4x_{A5,6}\, \frac{(15)(52)(36)(64)}{\hat{x}^2_{15}\hat{x}^2_{52}\hat{x}^2_{36}\hat{x}^2_{64}} \times \frac{1}{x^2_{A56}}(\theta^+_5 - (56^-)\theta^+_6)^4 \delta^{(-2,2)}(u_5, u_6)\,. \tag{8.99}$$

Here we have used the short-hand notation, e.g.,

$$(12) \equiv u^{+i}_1 u^+_{2i}\,, \qquad (12^-) \equiv u^{+i}_1 u^-_{2i}\,, \qquad \text{etc.} \tag{8.100}$$

The presence of Grassmann and harmonic delta functions in eq. (8.99) allows us to immediately carry out the integrations $\int du_6\, d^4\theta^+_6$. Further, remembering that we only want to compute the first component in the expansion of I, we can set $\theta_{1,2,3,4} = 0$. The result is

$$I_0 = \int du_5\, d^4\theta^+_5\, d^4x_{5,6}\, \frac{(15)(52)(35)(54)}{x^2_{56}} \times \left[x_{15} - 2i\frac{(15^-)}{(15)}\theta^+_5\sigma\bar{\theta}^+_5\right]^{-2} \left[x_{52} + 2i\frac{(5^-2)}{(52)}\theta^+_5\sigma\bar{\theta}^+_5\right]^{-2} \times \left[x_{36} - 2i\frac{(35^-)}{(35)}\theta^+_5\sigma\bar{\theta}^+_5\right]^{-2} \left[x_{64} + 2i\frac{(5^-4)}{(54)}\theta^+_5\sigma\bar{\theta}^+_5\right]^{-2}. \tag{8.101}$$

Note that we have dropped the index A of the analytic basis space-time coordinates. At the external points, setting $\theta_{1,2,3,4} = 0$ makes the difference between central and analytic basis irrelevant. At the interaction points 5 and 6, changing the basis amounts to a shift of the space-time variables, but the integral is translation invariant. We can profit from this freedom to shift the integration variables $x_{5,6}$ as follows:

$$x^m_5 \to x^m_5 - 2i\frac{(5^-2)}{(52)}\theta^+_5\sigma^m\bar{\theta}^+_5\,, \qquad x^m_6 \to x^m_6 - 2i\frac{(5^-4)}{(54)}\theta^+_5\sigma^m\bar{\theta}^+_5\,, \tag{8.102}$$

so that x_{52} and x_{64} in (8.101) loose their nilpotent extensions. With the help of the harmonic cyclic identity, e.g.,

$$(15)(25^-) - (25)(15^-) = (12) \tag{8.103}$$

eq. (8.101) becomes

$$I_0 = \int du_5\, d^4\theta^+_5\, d^4x_{5,6}\, \frac{(15)(52)(35)(54)}{x^2_{56}x^2_{52}x^2_{64}} \times \left[x_{15} + 2i\frac{(12)}{(15)(25)}\theta^+_5\sigma\bar{\theta}^+_5\right]^{-2} \left[x_{36} + 2i\frac{(34)}{(35)(45)}\theta^+_5\sigma\bar{\theta}^+_5\right]^{-2}. \tag{8.104}$$

It is now clear that the nilpotent terms in (8.104) can be rewritten as external shifts of the space-time integral at points 1 and 3:

$$I_0 = \int du_5\, d^4\theta_5^+ \exp\left[2i\left(\frac{(12)}{(15)(25)}\partial_1^m + \frac{(34)}{(35)(45)}\partial_3^m\right)\theta_5^+\sigma_m\bar\theta_5^+\right] \times (15)(52)(35)(54)\int \frac{d^4x_{5,6}}{x_{15}^2x_{25}^2x_{36}^2x_{46}^2x_{56}^2}. \qquad (8.105)$$

Obviously, only the quadratic term in the expansion of the exponential can contribute to the Grassmann integral, so one finds

$$I_0 = -(12)^2\frac{4i\pi^2}{x_{12}^2}g_2\int du_5\frac{(35)(45)}{(15)(25)} - (34)^2\frac{4i\pi^2}{x_{34}^2}g_4\int du_5\frac{(15)(25)}{(35)(45)} - (12)(34)\, 2\partial_1\cdot\partial_3 f. \qquad (8.106)$$

Here

$$f(x_1,x_2,x_3,x_4) = \int\frac{d^4x_5\, d^4x_6}{x_{15}^2x_{25}^2x_{36}^2x_{46}^2x_{56}^2} \qquad (8.107)$$

is a two-loop space-time integral and, e.g.,

$$g_2(x_1,x_3,x_4) = \frac{x_{12}^2}{4i\pi^2}\Box_1 f(x_1,x_2,x_3,x_4) = \int\frac{d^4x_5}{x_{15}^2x_{35}^2x_{45}^2} \qquad (8.108)$$

is a one-loop space-time integral.

The last step is to compute the harmonic integrals, for example,

$$\begin{aligned}\int du_5\frac{(35)(45)}{(15)(25)} &= \int du_5\frac{[D_5^{++}(35^-)](45)}{(15)(25)}\\ &= \int du_5(35^-)(45)\left[\frac{1}{(25)}\delta^{(-1,1)}(1,5) + \frac{1}{(15)}\delta^{(-1,1)}(2,5)\right]\\ &= -\frac{(31^-)(41)}{(12)} + \frac{(32^-)(42)}{(12)}. \end{aligned} \qquad (8.109)$$

Finally, the building block from Figure 8.8 becomes

$$\begin{aligned} I_0 &= (12)[(31^-)(41) - (32^-)(42)]\frac{4i\pi^2}{x_{12}^2}g_2\\ &\quad + (34)[(13^-)(23) - (14^-)(24)]\frac{4i\pi^2}{x_{34}^2}g_4\\ &\quad - (12)(34)\, 2\partial_1\cdot\partial_3 f. \end{aligned} \qquad (8.110)$$

The building block appearing in Figure 8.7b is obtained by identifying two end points (for instance, points 2 and 4 for the configuration (8.110)). To complete the graphs in Figure 8.7a,b one needs to multiply the building blocks by two free hypermultiplet propagators. Putting all terms together we arrive at the following final result:

$$\langle J^{++}(1)J^{++}(2)J^{++}(3)J^{++}(4)\rangle = (14)^2(23)^2A_1 + (12)^2(34)^2A_2 + (12)(23)(34)(41)A_3\,, \qquad (8.111)$$

where the coefficients are given by the one-loop integrals g_i and by particular combinations of derivatives of the two-loop integral f. In fact, they can be further simplified [E1, E2]:

$$A_1 = \frac{4\pi^4}{x_{14}^2x_{23}^2x_{12}^2x_{34}^2}\Phi^{(1)}\Big(\frac{x_{14}^2x_{23}^2}{x_{12}^2x_{34}^2}, \frac{x_{13}^2x_{24}^2}{x_{12}^2x_{34}^2}\Big), \qquad A_2 = A_1(2 \leftrightarrow 4)\,, \quad (8.112)$$

$$A_3 = 4\pi^4\frac{x_{13}^2x_{24}^2 - x_{14}^2x_{23}^2 - x_{12}^2x_{34}^2}{x_{12}^4x_{34}^4x_{23}^2x_{14}^2}\Phi^{(1)}\left(\frac{x_{14}^2x_{23}^2}{x_{12}^2x_{34}^2}, \frac{x_{13}^2x_{24}^2}{x_{12}^2x_{34}^2}\right). \qquad (8.113)$$

Here the function $\Phi^{(1)}$ corresponds to a well-known one-loop integral, the so-called scalar box [U1]:

$$\Phi^{(1)}(x, y) = \frac{1}{\lambda}\left\{2\Big(\mathrm{Li}_2(-\rho x) + \mathrm{Li}_2(-\rho y)\Big) + \ln\frac{y}{x}\ln\frac{1+\rho y}{1+\rho x} + \ln(\rho x)\ln(\rho y) + \frac{\pi^2}{3}\right\} \qquad (8.114)$$

and

$$\lambda(x, y) \equiv \sqrt{(1-x-y)^2 - 4xy}\,, \qquad \rho(x, y) \equiv 2(1-x-y+\lambda)^{-1}\,, \quad (8.115)$$

Li_2 denoting the Euler dilogarithm.

As expected, this correlator obviously satisfies the harmonic analyticity condition (8.98) and has the required conformal properties. In particular, the non-trivial function $\Phi^{(1)}$ is manifestly conformally invariant.

8.7 Ultraviolet finiteness of $N = 4$, $d = 2$ supersymmetric sigma models

The quantization scheme developed in this chapter provides us with a direct and simple proof of the ultraviolet finiteness of $N = 4$ supersymmetric sigma models in two space-time dimensions. As shown in Chapter 11, such sigma models are in one-to-one correspondence with the self-interacting theories of q^+ hypermultiplets. The general (4,4) $d = 2$ sigma model is described by the action

$$S = \int du\, d^2x\, d^4\theta^+\ \big[q_a^+ D^{++} q^{+a} + L^{+4}(q^+, u)\big]\,, \qquad (8.116)$$

where L^{+4} is an arbitrary function. This action is obtained by dimensional reduction from (5.82) (in the gauge $L_a^+ = q_a^+$). Therefore, we can use the Feynman rules developed earlier in this chapter. The propagator is the same as (8.57), while the vertices are obtained from the expansion of L^{+4} in powers of q^+:

$$L^{+4}(q^+, u) = \sum_k \frac{\lambda^k}{k!} L^{(4-k)}(u)(q^+)^k . \tag{8.117}$$

Here λ is the expansion parameter (the coupling constant).

All this allows us to perform a direct power-counting analysis of the ultraviolet properties of the sigma models. The crucial point in the whole argument is the fact that although the action (8.116) is an analytic superspace integral, the contribution of each graph in perturbation theory is given by a full superspace integral. To see this one should go through the standard D algebra described above. We briefly repeat the main steps. One has to remove the spinor derivatives from all but one of the Grassmann delta functions in a given loop. To this end one restores the full superspace integrals $\int d^8\theta$ at all the vertices using spinor derivatives from the manifestly analytic propagators. After that the remaining spinor derivatives may be moved around by integration by parts. In the process some of them may hit external lines, the rest go over to other internal lines. Once the delta function in a propagator has been cleared, the $d^8\theta$ integration can be done at one of the vertices that the propagator links. One continues in this way until all the delta functions in the loop have been used and the loop shrinks to a point in θ space. The same is repeated with the other loops, and one ends up with a graph contribution that is local in the θ coordinates under a single full $d^8\theta$ integral. The generic form of such a contribution is (suppressing the dimensionless harmonic integral)

$$\begin{aligned} \Gamma &= \lambda^k \int d^8\theta\, d^2p_1 \ldots d^2p_m\, \delta^2(p_1 + \cdots + p_m)(D)^n \\ &\quad \times [q^+(p_1) \ldots q^+(p_m)] I(p_1, \ldots, p_m) . \end{aligned} \tag{8.118}$$

Here $q^+(p_i)$, $i = 1, \ldots, m$ are m external-line superfields with two-dimensional momenta p_i. $(D)^n$ denotes n spinor derivatives originating from the D algebra and distributed in a certain way over the external lines. The coupling constant factor λ^k indicates the perturbation order of the graph. Finally, $I(p_1, \ldots, p_m)$ is the loop momentum integral which depends on the external momenta and which is potentially divergent.

The ultraviolet behavior of the loop integral $I(p_1, \ldots, p_m)$ is governed by its dimension ('power counting'). As a whole, the graph contribution Γ must be dimensionless, $[\Gamma] = 0$. The dimension of q^+ in the momentum representation is related to its dimension in the space-time representation:

$$[q^+(p)] = [q^+(x)] - 2 . \tag{8.119}$$

The latter can be determined from the kinetic term in the action (8.116). One gets

$$[q^+(x)] = 0 \quad \Rightarrow \quad [q^+(p)] = 2\,. \tag{8.120}$$

In two dimensions we also have $[\lambda] = 0$ (a necessary condition for the theory to be renormalizable). Finally, $[d\theta] = 1/2,\ [dp] = 1,\ [D] = 1/2,\ [\delta^2(p)] = -2$. Putting all this together, one finds

$$\begin{aligned} 0 &= k \cdot 0 + 8 \cdot \frac{1}{2} + m \cdot 2 - 2 + n \cdot \frac{1}{2} - m \cdot 2 + [I] \\ &\Rightarrow \quad [I] = -\frac{n}{2} - 2 < 0\,. \end{aligned} \tag{8.121}$$

We see that the loop integral always has negative dimension, so it is superficially convergent. Since the same argument applies to each of its subloops separately, we conclude that any graph in the theory will be ultraviolet finite.

This phenomenon is an example of the so-called non-renormalization theorems (see, e.g., [G26, G41, H16]). The essential assumption there is the existence of a formulation of the theory in terms of unconstrained superfields, such that by applying the Feynman rules one can always cast any graph in the form (8.118). In the case of $N = 4$, $d = 2$ sigma models* the only such formulation is the harmonic superspace one. For other proofs of ultraviolet finiteness of this class of $d = 2$ sigma models see refs. [A3, M4, S19].

At this point we switch our attention from two-dimensional sigma models and briefly discuss another class of theories, also known to be ultraviolet finite. These are the minimal couplings of $N = 2$ SYM to specific combinations of $N = 2$ matter in four dimensions discussed at the end of Section 8.5. In this case simple power counting leaves room for logarithmic divergences. Indeed, the contribution to the effective action has the form

$$\begin{aligned} \Gamma = g^k \int & d^8\theta\, d^4p_1 \dots d^4p_{m+n} \delta^4(p_1 + \cdots + p_{m+n}) \\ & \times (D)^r [V^{++}(p_1) \dots V^{++}(p_m) q^+(p_{m+1}) \dots q^+(p_{m+n})] I(p_1, \dots, p_{m+n})\,. \end{aligned} \tag{8.122}$$

Taking into account the dimension of the external superfields $[V] = -4,\ [q] = -3$, one finds $[I] = -n - 1/2\, r$. Therefore, graphs with at least one external matter line are superficially convergent whereas graphs with only SYM external lines may have logarithmic divergences. One concludes that $N = 2$ supersymmetry alone is not sufficient to explain the total cancellation of divergences in these four-dimensional theories. Howe, Stelle and Townsend [H16] have shown that invoking gauge-invariance arguments in addition to $N = 2$ supersymmetry one can prove the absence of ultraviolet divergences.

* Note that in d space-time dimensions $[q^+(x)] = 1/2\, d - 1$, and the coupling constant λ has dimension $[\lambda] = 1 - 1/2\, d$, which is negative for $d > 2$. So, the corresponding sigma models are non-renormalizable.

Finally, let us emphasize that the diagram technique presented here is pertinent to the case of harmonic superspace without central charges. Its generalization to the theories having $N = 2$ supersymmetry with central charges (induced, e.g., via the Scherk–Schwarz mechanism) has been given in [B14, E9, I8, O8, O9]. It was applied, e.g., for computing the low-energy quantum effective actions in the Coulomb branch of $N = 2$ SYM theory.

9
Conformal invariance in $N = 2$ harmonic superspace

In this chapter, using Cartan's techniques, we derive the realizations of the rigid $N = 2$ superconformal group $SU(2,2|2)$ in ordinary $N = 2$ superspace, its harmonic extension and the analytic subspace of the latter. To do this in a systematic way, we identify these superspaces with the appropriate coset spaces of the extended supergroup $SU(2,2|2) \times SU(2)_A$ where $SU(2)_A$ is the group of outer automorphisms of $SU(2,2|2)$. The superconformal properties of the basic $N = 2$ multiplets (including those used as supergravity compensators in Chapter 10) are examined.

9.1 Harmonic superspace for $SU(2,2|2)$

9.1.1 Cosets of $SU(2,2|2)$

The $N = 2$ conformal supersymmetry supergroup is $SU(2,2|2)$. Its algebra can be found in Chapter 2 as a particular case of the superalgebra $SU(2,2|N)$ given in eqs. (2.7), (2.8), (2.15), (2.16), (2.18) (with $Z = 0$) and (2.21). Here we recall just the structure of the part of that algebra which involves the conformal boosts (K_a), dilatation (D), R-symmetry (R), special supersymmetry ($S_{\alpha i}, \bar{S}^i_{\dot\alpha}$) and $SU(2)_C$ ($I^{\pm\pm}, I^0$):

$$
\begin{array}{lll}
[D,P] \propto P\,, & [D,K] \propto K\,, & [K,P] \propto L + D\,, \\
\{S,S\} \propto K\,, & [Q,K] \propto S\,, & [S,P] \propto Q\,, \\
\{S,K\} = 0\,, & \{Q,S\} \propto L + D + I + R\,. &
\end{array}
\qquad (9.1)
$$

The traditional superspace used for the realization of $SU(2,2|2)$ is given by the coset

$$
\mathbb{R}^{4|8} = \frac{SU(2,2|2)}{\{L, K, D, S, R, I\}} = (x^a, \theta^\alpha_i, \bar\theta^{\dot\alpha i})\,. \qquad (9.2)
$$

It coincides with the real superspace $\mathbb{R}^{4|8}$ (3.15) for the super-Poincaré group. The question now is how to define a *harmonic* superspace as some appropriate coset of $SU(2,2|2)$. It seems natural to try the direct analog of eq. (3.41):

$$\mathbb{HR}^{4+2|8} = \frac{\{L, P, Q, K, S, R, I^{\pm\pm}, I^0\}}{\{L, K, D, S, R, I^0\}} \,. \tag{9.3}$$

In this case the harmonic coset would be the standard one, $SU(2)_\mathrm{C}/U(1)_\mathrm{C}$. The next step should be to find a Grassmann analytic subspace similar to (3.51). To this end we have to put the + projections of the Q-supersymmetry generators (see (3.48)) in the denominator in (9.3). However, such a set of generators in the denominator would not be closed, since $\{S^+, Q^+\} \propto I^{++}$. One solution could be to add I^{++} to the denominator and thus to define a new (not self-conjugate) harmonic coset $SU(2)_\mathrm{C}/\{I^{++}, I^0\} \sim CP^1$ instead of $SU(2)_\mathrm{C}/\{I^0\} \sim S^2$. Such a coset would be parametrized by a single complex coordinate (see [K6, L3] where the so-called 'projective superspace' was introduced; a similar approach has been adopted in the recent review [H10]). However, in this case one looses the possibility to describe the sphere S^2 in a parametrization independent way by the harmonic variables $u^\pm_i$ as defined in Chapter 3. As a consequence, the manifest $SU(2)$ covariance of the harmonic expansion on the sphere S^2 is lacking.

An alternative approach is to use a different group $SU(2)_\mathrm{A} = \{T^{\pm\pm}, T^0\}$ in order to preserve the standard definition of the harmonic coset as $SU(2)_\mathrm{A}/U(1)_\mathrm{A}$. This new $SU(2)$ is the group of *external* automorphisms of the conformal superalgebra $SU(2,2|2)$. It uniformly rotates the $SU(2)$ indices of all the generators, including those of $SU(2)_\mathrm{C}$:

$$[T, I] \propto I \,. \tag{9.4}$$

The starting point is a bigger real harmonic superspace with *two independent* sets of harmonics:

$$\mathbb{HR}^{4+5|8} = \frac{\{L, P, Q, K, S, R, I^{\pm\pm}, I^0\}}{\{L, K, D, S, R\}} \times \frac{\{T^{\pm\pm}, T^0\}}{\{T^0\}} = (x^a, \theta^\alpha_i, \bar\theta^{\dot\alpha i}, v^i_j, u^\pm_i) \,. \tag{9.5}$$

The harmonics $u^\pm_i$ parametrize the coset $SU(2)_\mathrm{A}/U(1)_\mathrm{A}$. The new 'harmonics' v^j_i parametrize the group $SU(2)_\mathrm{C}$. Their rôle is purely auxiliary, they help us define a Grassmann analytic subspace of (9.5) in which both the superconformal group $SU(2,2|2)$ and $SU(2)_\mathrm{A}$ can be realized. As we explained earlier, putting the + projections of the Q-supersymmetry generators in the denominator forces us to add to them the generators I^{++} and I^0 of $SU(2)_\mathrm{C}$. The remaining generator I^{--} cannot be put in the denominator since it would convert $Q^+, \bar Q^+$ into $Q^-, \bar Q^-$, and so we would have to add the translation generator P_a as well ($\{\bar Q^+, Q^-\} \propto P$). This can be avoided by realizing that the generator T^{--} of $SU(2)_\mathrm{A}$ transforms $Q^+, \bar Q^+$ in the same way as I^{--}. Then the combination $\mathcal{I}^{--} = I^{--} - T^{--}$ commutes with $Q^+, \bar Q^+, I^{++}$ and I^0 and so can be put in the denominator without trouble. Thus we obtain the Grassmann analytic coset

$$\mathbb{HA}^{4+4|4} = \frac{SU(2,2|2) \ltimes SU(2)_{\rm A}}{\{L, K, D, S, R, Q^+, \bar{Q}^+, I^{++}, I^0, \mathcal{I}^{--}, T^0\}} . \quad (9.6)$$

This coset is complex not only because of the presence of $Q^+, \bar{Q}^+$ in the denominator, but also due to the fact that the generator $\mathcal{I}^{--}$ is not conjugate to I^{++}. As we shall see later on, for this reason the harmonic part of the coset (9.6) is parametrized by two complex (or four real) coordinates.* Thus, it is not precisely the desired sphere S^2. Nevertheless, this superspace can be made real with respect to the ~conjugation and thus can be considered as a subspace of the real superspace $\mathbb{HR}^{4+5|8}$.

9.1.2 *Structure of the analytic superspace*

Now we explain in detail how the coset (9.6) can be obtained from the real superspace (9.5). We start by choosing the following representative for the coset (9.5):

$$\Omega = \exp i\{-x^a P_a + \theta^\alpha_i Q^i_\alpha + \bar{\theta}^i_{\dot\alpha} \bar{Q}^{\dot\alpha}_i\} \exp\{i\eta^{ij} I_{ij}\} \exp i\{\xi T^{++} + \bar{\xi} T^{--}\} . \quad (9.7)$$

Here η^{ij} are coordinates for $SU(2)_{\rm C}$ related to the matrix v^i_j as follows:

$$v^j_i = \left(\exp\{i\eta^{kl}\tau_{kl}\}\right)^j_i . \quad (9.8)$$

The specific order of the exponents in (9.7) is chosen so that the Grassmann coordinates transform linearly and uniformly under the two $SU(2)$ groups. Now we start rearranging the various factors according to the coset structure (9.6). First we interchange the two $SU(2)$ factors using the relation

$$\begin{aligned} & e^{-i(\xi T^{++}+\bar{\xi}T^{--})} \exp\{i\eta^{ij} I_{ij}\} e^{i(\xi T^{++}+\bar{\xi}T^{--})} \\ &= \exp\left\{i\eta^{ij}\left[e^{-i(\xi T^{++}+\bar{\xi}T^{--})} I_{ij} e^{i(\xi T^{++}+\bar{\xi}T^{--})}\right]\right\} \\ &= \exp\left\{\eta^{ij}\left[u^+_i u^+_j I^{--} - u^+_i u^-_j I^0 - u^-_i u^-_j I^{++}\right]\right\} , \end{aligned} \quad (9.9)$$

where $u^\pm_i$ are defined as in (3.58). Next we have to get rid of the generators I^{++}, $I^0, \mathcal{I}^{--}$. This is achieved in two steps. First we factor out I^{++}, I^0:

$$\Omega \Rightarrow \Omega_{(1)} = \exp i\{-x^a P_a + \theta^\alpha_i Q^i_\alpha + \bar{\theta}^i_{\dot\alpha} \bar{Q}^{\dot\alpha}_i\} \exp i\{\xi T^{++} + \bar{\xi} T^{--}\} \exp\{z^{++} I^{--}\} , \quad (9.10)$$

* As an independent set of the $SU(2)$ generators in the denominator of the coset (9.6) one can choose $I^{++}, I^0, I^{--} - T^{--}, I^0 - T^0$. Then, taking into account that the generators $I, I - T$ form two commuting $SU(2)$ algebras, one can interpret the $SU(2)$ part of the coset as the product of two CP^1 manifolds, $CP^1 \times CP^1$. This product is not real in the sense of ordinary complex conjugation, but it can be made real under ~ conjugation.

where

$$z^{++} = \frac{v^{++}}{v^{-+}}, \qquad v^{++} = v_i^j u_j^+ u^{+i}, \qquad v^{-+} = -v_i^j u_j^+ u^{-i}, \qquad \widetilde{z^{++}} = z^{++}. \tag{9.11}$$

Then we write down

$$\exp\{z^{++} I^{--}\} = \exp\{z^{++}(\mathcal{I}^{--} + T^{--})\}, \tag{9.12}$$

and using the fact that $[\mathcal{I}^{--}, T^{--}] = 0$, we can factor out $\mathcal{I}^{--}$ as well:*

$$\Omega_{(1)} \Rightarrow \Omega_{(2)} = \exp i\{-x^a P_a + \theta_i^\alpha Q_\alpha^i + \bar\theta_{\dot\alpha}^i \bar Q_i^{\dot\alpha}\} \exp i\{\xi T^{++} + \bar\xi T^{--}\} \times \exp\{z^{++} T^{--}\}. \tag{9.13}$$

At this stage we have obtained a harmonic superspace the harmonic part of which coincides with that of the coset (9.6). Note that the harmonic factor contains only the generators of the automorphism group $SU(2)_{\rm A}$. However, comparing (9.13) to the parametrization (3.43) of the standard harmonic superspace $\mathbb{HR}^{4+2|8}$ (3.42), we observe an important difference: there is an extra $SU(2)$ factor in (9.13) containing the dependence on the $SU(2)_{\rm C}$ parameters v_i^j. This will result in a non-unitary set of harmonic variables, as we shall see shortly.

The next step towards the analytic subspace is to move the exponent containing (x, θ) in (9.13) to the right. This step is the same as when we passed from the central basis parametrization of the ordinary harmonic superspace, eq. (3.43), to the analytic basis parametrization (3.52):

$$\Omega_{(2)} = \exp i\left\{\xi T^{++} + \bar\xi T^{--}\right\} \exp\{z^{++} T^{--}\} \times \exp i\{-x_A^a P_a - \theta_A^{+\alpha} Q_\alpha^- - \bar\theta_A^{+\dot\alpha} \bar Q_{\dot\alpha}^-\} \exp i\{\theta_A^{-\alpha} Q_\alpha^+ + \bar\theta_A^{-\dot\alpha} \bar Q_{\dot\alpha}^+\}. \tag{9.14}$$

Here

$$x_A^a = x^a - 2i\theta^{(i}\sigma^a\bar\theta^{j)} w_i^+ w_j^-, \qquad \theta_{\alpha,\dot\alpha}^{\pm} = \theta_{\alpha,\dot\alpha}^i w_i^{\pm} \tag{9.15}$$

and the new conformal harmonics are defined by

$$\| w \| = e^{i(\xi\tau + \bar\xi\tau)} e^{z^{++}\tau^{--}} = \| u \| \begin{pmatrix} 1 & 0 \\ z^{++} & 1 \end{pmatrix}, \tag{9.16}$$

or, in detail,

$$w_i^+ = u_i^+ + z^{++} u_i^- = \frac{v_i^j}{v^{-+}} u_j^+, \qquad w_i^- = u_i^-. \tag{9.17}$$

* The steps (9.9)–(9.13) are most easily done using the τ-matrix representation of the generators I and T.

Note that these harmonics have the usual property

$$w^{+i}w_i^- = 1\,. \tag{9.18}$$

It is worth noting that the coordinate z^{++}, being expressed in terms of harmonics $w^{\pm i}$, satisfies the equation

$$\partial_w^{++}z^{++} + (z^{++})^2 = 0\,, \tag{9.19}$$

which coincides with the defining constraint (6.43) of the analytic non-linear multiplet $N^{++}(\zeta, u)$ which was considered in Chapter 6. We see soon that this resemblance is not accidental.

Finally, the coset representative of the analytic subspace $\mathbb{HA}^{4+4|4}$ (9.6) is obtained by dropping the factor containing the generators Q^+ in (9.14):

$$\begin{aligned}\Omega_A \;=\;& \exp i\left\{\xi T^{++} + \bar{\xi}T^{--}\right\}\exp\{z^{++}T^{--}\}\\ &\times \exp i\left\{-x_A^a P_a - \theta_A^{+\alpha}Q_\alpha^- - \bar{\theta}_A^{+\dot{\alpha}}\bar{Q}_{\dot{\alpha}}^-\right\}.\end{aligned} \tag{9.20}$$

Thus, we defined an analytic superspace in which the superconformal group $SU(2,2|2)$ is realized by left multiplication. It is spanned by the following set of coordinates:

$$\mathbb{HA}^{4+4|4} = (x_A^a, \theta_\alpha^+, \bar{\theta}_{\dot{\alpha}}^+, w_i^{\pm})\,. \tag{9.21}$$

Clearly, the matrix $\|\ w\ \|$ of the new harmonic variables is not unitary, unlike the matrix $\|\ u\ \|$ considered in Chapter 3. It corresponds to two complex degrees of freedom (one from the unitary harmonics $u^{\pm}$ and a second from z^{++}). Nevertheless, the special conjugation (3.102) can be applied to w as well:

$$\widetilde{w_i^+} = \widetilde{u_i^+} + \widetilde{z^{++}}\widetilde{u_i^-} = w^{+i}, \qquad \widetilde{w_i^-} = w^{-i}\,. \tag{9.22}$$

This property will allow us to have real analytic superspace and superfields as before.

9.1.3 *Transformation properties of the analytic superspace coordinates*

The reason for considering the new non-unitary harmonic variables $w_i^{\pm}$ is the desire to simultaneously have a manifest realization of the automorphism group $SU(2)_{\mathrm{A}}$ and a sufficiently simple realization of the $SU(2)_{\mathrm{C}}$ subgroup of $SU(2,2|2)$. The harmonics $w_i^{\pm}$ undergo the same $SU(2)_{\mathrm{A}}$ rotations in the index i as the old unitary harmonics $u_i^{\pm}$. To find out how they transform under $SU(2)_{\mathrm{C}}$, we can use the definitions (9.17), (9.11). According to these definitions, the unitary harmonics $u_i^{\pm}$ do not transform under $SU(2)_{\mathrm{C}}$ whereas the matrices v_i^j (9.8) have the transformation law

$$\delta v_i^j = -\lambda_i^k v_k^j\,. \tag{9.23}$$

Here λ_i^k are the parameters of an element of $SU(2)_{\mathrm{C}}$

$$\exp\{i\lambda^{kj} I_{kj}\}$$

multiplying the coset representative (9.7) from the left (this has to be compared with the action of the group $SU(2)_{\mathrm{A}}$ which uniformly rotates *both* indices of v_i^j). Using (9.11), (9.23) and the completeness relation for $u^\pm$, it is easy to find the transformation law of z^{++}

$$\delta z^{++} = \lambda^{ij} w_i^+ w_j^+ \equiv \lambda^{++} = \widetilde{\lambda^{++}}. \tag{9.24}$$

Thus, for $w^\pm$ we obtain the following *asymmetric* transformation laws under $SU(2)_{\mathrm{C}}$:

$$\begin{aligned} \delta w_i^+ &= \lambda^{++} w_i^- , \\ \delta w_i^- &= 0, \qquad \delta\partial_w^{++} = -\lambda^{++}\partial_w^0 , \end{aligned} \tag{9.25}$$

It is this asymmetry which causes the non-unitarity of the conformal harmonics mentioned above.

In order to better understand the difference between the realization of the two $SU(2)$ groups ($SU(2)_{\mathrm{A}}$ and $SU(2)_{\mathrm{C}}$) on the harmonic variables, it is instructive to make the following comparison. Let us take the unitary harmonics $u_i^\pm$ in the stereographic parametrization (4.3). It is not hard to find out that $SU(2)_{\mathrm{A}}$ acting as left multiplications of the matrix (4.3) is realized on $t = u_2^+/u_1^+$ as a fractional linear (Möbius) transformation:

$$t' = \frac{dt + c}{bt + a}, \qquad \bar{t}' = \overline{(t')}, \qquad \text{where} \qquad \begin{pmatrix} a & b \\ c & d \end{pmatrix} \in SU(2)_{\mathrm{A}}. \tag{9.26}$$

It is accompanied by a transformation of the phase factor ($e^{2i\psi} = u_1^+/u_2^-$):

$$e^{2i\psi'} = e^{2i\psi}\frac{a + bt}{d - c\bar{t}} . \tag{9.27}$$

Now, the harmonics $u^\pm$ can be identified with $w^\pm$ at the origin of the group $SU(2)_{\mathrm{C}}$, i.e., at the point $v_i^j = \delta_i^j \;\rightarrow\; z^{++} = 0$. The transformation group $SU(2)_{\mathrm{C}}$ acts transitively on the space of its parameters. Thus, one can reach any $w^\pm$ (i.e., any point $v_i^j \neq \delta_i^j$) by performing a finite $SU(2)_{\mathrm{C}}$ transformation on $u^\pm$. The infinitesimal transformations (9.25), with $w^\pm$ replaced by $u^\pm$ in the stereographic parametrization (4.3), are easily integrated to the following finite form

$$t_w = \frac{\delta t + \gamma}{\beta t + \alpha}, \qquad \bar{t}_w = \bar{t}, \qquad \text{where} \qquad \begin{pmatrix} \alpha & \beta \\ \gamma & \delta \end{pmatrix} \in SU(2)_{\mathrm{C}} \tag{9.28}$$

and

$$e^{2i\psi_w} = e^{2i\psi}\frac{1 + t\bar{t}}{1 + t_w\bar{t}} . \tag{9.29}$$

We immediately see that t_w is not conjugated to $\bar{t}_w$ and, in addition, the phase ψ_w becomes complex. This clearly shows how $SU(2)_{\rm C}$ breaks the unitarity of the harmonics. The implications of this will be discussed in the next section.

In fact, (4.3) with t and ψ replaced by t_w and ψ_w from eqs. (9.28), (9.29) provides an explicit six-parameter realization of the harmonics $w^\pm$ (9.17): three parameters come from $SU(2)_{\rm A}$ $(t, \bar{t}, \psi)$ and another three from the $SU(2)_{\rm C}$ matrix with elements $\alpha, \beta, \gamma, \delta$. Note that in the coset construction in the preceding subsection we assumed that the parameters associated with the stability subgroup generators T^0, I^0 were factored out, which amounts to a four-dimensional harmonic coset.

The full realization of $SU(2,2|2)$ on the analytic basis coordinates x_A, θ^+, θ^-, $w^\pm$ is obtained by left multiplications of the coset representative (9.14) (we omit the evident super-Poincaré part):

$$\begin{aligned} \delta w_i^+ &= \Lambda^{++} w_i^- \, , \\ \delta w_i^- &= 0 \, , \end{aligned} \tag{9.30}$$

where

$$\Lambda^{++} = \lambda^{ij} w_i^+ w_j^+ + 4i\theta^{+\alpha} k_{\alpha\dot\alpha} \bar\theta^{+\dot\alpha} + 4i(\theta^{+\alpha}\eta^i_\alpha + \bar\eta^i_{\dot\alpha}\bar\theta^{+\dot\alpha}) w_i^+ \, , \tag{9.31}$$

$$\widetilde{\Lambda^{++}} = \Lambda^{++} \, .$$

Further,

$$\begin{aligned} \delta x_A^{\dot\alpha\alpha} &= x_A^{\dot\alpha\beta} k_{\beta\dot\beta} x_A^{\dot\beta\alpha} + 4i(\bar\theta^{+\dot\alpha}\bar\eta^i_{\dot\beta} x_A^{\dot\beta\alpha} + x_A^{\dot\alpha\beta}\eta^i_\beta\theta_A^{+\alpha}) w_i^- \\ &\quad + 4i\lambda^{ij} w_i^- w_j^- \bar\theta^{+\dot\alpha}\theta^{+\alpha} + b x_A^{\dot\alpha\alpha} \, , \\ \delta\theta^{+\alpha} &= \theta^{+\beta} k_{\beta\dot\beta} x_A^{\dot\beta\alpha} - 2i(\theta^+)^2 \eta^{\alpha i} w_i^- + \bar\eta^i_{\dot\beta} w_i^+ x_A^{\dot\beta\alpha} \\ &\quad + \lambda^{ij} w_i^+ w_j^- \theta^{+\alpha} + \frac{1}{2} b\theta^{+\alpha} + \frac{1}{2} i\gamma\theta^{+\alpha} \, , \\ \delta\theta^{-\alpha} &= \theta^{-\beta} k_{\beta\dot\beta} x_A^{\dot\beta\alpha} - 2i(\theta^-)^2 \bar\theta^+_{\dot\beta} k^{\dot\beta\alpha} + 4i\eta^i_\beta \theta^{-\beta}(\theta^{-\alpha} w_i^+ - \theta^{+\alpha} w_i^-) \\ &\quad + \bar\eta^i_{\dot\beta}(x_A^{\dot\beta\alpha} - 4i\bar\theta^{+\dot\beta}\theta^{-\alpha}) w_i^- + \lambda^{ij} w_i^-(w_j^-\theta^{+\alpha} - w_j^+\theta^{-\alpha}) \\ &\quad + \frac{1}{2} b\theta^{-\alpha} + \frac{1}{2} i\gamma\theta^{-\alpha} \, , \end{aligned} \tag{9.32}$$

where the parameters λ^{ij} correspond to $SU(2)_{\rm C}$, $k_{\alpha\dot\alpha}$ to conformal boosts, η^i_α, $\bar\eta_{\dot\alpha i} \equiv \overline{(\eta^i_\alpha)}$ to special conformal supersymmetry, b and γ to dilatations and γ_5 transformations. By construction, these transformations leave the analytic subspace $(x_A, \theta^+, \bar\theta^+, w)$ invariant.

The covariant derivatives D^{++}, D^{--}, D^+_α, D^-_α have the same form in the analytic basis (9.14) as before (3.84), (3.53), with $u^\pm_i$ replaced by $w^\pm_i$. They transform as follows:

$$\begin{aligned} \delta D^{++} &= -\Lambda^{++} D^0 , \\ \delta D^{--} &= -(D^{--}\Lambda^{++}) D^{--}, \\ \delta D^+_\alpha &= -D^+_\alpha(\delta\theta^{-\beta}) D^+_\beta \end{aligned} \tag{9.33}$$

(we do not need the transformation law of D^-).

It will be useful for future purposes to compute the Berezinian of an infinitesimal $SU(2,2|2)$ transformation of the analytic superspace coordinates:

$$\begin{aligned} \mathrm{Ber}\left(\frac{\partial(\zeta', w')}{\partial(\zeta, w)}\right) &\simeq 1 + \frac{\partial \delta x^a_A}{\partial x^a_A} - \frac{\partial \delta\theta^{+\alpha}}{\partial\theta^{+\alpha}} - \frac{\partial\delta\bar\theta^{+\dot\alpha}}{\partial\bar\theta^{+\dot\alpha}} + \partial^{--}\Lambda^{++} \\ &\equiv 1 - 2\Lambda , \end{aligned} \tag{9.34}$$

$$\Lambda = -b + \lambda^{ij} w^+_i w^-_j + 4i(\theta^{+\alpha}\eta^i_\alpha + \bar\eta^i_{\dot\alpha}\bar\theta^{+\dot\alpha}) w^-_i - k_{\alpha\dot\alpha} x^{\dot\alpha\alpha}_A .$$

Note the following properties of Λ:

$$D^{++}\Lambda = \Lambda^{++}, \qquad D^{++}\Lambda^{++} = 0 . \tag{9.35}$$

9.1.4 Superfield representations of $SU(2,2|2)$

As we saw above, the harmonics $w^\pm$ suitable for representing the superconformal group $SU(2,2|2)$ are non-unitary. This might ruin our rules of harmonic calculus developed in Chapter 4. The way out of this situation is to interpret the harmonics $w^\pm$ just as an auxiliary notion which helps us understand how $SU(2,2|2)$ acts in the analytic superspace. The superfields defined on this superspace are still to be considered as functions of the old, unitary harmonics $u^\pm$ and all the rules of harmonic calculus are to be applied to them. Effectively, we thus restrict the analytic harmonic superspace $\mathbb{HA}^{4+4|4}$ (9.21) to $\mathbb{HA}^{4+2|4}$ by putting $w^\pm = u^\pm$:

$$\mathbb{HA}^{4+2|4} = (x^a_A, \theta^+_\alpha, \bar\theta^+_{\dot\alpha}, u^\pm_i) \equiv (\zeta_A, u^\pm_i) . \tag{9.36}$$

Then the variables v^j_i in eq. (9.17) are to be viewed as the matrix of a finite $SU(2)_{\mathrm{C}}$ non-linear transformation of the harmonics u. In accordance with the coset structure (9.6), a harmonic function $f^{(q,p)}(u)$ can now carry an $SU(2)_{\mathrm{C}}$ weight p (the matrix part of the operator I^0) in addition to the $U(1)_{\mathrm{A}}$ charge q. Then we write down the transformation law of $f^{(q,p)}(u)$ in its active form

$$f'^{(q,p)}(u) = (\Lambda^{-+})^p f^{(q,p)}(u') , \tag{9.37}$$

where

$$(u^+_i)' = \frac{\Lambda^j_i u^+_j}{\Lambda^{-+}}, \qquad (u^-_i)' = u^-_i , \qquad \Lambda^{-+} = -\Lambda^l_k u^{-k} u^+_l , \qquad \Lambda^j_i \in SU(2)_{\mathrm{C}} . \tag{9.38}$$

The infinitesimal form of (9.37) with $\Lambda_i^j = \delta_i^j + \lambda_i^j$, $\Lambda^{-+} = 1 + \lambda^{ij} u_i^- u_j^+$ is as follows:

$$\begin{aligned}
\delta^* f^{(q,p)}(u) &\simeq f'^{(q,p)}(u) - f^{(q,p)}(u) \\
&= -(\lambda^{ij} u_i^+ u_j^+)\partial^{--} f^{(q,p)}(u) + p(\lambda^{ij} u_i^- u_j^+) f^{(q,p)}(u) \\
&= (p-q)(\lambda^{ij} u_i^- u_j^+) f^{(q,p)}(u) \qquad (9.39)\\
&\quad + \lambda_i^j \left(u^+_{(j} \frac{\partial}{\partial u^+_{i)}} + u^-_{(j} \frac{\partial}{\partial u^-_{i)}} \right) f^{(q,p)}(u) - (\lambda^{ij} u_i^- u_j^-)\partial^{++} f^{(q,p)}(u).
\end{aligned}$$

The use of the active form (9.39) of the $SU(2)_C$ transformations allows us to keep the old definition of the harmonic integral over the unitary harmonics u, i.e., on the sphere S^2. Let us investigate the conditions for invariance under $SU(2)_C$. Consider the harmonic integral

$$I = \int du \; f^{(0,p)}(u). \qquad (9.40)$$

The integrand $f^{(0,p)}(u)$ must have zero $U(1)_A$ charge, according to our rule (4.10). Now, perform an active infinitesimal transformation (9.39) on $f^{(0,p)}(u)$ (the measure does not transform under active transformations):

$$\begin{aligned}
\delta^* I &= \int du \; \delta^* f^{(0,p)}(u) \\
&= \int du \; [-(\lambda^{ij} u_i^+ u_j^+)\partial^{--} f^{(0,p)}(u) + p(\lambda^{ij} u_i^- u_j^+) f^{(0,p)}(u)]. \qquad (9.41)
\end{aligned}$$

Integrating by parts in the first term, we obtain

$$\delta^* I = (p+2) \int du \; (\lambda^{ij} u_i^- u_j^+) f^{(0,p)}(u). \qquad (9.42)$$

Then it is clear that the condition for invariance is

$$\delta^* I = 0 \quad \to \quad p = -2. \qquad (9.43)$$

One has to realize that the above $SU(2)_C$ transformations, in general, take the function $f(u)$ out of the class of functions admitting a harmonic expansion on the sphere. The sufficient condition for a function to have an expansion on S^2 assumed in Chapter 4 was square-integrability. It is easy to find examples where a function $f(u)$ ceases to be square-integrable after a transformation (9.37) (or even after an infinitesimal transformation (9.39)). The only functions which are undoubtedly 'safe' are those satisfying the Grassmann analyticity condition $\partial^{++} f^{(q,p)}(u) = 0$ and having equal conformal weight and $U(1)_A$ charge, $p = q$. Indeed, for them (9.39) is reduced to the 'good' $SU(2)_A$ transformation. What concerns general harmonic functions is whether the restrictions imposed by the conformal transformations create a problem for us or not. We may say that the conformal properties are important in certain applications in field theory. There, in principle, we may restrict the class of harmonic functions to those still admitting an expansion on S^2 after a conformal transformation. On the other hand, in applications like hyper-Kähler geometry (see Chapter 11) where it is important to keep the class of harmonic functions as large as possible, we do not investigate the conformal properties.

The above consideration directly generalizes to superfields. The $N = 2$ conformal supergroup is rather restrictive concerning the possibilities of defining analytic superfields. In Cartan's framework the superfields defined on the coset (9.6) are assumed to be transformed homogeneously by the generators in the denominator. Another standard assumption is that the translation-like generators $K, S, Q^+, \bar{Q}^+, I^{++}, \mathcal{I}^{--}$ do not have matrix parts, i.e., the superfields behave as scalars under such transformations. The rest of the generators (L_{ab}, D, R, I^0, T^0) can, in principle, have matrix parts, i.e., the superfield $\Phi(\zeta_A, u)$ can carry Lorentz indices, can have dilatation and R weights and two $U(1)$ charges (for I^0 and T^0). However, the algebra (2.8), (2.21) tells us that the assumptions

$$\hat{K}\Phi = \hat{S}\Phi = \hat{Q}^+\Phi = \hat{\bar{Q}}^+\Phi = \hat{I}^{++}\Phi = \hat{\mathcal{I}}^{--}\Phi = 0 \qquad (9.44)$$

($\hat{}$ means the matrix part only) imply

$$\hat{L}_{ab}\Phi = \hat{R}\Phi = (\hat{I}^0 + i\hat{D})\Phi = 0\,. \qquad (9.45)$$

This means that analytic superfields covariant under $SU(2,2|2)$ can have neither Lorentz indices nor R weight, and their dilatation and I^0 weights have to be related according to (9.45):

$$-i\hat{D}\Phi = \hat{I}^0\Phi \quad \rightarrow \quad d = p\,. \qquad (9.46)$$

The T^0 charge remains arbitrary. The conclusion is that an analytic superfield $\Phi^{(q)}$ transforms as follows:

$$\delta^*\Phi^{(q)}(x,\theta^+,u) = -\lambda\cdot\partial\Phi^{(q)}(x,\theta^+,u) + p\Lambda\Phi^{(q)}(x,\theta^+,u)\,, \qquad (9.47)$$

where $\lambda\cdot\partial \equiv \lambda^M\partial_M + \Lambda^{++}\partial^{--}$, $\lambda^M = \delta\zeta^M$ and Λ is defined in (9.34). This transformation law is derived by considering the left action of D, $SU(2)_{\mathrm{C}}$, K and S on the coset element Ω (9.20), and using (9.44) and (9.45).

It is instructive to compare these restrictions with those arising for chiral $N = 2$ superfields, say the right-chiral ones $\Phi(x,\bar{\theta})$. For such superfields an analog of the coset (9.6) is the following:

$$\mathbb{C}^{4|4} = \frac{SU(2,2|2)}{\{L,K,D,S,\bar{S},R,Q,I\}}\,. \qquad (9.48)$$

By inspecting the little group algebra in this case, one immediately concludes that the right-chiral superfields can carry dotted Lorentz indices, should be $SU(2)_C$ singlets and their dilatation and R weights should be related by

$$(\hat{D} + i\hat{R})\Phi = 0 \quad \rightarrow \quad d = -r\,. \tag{9.49}$$

An example of such an $SU(2, 2|2)$ covariant chiral superfield is supplied by the $N = 2$ Maxwell covariant superfield strength W.

Finally, note that the harmonic derivative D^{++} of an analytic superfield with $U(1)_A$ charge q and conformal weight p transforms, in general, non-homogeneously:

$$\delta(D^{++}\Phi^{(q,p)}) = (q - p)\Lambda^{++}\Phi^{(q,p)} + p\Lambda D^{++}\Phi^{(q,p)} \tag{9.50}$$

(here we use the passive form of the transformation; see (9.33)–(9.35)). Only in the special case $q = p$ does the transformation law (9.50) become homogeneous. This circumstance will play an important rôle in the next section.

9.2 Conformal invariance of the basic $N = 2$ multiplets

In this section we examine the conformal properties of the $N = 2$ matter and SYM multiplets. It turns out that some of them have conformally invariant actions (q^+, the tensor and non-linear multiplets, the SYM multiplet), while others do not (e.g., the relaxed hypermultiplet). Only the former can be coupled to $N = 2$ conformal supergravity and are of interest for the construction of Einstein supergravity (see Chapter 10).

9.2.1 Hypermultiplet

We begin with the q^+ hypermultiplet. Its free action

$$S = \frac{1}{2}\int du\, d\zeta^{(-4)} q^+_a D^{++} q^{+a} \tag{9.51}$$

is obviously invariant under the superconformal transformations (9.47) provided q^+ transforms with weight +1 (see (9.50)):

$$\delta^* q^+ = -\lambda \cdot \partial q^+ + \Lambda q^+. \tag{9.52}$$

For a single hypermultiplet ($a = 1, 2$) (9.51) is the only possible conformally invariant action. For several hypermultiplets the action

$$S = \frac{1}{\xi^2}\int du\, d\zeta^{(-4)} \left[q^+_a D^{++} q^{+a} + L^{+4}(q, u)\right] \tag{9.53}$$

is invariant if L^{+4} has weight +2 and does not depend on u^+_i:

$$\partial^{--} L^{+4} = 0\,, \qquad q^+ \frac{\partial}{\partial q^+} L^{+4} = 2L^{+4}\,. \tag{9.54}$$

This is a direct consequence of (9.52).

9.2.2 Tensor multiplet

The defining constraint (6.11) for the tensor multiplet $L^{++}(\zeta, u)$

$$D^{++}L^{++} = 0 \tag{9.55}$$

is conformally covariant provided L^{++} has weight +2 (see (9.50)):

$$\delta^* L^{++} = -\lambda \cdot \partial L^{++} + 2\Lambda L^{++} \,. \tag{9.56}$$

It is clear that the quadratic action (6.14)

$$S_{\text{quadratic}} = \int du\, d\zeta^{(-4)} (L^{++})^2 \tag{9.57}$$

is not invariant. The simplest way to find a conformally invariant action for L^{++} is to start from the free q^+ action (9.51) and then to perform in it the duality transformation discussed in Chapter 6. The conformally invariant (improved) L^{++} action reads [G10] (recall eq. (6.17))

$$S_{\text{improved}} = \frac{1}{\gamma^2} \int du\, d\zeta^{(-4)} \left(\frac{\ell^{++}}{1 + \sqrt{1 + \ell^{++} c^{--}}} \right)^2 \equiv \frac{1}{\gamma^2} \int du\, d\zeta^{(-4)}\, (g^{++})^2 \,, \tag{9.58}$$

where

$$\ell^{++} \equiv L^{++} - c^{++}, \qquad c_{ij} = c_{ji} = \text{const}, \qquad c^{\pm\pm} = c^{ij} u_i^{\pm} u_j^{\pm} \,, \qquad c^{ij} c_{ij} = 2 \,. \tag{9.59}$$

It can be checked that the integrand is shifted by a full harmonic derivative under the transformation (9.56) which, being rewritten through ℓ^{++}, looks as

$$\delta^* \ell^{++} = -\lambda \cdot \partial \ell^{++} + 2\Lambda(\ell^{++} + c^{++}) - 2\Lambda^{++} c^0, \qquad c^0 = c^{ik} u_i^+ u_k^- \,. \tag{9.60}$$

Note that the presence of an isotriplet constant c^{ik} in the superfield Lagrangian density in (9.58) reflects the topological non-triviality of the action of the improved $N = 2$ tensor multiplet. It was discovered at the component level in [D17] and means that the Lagrangian density, when written in terms of the notoph field strength V_m, $\partial^m V_m = 0$, contains a Dirac string of singularities parametrized by c^{ik}. The dependence on c^{ik} disappears after rewriting the Lagrangian through the notoph potential $E_{[mn]}$. The superfield action (9.58), in contrast to the Lagrangian density, actually does not depend on the particular choice of this constant. It can be absorbed into some special superconformal transformation of ℓ^{++} (the only restriction is $c^{ik} c_{ik} \neq 0$).

9.2.3 Non-linear multiplet

The non-linear multiplet N^{++} is defined by the constraint (6.43)

$$D^{++} N^{++} + (N^{++})^2 = 0 \,. \tag{9.61}$$

It is $SU(2, 2|2)$ covariant if N^{++} transforms in the following unusual way:

$$\delta^* N^{++} = -\lambda \cdot \partial N^{++} + \Lambda^{++} . \tag{9.62}$$

To check this one should use (9.33), (9.35). Equation (9.62) follows also from the duality transformation (6.79), which expresses N^{++} in terms of q^+. Note the analogy of eqs. (9.62), (9.61) with the transformation law (9.24) of the auxiliary variable z^{++} and the equation (9.19) that it satisfies. This analogy suggests that the superfield N^{++} itself can be viewed as parametrizing the whole group $SU(2)_C$, i.e., as a Goldstone superfield accomplishing its full spontaneous breakdown (or compensation in the case of $N = 2$ supergravity, see Chapter 10).

9.2.4 Relaxed hypermultiplet

Not every off-shell $N = 2$ matter multiplet possesses a conformally invariant action. Take, for instance, the relaxed hypermultiplet, described by a constrained analytic superfield L^{++} (6.19) and a gauge one V (6.20):

$$(D^{++})^2 L^{++} = 0 , \qquad \delta_{\text{gauge}} V(\zeta, u) = D^{++} \Lambda^{--}(\zeta, u) . \tag{9.63}$$

Because of the gauge invariance the gauge superfield V can enter the action only coupled to $D^{++} L^{++}$, so one of the equations of motion must be

$$D^{++} L^{++} = 0 . \tag{9.64}$$

Now, the constraint in (9.63) is conformally invariant only if L^{++} has weight +3:

$$\delta^* L^{++} = -\lambda \cdot \partial L^{++} + 3\Lambda L^{++} . \tag{9.65}$$

However, the equation of motion (9.64) is not consistent with (9.65). Hence, there does not exist a conformally invariant action for the relaxed hypermultiplet. The reader can easily generalize the above arguments to the further relaxed hypermultiplets of Section 6.4.

9.2.5 Yang–Mills multiplet

Finally, we consider $N = 2$ super Yang–Mills theory, which must be conformally invariant in four dimensions. The geometric meaning of the prepotential V^{++} (see Chapter 7) as the connection for the covariant derivative D^{++} implies that V^{++} should have vanishing dilatation weight $d = 0$ and, according to eq. (9.46), $p = 0$. Its variation written down in the central basis of $\mathbb{HR}^{4+2|8}$ is

$$\delta^* V^{++} = -(\lambda^M \partial_M + \Lambda^{++} D^{--}) V^{++} , \tag{9.66}$$

where λ^M are now the conformal shifts of the coordinates $x^a, \theta_i^\alpha, \bar\theta^{\dot\alpha i}$ of $\mathbb{R}^{4|8}$ and $D^{--} = \partial^{--}$ in the central basis. The corresponding variation of the action is (see (7.99)):

$$\delta^* S = \frac{1}{2}\mathrm{Tr}\int d^{12}X\, du\; \delta^* V^{++}(X,u) V^{--}(X,u)\,. \tag{9.67}$$

The term containing λ_M is a total derivative. Indeed, inserting the expression (7.80) for V^{--} in terms of V^{++} into (9.67), we obtain

$$\begin{aligned} &\mathrm{Tr}\int du\, (\lambda^M \partial_M V^{++}) V^{--} \qquad (9.68)\\ &= \mathrm{Tr}\int du\, (\lambda^M\partial_M V^{++}) \sum_{n=1}^{\infty}\int du_1 \ldots du_n \frac{(-i)^{n+1} V^{++}(1)\ldots V^{++}(n)}{(u^+u_1^+)\ldots(u_n^+u_1^+)} \\ &= \lambda^M \partial_M \left[\sum_{n=1}^{\infty}\frac{(-i)^{n+1}}{n+1}\mathrm{Tr}\int du\, du_1\ldots du_n \frac{V^{++}(X,u)\ldots V^{++}(X,u_n)}{(u^+u_1^+)\ldots(u_n^+u_1^+)}\right]. \end{aligned}$$

Further, the supervolume of $\mathbb{R}^{4|8}$ is invariant under $SU(2,2|2)$ [F6], therefore

$$(-1)^M \partial_M \lambda^M = 0\,. \tag{9.69}$$

The term proportional to Λ^{++} also leads to a total derivative:

$$\begin{aligned} \int du\, dX\; \mathrm{Tr}\left[\Lambda^{++}(D^{--}V^{++})V^{--}\right] &= \int du\, dX\; \mathrm{Tr}\left[\Lambda^{++}(\mathcal{D}^{++}V^{--})V^{--}\right] \\ &= \frac{1}{2}\int du\, dX\; \Lambda^{++} D^{++} \mathrm{Tr}(V^{--})^2 = 0\,. \end{aligned} \tag{9.70}$$

Here we have used eq. (7.79) in the form $D^{--}V^{++} = \mathcal{D}^{++}V^{--}$ and the second relation in (9.35).

Note that the minimal coupling of a hypermultiplet to Yang–Mills,

$$i\int du\, d\zeta^{(-4)}\; \tilde{q}^+ V^{++} q^+\,, \tag{9.71}$$

is manifestly conformally invariant.

10
Supergravity

In this chapter we define the gauge group and prepotentials of conformal $N = 2$ supergravity in harmonic superspace. The gauge group is a local version of the analytic superspace realization of the rigid superconformal group $SU(2, 2|2)$ discussed in Chapter 9. The prepotentials arise as the Grassmann analytic vielbeins covariantizing the harmonic derivative D^{++}. Then we construct the harmonic superspace action of Einstein $N = 2$ supergravity. Its first basic ingredient is the action for a Maxwell compensating superfield in the background of conformal supergravity. As a second compensator we take a q^+ hypermultiplet. This gives rise to an essentially new, 'principal' version of Einstein $N = 2$ supergravity. It contains an infinite number of auxiliary fields (coming from the q^+ compensator) and is the only one allowing for the most general matter couplings. The previously known versions correspond to choosing as compensators $N = 2$ matter superfields with finite sets of auxiliary fields. They can be reduced to the principal version by duality transformations.

10.1 From conformal to Einstein gravity

In the formulation of supergravity we are going to follow the method of compensation of conformal supergravity [D17, D18, D21, K1, S7]. The idea of the method is to start with a graviton field belonging to an irreducible representation of the Poincaré group with spin 2 (conformal gravity). In the case of ordinary ($N = 0$) gravity this means that the gravitation field g_{mn} has an additional gauge invariance with a scalar parameter (Weyl invariance), which, roughly speaking, carries away the determinant of g. Then one may add a matter scalar field transforming under the same Weyl group as a density, thus restoring the missing degree of freedom in the graviton field. The result is the familiar Einstein theory, in which the graviton field is in a reducible representation of the Poincaré group (spins 2,0).

In more detail, the procedure of compensation of the Weyl group in the case $N = 0$ is as follows. Consider the action of Einstein gravity with a cosmological term:

$$S_{\text{Einstein}} = \int d^4x \sqrt{-g}\left(-\frac{1}{2\kappa^2}R + \frac{\xi^2}{\kappa^4}\right). \tag{10.1}$$

Here κ is Newton's constant ($[\kappa] = -1$) and $\kappa^{-4}\xi^2$ is a cosmological constant ($[\xi] = 0$). It is invariant under the group of diffeomorphisms of M^4:

$$\delta x^m = \tau^m(x) \quad \rightarrow \quad \delta g_{mn}(x) \simeq g_{mn}{}'(x') - g_{mn}(x) = \partial_m \tau^k g_{kn} + \partial_n \tau^k g_{km} \,. \tag{10.2}$$

Off shell the symmetric tensor field $g_{mn}(x)$ carries Poincaré spins 2,1,0,0. The gauge parameter $\tau^m(x)$ in (10.2) takes away the spins 1,0, so we are left with the spins 2, 0 in $g_{mn}(x)$.

Now the trick is to redefine the metric tensor:

$$g_{mn} = (\kappa\phi)^2 \hat{g}_{mn} \,, \tag{10.3}$$

where ϕ is a non-vanishing scalar field of physical dimension $[\phi] = 1$. Clearly, this splitting allows for local dilatations of ϕ and $\hat{g}_{mn}$:

$$\delta \hat{g}_{mn}(x) = 2a(x)\hat{g}_{mn}(x) \,, \qquad \delta\phi(x) = -a(x)\phi(x) \,. \tag{10.4}$$

The new independent parameter $a(x)$ gauges away the second spin 0, so $\hat{g}_{mn}$ describes only spin 2 off shell. In other words, this is the conformal graviton field. One might think that gravity should be the theory of pure spin 2 (recall Yang–Mills theory, which is the theory of pure spin 1). However, it can be shown that there does not exist a second-order equation of motion for $\hat{g}_{mn}$ with such a wide gauge invariance (the equations of conformal gravity are of fourth order) [F14]. So, in order to have a physical theory of gravity, one must allow for an extra degree of freedom (spin 0).

Let us now insert (10.3) into the action (10.1) and rewrite the latter as follows:

$$S_{\text{Einstein}} = \int d^4x \sqrt{-\hat{g}} \left[3\phi \left(\hat{\Box} - \frac{1}{6}\hat{R} \right) \phi + \xi^2 \phi^4 \right], \tag{10.5}$$

where

$$\hat{\Box} = \hat{\mathcal{D}}^m \partial_m \,, \qquad \hat{R} = R(\hat{g}) \,. \tag{10.6}$$

We can interpret (10.5) as the action for the scalar field $\phi(x)$ in a conformal gravity background. In particular, it is invariant under the Weyl transformations (10.4). This freedom can be used to fix the gauge $\phi(x) = \kappa^{-1}$, after which one recovers the original Einstein action (10.1).

On the other hand, one can switch off the conformal gravity field $\hat{g}_{mn}$ in (10.5) by putting it equal to its flat limit η_{mn}. Then one obtains the flat-space scalar field action

$$S_\phi = \int d^4x (3\phi\Box\phi + \xi^2\phi^4) \,. \tag{10.7}$$

We stress that the kinetic term has the wrong sign. This is a common feature of the compensating matter multiplets. The reason is that the R term in (10.5) appears with a sign determined by its rôle as a Weyl gauge connection for the covariant $\Box$. Note also that the kinetic term is related to the pure Einstein action, and the conformally invariant self-interaction to the cosmological term.

The action (10.7) is invariant under those transformations (10.2), (10.4) which preserve the flat limit $\hat{g}_{mn} = \eta_{mn}$. They constitute the rigid conformal group* $SO(2,4)$:

$$x^{m\prime} \simeq x^m + ax^m - k^m x^2 + 2k^n x_n x^m \,, \tag{10.8}$$

$$\begin{aligned}\phi'(x') &= \left|\frac{\partial x'}{\partial x}\right|^{-1/4} \phi(x) \\ &\simeq (1 - a - 2k^m x_m)\phi(x)\,.\end{aligned} \tag{10.9}$$

Now one may reverse the whole argument. Starting from the rigid conformally invariant action (10.7), one generalizes (10.8) to the diffeomorphism group and (10.9) to an independent Weyl group. Then one introduces the gauge field $\hat{g}_{mn}$, covariantizes (10.7) and obtains the action (10.5) for a scalar field in a conformal gravity background. In fact, the Weyl invariance is in a sense fake, i.e., it is *compensated* by the scalar field ϕ. This means that there exists the gauge $\phi = \kappa^{-1}$, in which the action (10.5) becomes the action of Einstein gravity that is invariant under the diffeomorphism group only. Note that this 'compensation' mechanism is based on the assumption that the scalar field has a non-vanishing constant part (or vacuum expectation value). In this it resembles the metric g_{mn} which must have a non-vanishing determinant.

The above approach can be generalized to $N = 1$ and $N = 2$ supergravities. There the conformal graviton is a member of an irreducible supermultiplet (superspin 3/2 for $N = 1$ and 1 for $N = 2$). The compensating scalar field gives rise to a separate conformal matter supermultiplet. Unlike the case $N = 0$ where the choice of the compensator is unique, in the supersymmetric case there are several different choices of the compensating off-shell matter multiplets. This results in different off-shell versions of Einstein supergravity.†

The main issue for us is $N = 2$ supergravity. However, before turning to it we review the case $N = 1$. We demonstrate that the concept underlying $N = 1$ conformal supergravity is the preservation of Grassmann analyticity (chirality). Later on we shall see that the preservation of the harmonic version of Grassmann analyticity is the key to the formulation of $N = 2$ conformal supergravity too.

* Recall that the conformal group leaves invariant the flat space interval (2.2) modulo an overall scale factor.

† With the auxiliary fields eliminated and in the absence of matter couplings they all are reduced to the well-known on-shell $N = 1$ [D11, F16] and $N = 2$ [F10] Einstein supergravity theories.

10.2 $N = 1$ supergravity

According to the compensation procedure above, we start with the most general conformally invariant action* for a chiral matter superfield $\Phi(x_L, \theta)$ (other choices of matter multiplets are also possible, see the discussion below):

$$S = -\int d^4x\, d^2\theta\, d^2\bar{\theta}\; \bar{\Phi}(x_L, \theta)\Phi(x_R, \bar{\theta}) + \xi \left[\int d^4x_L\, d^2\theta\; \Phi^3(x_L, \theta) + \text{c.c.}\right], \tag{10.10}$$

where

$$x_L^m = x^m + i\theta\sigma^m\bar{\theta}\,, \qquad x_R^m = x^m - i\theta\sigma^m\bar{\theta} = \overline{x_L^m}\,. \tag{10.11}$$

The action of the conformal supergroup $SU(2, 2|1)$ is given by

$$\begin{aligned} \delta x_L^{\dot{\mu}\mu} &= a x_L^{\dot{\mu}\mu} + x_L^{\dot{\mu}\nu} k_{\nu\dot{\nu}} x_L^{\dot{\nu}\mu} - 4i x_L^{\dot{\mu}\nu} \eta_\nu \theta^\mu\,, \\ \delta\theta^\mu &= \left(\frac{a}{2} + ib\right)\theta^\mu + \theta^\nu k_{\nu\dot{\nu}} x_L^{\dot{\nu}\mu} + 2i\theta^2\eta^\mu + \bar{\eta}_{\dot{\nu}} x_L^{\dot{\nu}\mu}\,, \end{aligned} \tag{10.12}$$

where a, b, k and η are the parameters of dilatation, γ_5 (R-symmetry) transformations, conformal boosts and conformal supersymmetry, respectively (the super Poincaré part of $SU(2, 2|1)$ is not shown). Equations (10.12) are obtained by implementing $SU(2, 2|1)$ as left shifts in the coset

$$\mathbb{C}^{4|2} = \frac{SU(2, 2|1)}{\{L, D, K, S, \bar{S}, R, \bar{Q}\}} = (x_L^m, \theta^\mu)\,. \tag{10.13}$$

The chiral superfields defined on this coset can have equal dilatation and R weights:

$$\Phi'(x_L', \theta') = \left(\text{Ber}\left|\frac{\partial\zeta_L'}{\partial\zeta_L}\right|\right)^d \Phi(x_L, \theta)\,, \tag{10.14}$$

or, infinitesimally,

$$\delta\Phi(x_L, \theta) = 3d\left(a - \frac{2}{3}ib + k_{\mu\dot{\mu}} x_L^{\mu\dot{\mu}} + 4i\theta^\mu\eta_\mu\right)\Phi(x_L, \theta)\,. \tag{10.15}$$

The kinetic term in (10.10) is conformally invariant if the Weyl weight of Φ is $d = -1/3$. To see this one uses the identity

$$\text{Ber}\left|\frac{\partial z'}{\partial z}\right| = \left(\text{Ber}\left|\frac{\partial\zeta_L'}{\partial\zeta_L}\right| \text{Ber}\left|\frac{\partial\zeta_R'}{\partial\zeta_R}\right|\right)^{1/3}, \tag{10.16}$$

which is derived from (10.11) and (10.12). The invariance of the interaction terms in (10.10) is then evident.

---

* As before, the kinetic term should have the wrong sign.

The extension of this rigid picture to a full conformal supergravity background goes through several steps. First of all, we promote the coordinate transformations (10.12) to the full diffeomorphism group that leaves the chiral superspace $\mathbb{C}^{4|2}$ invariant:

$$\begin{aligned} \delta x_L^m &= \lambda^m(x_L, \theta)\,, \\ \delta\theta^\mu &= \lambda^\mu(x_L, \theta)\,. \end{aligned} \tag{10.17}$$

The second, most non-trivial step is related to the question: What do x_L^m and x^m mean now? The rigid superspace relations (10.11) suggest the following definition:

$$x^m = \frac{1}{2}(x_L^m + x_R^m)\,, \qquad x_R^m = \overline{x_L^m}\,. \tag{10.18}$$

At the same time, the imaginary part of x_L ($i\theta\bar{\theta}$ in (10.11)) should become an arbitrary function of $x, \theta, \bar{\theta}$:

$$\frac{1}{2i}(x_L^m - x_R^m) = H^m(x, \theta, \bar{\theta})\,. \tag{10.19}$$

In other words, we define the real superspace $\mathbb{R}^{4|4}$ as an arbitrary hypersurface in $\mathbb{C}^{4|2}$, the shape of which is determined by the superfield H^m. The transformation law of H^m follows from (10.17)–(10.19):

$$H^{m\prime}(x', \theta', \bar{\theta}') \simeq H^m(x, \theta, \bar{\theta}) + \frac{1}{2i}\left(\lambda^m(x_L, \theta) - \bar{\lambda}^m(x_R, \bar{\theta})\right). \tag{10.20}$$

The crucial observation now is that the newly introduced object H^m is the basic gauge superfield for $N = 1$ conformal supergravity [O6]. One way to see this is to do a superspin counting. The real superfield H^m contains superspins 3/2, $(1)^2$, $(1/2)^2$, $(0)^2$ (see Section 2.3.2), the chiral parameters λ^m and λ^μ contain $(1)^2$, $(0)^2$ and $(1/2)^2$, respectively. So, the only supermultiplet in H^m, which is not a pure gauge, has superspin 3/2, exactly as expected from the $N = 1$ conformal supergravity multiplet. An alternative proof is to go to the Wess–Zumino gauge with the help of (10.17) and (10.20):

$$H^m = \theta\sigma^a\bar{\theta}e_a^m + \bar{\theta}\bar{\theta}\theta^\mu\psi_\mu^m + \theta\theta\bar{\theta}_{\dot{\mu}}\bar{\psi}^{m\dot{\mu}} + \theta\theta\bar{\theta}\bar{\theta}A^m\,. \tag{10.21}$$

Here one finds the vierbein e_a^m containing the conformal graviton (gauge-independent spin 2 off-shell), the gravitino ψ_μ^m (spins $(3/2)^2$) and the R-symmetry gauge field A^m (spin 1), which form the conformal supergravity multiplet (superspin 3/2). Note that the vierbein $e_a^m(x)$ undergoes diffeomorphisms and Weyl transformations with parameters $\lambda^m(x)$ and $a(x)$ contained in the superdiffeomorphism parameters:

$$\lambda^m(x, \theta) = \lambda^m(x) + \cdots\,, \qquad \lambda^\mu(x, \theta) = \cdots + \theta^\mu\frac{1}{2}a(x) + \cdots\,. \tag{10.22}$$

The third step is to generalize the rigid-space transformation law (10.14), (10.15) for the chiral superfield Φ. In this case it is sufficient to insert the general Berezinian of the transformations (10.17) in (10.14). Unlike the case $N = 0$, no new Weyl factor (cf. a in (10.4)) is needed, since an independent Weyl parameter is already contained in $\lambda^\mu(x_L, \theta)$ (see (10.22)). Once again, choosing the weight $d = -1/3$ for Φ, we achieve invariance of the potential term in (10.10).

The final step is the covariantization of the kinetic term in (10.10). This is not so easy now. The $SU(2, 2|1)$ relation (10.16) does not hold in the context of the $\mathbb{C}^{4|2}$ diffeomorphism group (10.17), so one needs a density E with the following transformation law:

$$E' = \left(\mathrm{Ber}\left|\frac{\partial z'}{\partial z}\right|\right)^{-1}\left(\mathrm{Ber}\left|\frac{\partial \zeta_L'}{\partial \zeta_L}\right|\mathrm{Ber}\left|\frac{\partial \zeta_R'}{\partial \zeta_R}\right|\right)^{1/3} E\,. \tag{10.23}$$

In principle, one could construct this object as the Berezinian of the supervielbeins in a suitable differential geometry scheme based on the superfield H^m [O7]. Instead, it is much easier to directly construct E from the following two building blocks:

$$\det\left(\delta_m^n + i\partial_m H^n\right) \equiv \det\left(1 + i\partial H\right),$$
$$\det\left(\sigma_a^{\mu\dot\mu}[\Delta_\mu, \bar\Delta_{\dot\mu}]H^m\right) \equiv \det\left([\Delta\sigma\bar\Delta]H\right),$$

where

$$\Delta_\mu = \partial_\mu + i\partial_\mu H^n(1 + i\partial H)_n^{-1m}\partial_m, \qquad \bar\Delta_{\dot\mu} = \overline{(\Delta_\mu)}\,.$$

These blocks have simple transformation laws:

$$\delta\ln\det\left(1 + i\partial H\right) = \bar\Delta_{\dot\mu}\bar\lambda^{\dot\mu} + \frac{1}{2}\partial_m(\lambda^m + \bar\lambda^m) - \partial_\mu\lambda^\mu - \bar\partial_{\dot\mu}\bar\lambda^{\dot\mu}\,,$$
$$\delta\ln\det\left([\Delta\sigma\bar\Delta]H\right) = \Delta^\mu\lambda_\mu + \bar\Delta_{\dot\mu}\bar\lambda^{\dot\mu} + \frac{1}{2}\partial_m(\lambda^m + \bar\lambda^m) - \partial_\mu\lambda^\mu - \bar\partial_{\dot\mu}\bar\lambda^{\dot\mu}\,.$$

Then the desired expression for E is

$$E = \det\left([\Delta\sigma\bar\Delta]H\right)^{-1/6}\left(\det\left(1 + i\partial H\right)\det\left(1 - i\partial H\right)\right)^{1/3}\,. \tag{10.24}$$

Now we are able to write down the action of a chiral superfield in a superconformal background:

$$\begin{aligned} S &= -\frac{1}{\kappa^2}\int d^4x\, d^2\theta\, d^2\bar\theta\; E\bar\Phi(x_L, \theta)\Phi(x_R, \bar\theta) \\ &\quad + \xi\left(\int d^4x_L\, d^2\theta\; \Phi^3(x_L, \theta) + \text{c.c.}\right). \end{aligned} \tag{10.25}$$

As in the case $N = 0$, the superfield $\Phi(x_L, \theta)$,

$$\Phi = f + ig + \theta^\mu\chi_\mu + \theta^2(S + iP) \tag{10.26}$$

with non-vanishing vacuum expectation value plays the rôle of a compensator. Its components compensate the part of the superdiffeomorphisms (10.17) corresponding to dilatations (the field f in (10.26)), R transformations (field g) and conformal supersymmetry (field χ_μ). The fields S and P in (10.26) are auxiliary fields. The component A^m of H^m (10.21), which used to be the R-symmetry gauge field, now becomes an auxiliary field too. Thus one obtains the so-called minimal version [F11, S22] of the off-shell action for $N = 1$ Einstein supergravity [D11, F16] with a cosmological term:

$$
\begin{aligned}
S &= -\frac{1}{2\kappa^2}\int d^4x\, e\Big[R + \epsilon^{mnkl}\bar{\psi}_m\gamma_5\gamma_n\mathcal{D}_k\psi_l + A^m A_m \\
&\quad + S^2 + P^2 + \frac{\xi}{\kappa}(S + \bar{\psi}^m\sigma_{mn}\psi^n)\Big].
\end{aligned} \tag{10.27}
$$

An important feature of this version is the possibility to couple supergravity to chiral matter with an arbitrary potential term. For this purpose it is sufficient to use the compensator Φ^3 to cancel out the transformations of the chiral measure:

$$
\int d^4x_L\, d^2\theta\ \Phi^3 P(\chi) + \text{c.c.}\,, \tag{10.28}
$$

where $\chi(x_L, \theta)$ is a weightless matter chiral superfield and P is an arbitrary holomorphic function.

There exist other off-shell versions of $N = 1$ Einstein supergravity. They can be obtained by covariantizing the action of a complex or real tensor multiplet instead of the chiral one, or by making a duality transformation from the chiral superfield action (10.25) to a tensor multiplet action (the procedure is analogous to the flat space one, see Chapter 6). In these alternative versions there is no natural chiral density, so coupling to matter is rather restrictive.

10.3 $N = 2$ supergravity

In this section we construct the principal version of $N = 2$ supergravity, which is in a sense the analog of the minimal version of $N = 1$ supergravity. To this end we first find the gauge group and the unconstrained prepotentials of $N = 2$ conformal supergravity. The action of $N = 2$ Einstein supergravity is obtained by putting a q^+ hypermultiplet and a Maxwell compensating multiplet in the background of conformal supergravity.

10.3.1 $N = 2$ conformal supergravity: Gauge group and prepotentials

Our starting point in the formulation of $N = 2$ supergravity is the conformally invariant action of a hypermultiplet (see Chapter 9):

$$
S = \frac{1}{2}\int du\, d\zeta^{(-4)}\ q^{+r} D^{++} q^+_r\,, \qquad r = 1, 2. \tag{10.29}
$$

Note the wrong sign of the kinetic term. The rigid conformal supergroup $SU(2,2|2)$ (9.30)–(9.32) leaves the analytic superspace $(x, \theta^+, \bar{\theta}^+, u)$ invariant. Therefore, we generalize it to a diffeomorphism supergroup with the same property [G16]:

$$\begin{aligned}
\delta x_A^m &= \lambda^m(x, \theta^+, u)\,, && (10.30)\\
\delta\theta^{\hat{\mu}+} &= \lambda^{\hat{\mu}+}(x, \theta^+, u)\,, \qquad \hat{\mu} \equiv (\mu, \dot{\mu})\,, && (10.31)\\
\delta u_i^+ &= \lambda^{++}(x, \theta^+, u)u_i^-\,, && (10.32)\\
\delta u_i^- &= 0\,, && (10.33)\\
\delta\theta^{\hat{\mu}-} &= \lambda^{\hat{\mu}-}(x, \theta^+, \theta^-, u)\,. && (10.34)
\end{aligned}$$

The most unusual part of this group is the transformation of the harmonics u (10.32), (10.33). Although it follows the pattern of the rigid conformal case, one has to make sure that the transformations (10.30)–(10.34) form a group. In particular,*

$$[\delta_1, \delta_2]u_i^+ = (\lambda_1^m \partial_m + \lambda_1^{\hat{\mu}+}\partial_{\hat{\mu}+} + \lambda_1^{++}\partial^{--})\lambda_2^{++}u_i^- - (1 \leftrightarrow 2)\,. \tag{10.35}$$

We see that the commutator produces a transformation of u_i^+ of the same type as in (10.32) with a new analytic parameter.

Further, we make q_r^+ transform under the group (10.30)–(10.34) as in the rigid case (9.52):

$$\begin{aligned}
q^{+'}{}_r(\zeta', u') &= \left(\mathrm{Ber}\left|\frac{\partial(\zeta', u')}{\partial(\zeta, u)}\right|\right)^{-1/2} q_r^+(\zeta, u)\\
&\simeq \left[1 - \frac{1}{2}(\partial_m\lambda^m - \partial_{\hat{\mu}+}\lambda^{\hat{\mu}+} + \partial^{--}\lambda^{++})\right] q_r^+(\zeta, u)\\
&\equiv (1 + \Lambda)\, q_r^+\,. && (10.36)
\end{aligned}$$

Note that the weight superparameter Λ does not satisfy any constraints, unlike the rigid conformal case (see (9.35)). As in the case $N = 1$, there is no need to make an independent Weyl rescaling of q_r^+, because the superparameter $\lambda^{\hat{\mu}+} = \cdots + \theta^{\hat{\mu}+}a(x) + \cdots$ already contains such a parameter.

The last step is to replace the flat harmonic derivative D^{++} in (10.29) by the properly covariantized one $\mathcal{D}^{++}$ and to ascribe a certain transformation law to $\mathcal{D}^{++}$. It is once again suggested by the rigid case (9.33):

$$\delta\mathcal{D}^{++} = -\lambda^{++}D^0\,. \tag{10.37}$$

In addition to (10.37) we postulate

$$\delta D^0 = 0\,, \tag{10.38}$$

* Here we use the notation $\partial_{\hat{\mu}\pm} = \partial/\partial\theta^{\hat{\mu}\pm}$.

i.e., all the parameters λ in (10.30)–(10.34) are assumed to have a definite $U(1)$ charge. Thus D^0 is still given by its flat space expression (3.84). This, together with (10.35) guarantees that the transformation law (10.37) for $\mathcal{D}^{++}$ is a group law.

Making use of the transformation laws (10.30)–(10.34), (10.36), (10.37), one can easily check the invariance of the action

$$S_q^{\text{curved}} = \frac{1}{2} \int du\, d\zeta^{(-4)}\; q^{+r} \mathcal{D}^{++} q_r^+ \,. \tag{10.39}$$

In particular, in $\delta S_q^{\text{curved}}$ the inhomogeneous terms

$$q^{+r} \left[\mathcal{D}^{++} (-1)^M \partial_M \lambda^M \right] q_r^+ - q^{+r} \lambda^{++} D^0 q_r^+ \tag{10.40}$$

($M = (++, m, \hat{\mu}+)$) vanish because $q^{+r} q_r^+ \equiv \epsilon^{rs} q_r^+ q_s^+ = 0$.

In order to realize the non-trivial transformation law (10.37) we have to introduce new objects – harmonic vielbeins:

$$\mathcal{D}^{++} = \partial^{++} + H^{++++} \partial^{--} + H^{++m} \partial_m + H^{++\hat{\mu}+} \partial_{\hat{\mu}+} + H^{++\hat{\mu}-} \partial_{\hat{\mu}-} \,. \tag{10.41}$$

In fact, (10.41) is a generalization of the flat space expression for D^{++} in the analytic basis (3.84):*

$$\underline{\text{Flat limit:}}\; H^{++++} = 0,\; H^{++m} = -2i\theta^+ \sigma^m \bar{\theta}^+,\; H^{++\hat{\mu}+} = 0,\; H^{++\hat{\mu}-} = \theta^{\hat{\mu}+} \,. \tag{10.42}$$

The transformation laws for these objects follow from eqs. (10.30)–(10.34), (10.37) and the explicit form of D^0 (3.84):

$$\begin{aligned} \delta H^{++++} &= \mathcal{D}^{++} \lambda^{++} \,, \\ \delta H^{++m} &= \mathcal{D}^{++} \lambda^m \,, \\ \delta H^{++\hat{\mu}\pm} &= \mathcal{D}^{++} \lambda^{\hat{\mu}\pm} \mp \lambda^{++} \theta^{\hat{\mu}\pm} \,. \end{aligned} \tag{10.43}$$

Note that the term ∂^{++} in (10.41) does not need a vielbein, because the coordinate u_i^- is inert under the diffeomorphisms (10.30)–(10.34).

The vielbeins H^{++++}, H^{++m}, $H^{++\hat{\mu}+}$ are present in the action (10.39), so we have to assume that they are analytic, as is q_r^+ itself:

$$\partial_{\hat{\nu}-} H^{++++} = \partial_{\hat{\nu}-} H^{++m} = \partial_{\hat{\nu}-} H^{++\hat{\mu}+} = 0 \,. \tag{10.44}$$

This does not apply to $H^{++\hat{\mu}-}$ which drops out from the action (10.39).

We remark that the covariant equation of motion for q^+ following from the action (10.39) is

$$\left(\mathcal{D}^{++} + \frac{1}{2} \Gamma^{++} \right) q_r^+ = 0 \,. \tag{10.45}$$

* Note that in the flat limit (10.42) the residual transformations (10.30)–(10.34) constitute precisely the conformal supergroup $SU(2, 2|2)$.

It contains a connection term

$$\Gamma^{++} = (-1)^M \partial_M H^{++M}, \qquad \delta\Gamma^{++} = 2(\lambda^{++} - \mathcal{D}^{++}\Lambda), \tag{10.46}$$

needed to compensate the conformal weight of q^+.

So far we have succeeded in covariantizing the hypermultiplet action by putting it in a background which generalizes the rigid superconformal group. The question now is: What do the newly introduced gauge superfields H^{++M} describe? To answer it we should analyze the superspin-superisospin content of the vielbeins H^{++M} and the parameters λ^M in (10.43) (recall (3.71)). The analytic vielbeins contain the following superspins Y and superisospins I:

$$\begin{array}{rlll} H^{++++}: & Y = 0; & I = 1,2,3,\ldots \\ H^{++m}: & Y = 1,0; & I = 0,1,2,\ldots \\ H^{++\hat{\mu}+}: & Y = 1/2; & I = 1/2,3/2,5/2,\ldots \end{array} \tag{10.47}$$

On the other hand, the analytic parameters contain:

$$\begin{array}{rlll} \lambda^{++}: & Y = 0; & I = 1,2,3,\ldots \\ \lambda^{m}: & Y = 1,0; & I = 1,2,3,\ldots \\ \lambda^{\hat{\mu}+}: & Y = 1/2; & I = 1/2,3/2,5/2,\ldots \end{array} \tag{10.48}$$

Comparing (10.47) with (10.48), one sees that the gauge-independent part of the analytic supervielbeins H^{++M} is a superspin 1, superisospin 0 *irreducible* $N = 2$ multiplet. This is exactly the expected content of the multiplet of $N = 2$ conformal supergravity.

The non-analytic vielbein $H^{++\hat{\mu}-}$ and the non-analytic parameter $\lambda^{\hat{\mu}-}$ have exactly the same component content. Therefore, this vielbein can be entirely gauged away. A convenient gauge is the one in which $H^{++\hat{\mu}-}$ equals its flat space value:

$$H^{++\hat{\mu}-} = \theta^{\hat{\mu}+} \quad \rightarrow \quad \mathcal{D}^{++}\lambda^{\hat{\mu}-} = \lambda^{\hat{\mu}+} - \theta^{\hat{\mu}-}\lambda^{++}. \tag{10.49}$$

One can display the field content of the vielbeins in the Wess–Zumino gauge. To do this it is sufficient to consider the linearized approximation of the transformations (10.43), i.e., to study small deviations from the flat limit (10.42):

$$\begin{gathered} H^{++m} = -2i\theta^+\sigma^m\bar{\theta}^+ + h^{++m}(\zeta_A, u), \\ H^{++\hat{\mu}+} = h^{++\hat{\mu}+}, \quad H^{++++} = h^{++++}; \end{gathered} \tag{10.50}$$

$$\begin{aligned} \delta h^{++m} &= D^{++}\lambda^m + 2i(\lambda^+\sigma^m\bar{\theta}^+ + \theta^+\sigma^m\bar{\lambda}^+) + O(h), \\ \delta h^{++\hat{\mu}+} &= D^{++}\lambda^{\hat{\mu}+} - \theta^{\hat{\mu}+}\lambda^{++} + O(h), \\ \delta h^{++++} &= D^{++}\lambda^{++} + O(h). \end{aligned} \tag{10.51}$$

Thus one finds the following Wess–Zumino gauge:

$$
\begin{aligned}
H^{++m}(\zeta_A, u) &= -2i\theta^+\sigma^a\bar{\theta}^+ e_a^m(x) + (\bar{\theta}^+)^2\theta^{\mu+}\psi^m_{\mu i}(x)u^{-i} \\
&\quad + (\theta^+)^2\bar{\theta}^+_{\dot{\mu}}\bar{\psi}_i^{m\dot{\mu}}(x)u^{-i} + (\theta^+)^2(\bar{\theta}^+)^2 V^m_{ij}u^{-i}u^{-j}\,, \\
H^{++\mu+}(\zeta_A, u) &= (\theta^+)^2\bar{\theta}^+_{\dot{\mu}} A^{\mu\dot{\mu}}(x) + (\bar{\theta}^+)^2\theta^+_\nu t^{(\nu\mu)}(x) \\
&\quad + (\theta^+)^2(\bar{\theta}^+)^2\chi_i^\mu(x)u^{-i}\,, \\
H^{++\dot{\mu}+}(\zeta_A, u) &= \widetilde{H^{++\mu+}}\,, \\
H^{++++}(\zeta_A, u) &= (\theta^+)^2(\bar{\theta}^+)^2 D(x)\,,
\end{aligned}
\tag{10.52}
$$

which is just the content of the $N = 2$ Weyl multiplet [B8, D16, D20]. The remaining gauge freedom in the parameters λ is

$$
\begin{aligned}
\lambda^m &= a^m(x) + \cdots\,, \\
\lambda^{\mu+} &= \epsilon^{\mu i}(x)u_i^+ + \theta^{\nu+}\left[(a(x) + ib(x) + \lambda^{(ij)}(x)u_i^+u_j^-)\delta_\nu^\mu + \ell_\nu^\mu(x)\right] \\
&\quad + (\theta^+)^2\eta^{\mu i}(x)u_i^- + \cdots\,, \\
\bar{\lambda}^{\dot{\mu}+} &= \widetilde{\lambda^{\mu+}}\,, \\
\lambda^{++} &= \lambda^{ij}(x)u_i^+u_j^+ + \cdots\,.
\end{aligned}
\tag{10.53}
$$

Here $a^m(x)$, $\epsilon^{\mu i}(x)$, $a(x)$, $b(x)$, $\ell^{(\mu\nu)}(x)$, $\eta^{\mu i}(x)$ and $\lambda^{ij}(x)$ are the space-time diffeomorphism, local supersymmetry, Weyl, R, local Lorentz, local conformal supersymmetry and local $SU(2)$ transformation parameters, respectively (the dots denote field-dependent terms and terms with derivatives on the parameters).

We would like to make a comment on the gauge-fixing procedure above. Looking at the multiplet content of H^{++++} in (10.47) and the corresponding parameter λ^{++} (10.48), one might expect that H^{++++} could be completely gauged away, like we did with $H^{++\hat{\mu}-}$. However, this is not possible because the last component in the θ expansion of H^{++++} ($D(x)$ in (10.52)) is shifted by the divergence of the vector parameter ρ^m ($\lambda^{++} = \cdots + i\theta^+\sigma_m\bar{\theta}^+\rho^m + \cdots$), $\delta D(x) = \partial_m\rho^m$. Thus the gauge $D = 0$ can be achieved only locally, because the boundary conditions for $\rho^m(x)$ do not allow one to gauge away $D(x)$ at infinity. Note at the same time that the gauge (10.49) is perfectly legal, because there both the vielbein and the parameter are general (non-analytic) superfields.

We can summarize this subsection by saying that we have found the prepotentials of $N = 2$ conformal supergravity. They are the unconstrained analytic superfields H^{++++}, H^{++m}, $H^{++\hat{\mu}+}$. We also constructed the action of a matter hypermultiplet (10.39) in the conformal background. In the case $N = 1$ similar steps lead to the compensation of all the extra local symmetries (Weyl, R and conformal supersymmetry), so we automatically obtained an action for $N = 1$ Einstein supergravity. The case $N = 2$ is different. Indeed, now one has five scalar parameters of extra local symmetries to worry about (Weyl, R and

$SU(2)$). In addition, the scalar field $D(x)$ of the Weyl multiplet (10.52) has dimension 2 ($[H^{++++}] = 0 \to [D] = 2$), so it requires a dimensionless scalar field $A(x)$ to form a Lagrange pair in an Einstein action:

$$S = \frac{1}{\kappa^2} \int d^4x \, (\cdots + D(x)A(x)) \,. \tag{10.54}$$

Altogether one needs six scalar fields from the compensating multiplets. The q^+ multiplet supplies only four of them (see Section 5.2.1), so two more are to be found elsewhere. The only $N = 2$ supermultiplet which describes two scalars is the Maxwell multiplet (see Chapter 7). Besides, it also supplies the spin 1 gauge field (the 'graviphoton') known to be a member of the $N = 2$ Einstein supergravity multiplet (2, 3/2, 3/2, 1).

We conclude that one possibility of constructing an $N = 2$ Einstein supergravity off-shell action is to consider a Maxwell and a hypermultiplet action in the conformal supergravity background.* In other words, our starting point now becomes the rigid conformally invariant action of a hypermultiplet minimally coupled to a Maxwell superfield V^{++} (see Chapters 7, 9):

$$\begin{aligned} S &= \frac{1}{2} \int du \, d\zeta^{(-4)} \, q^{+r} D^{++} q^+_r + \frac{i\xi}{2} \int du \, d\zeta^{(-4)} \, \tilde{q}^+ V^{++} q^+ \\ &\quad - \frac{1}{4} \int du \, d^{12}X \, V^{++} V^{--} \,. \end{aligned} \tag{10.55}$$

Here ξ is a dimensionless coupling constant. Note the wrong sign of the kinetic terms for q^+ and V^{++} which is explained by their future rôle as compensating superfields (cf. (10.7) and (10.10)). So far we have succeeded in covariantizing the q^+ term. It remains to find a way to covariantize the Maxwell and the Maxwell-hypermultiplet terms.

10.3.2 Central charge vielbeins

The compensating Maxwell superfield V^{++} in (10.55) admits the following natural geometric interpretation. The action (10.55) is invariant under Abelian gauge transformations of the analytic superfields q^+ and $H^{++5} \equiv \kappa V^{++}$:

$$\delta q^+ = -\frac{i\xi}{2\kappa} \lambda^5 q^+ , \tag{10.56}$$

$$\delta H^{++5} = D^{++} \lambda^5 \tag{10.57}$$

with an analytic gauge parameter $\lambda^5(\zeta, u)$. Following Kaluza and Klein, one can consider the transformation of q^+ as a shift of an extra coordinate x^5 with dimension of length:

$$\delta x^5 = \lambda^5(\zeta, u) \,. \tag{10.58}$$

* Actually, the choice of the second compensator is not unique. Instead of q^+ we could take a tensor or a non-linear or a central-charged multiplet (see the discussion below).

The dependence on x^5 is assumed almost trivial:

$$\hat{q}^+(\zeta, u, x^5) = \exp\left(\frac{i\xi x^5}{2\kappa}\right) q^+(\zeta, u)\,,$$
$$\delta\hat{q}^+ \simeq \hat{q}^+(\zeta, u, x^{5'}) - \hat{q}^+(\zeta, u, x^5) = 0\,. \tag{10.59}$$

In this picture the gauge superfield H^{++5} plays the rôle of a vielbein in the covariant derivative D^{++}:

$$D^{++} \quad \to \quad D^{++} + H^{++5}\frac{\partial}{\partial x^5}\,. \tag{10.60}$$

Note that such a coordinate has already appeared in the discussion of with central charges (see Section 3.6). There H^{++5} H^{++5} was represented by the constant term $i(\theta^+)^2 - i(\bar{\theta}^+)^2$ (see (3.98), (3.100)). Following this, we assume that H^{++5} has a non-vanishing flat limit:

$$H^{++5} = i(\theta^+)^2 - i(\bar{\theta}^+)^2\,. \tag{10.61}$$

This will allow us to interpret some components of H^{++5} as compensators for the Weyl and R transformations.

The advantage of the above interpretation of H^{++5} as a vielbein associated with the new central charge coordinate x^5 is that in curved superspace one can treat it on an equal footing with the prepotentials of conformal supergravity. In other words, we add the transformation law (10.58) to the diffeomorphisms (10.30)–(10.34) and extend the covariant harmonic derivative (10.41):

$$\begin{aligned} \mathcal{D}^{++} \quad \to \quad \mathcal{D}^{++} &= \partial^{++} + H^{++++}\partial^{--} + H^{++m}\partial_m + H^{++5}\partial_5 \\ &\quad + H^{++\hat{\mu}+}\partial_{\hat{\mu}+} + H^{++\hat{\mu}-}\partial_{\hat{\mu}-}\,. \end{aligned} \tag{10.62}$$

It is important to realize that neither the gauge parameter λ^5 nor the vielbeins H^{++} depend on the central charge coordinate x^5. Only matter superfields with central charge are allowed to depend on x^5.

The transformation law of the new $\mathcal{D}^{++}$ under the extended diffeomorphisms (10.30)–(10.34), (10.58) is the same as before (see (10.37)). This means that the new vielbein H^{++5} transforms as follows:

$$\delta H^{++5} = \mathcal{D}^{++}\lambda^5\,. \tag{10.63}$$

As in the case of $N = 2$ SYM theory, the transformations (10.63) allow one to fix a Wess–Zumino gauge, in which H^{++5} contains only a finite set of ordinary fields:

$$\begin{aligned} H^{++5}_{\rm WZ} &= (\theta^+)^2 M + (\bar{\theta}^+)^2\bar{M} + i\theta^{\mu+}\bar{\theta}^{\dot{\mu}+}V_{\mu\dot{\mu}} \\ &\quad + (\theta^+)^2\bar{\theta}^+_{\dot{\mu}}\bar{\rho}^{\dot{\mu}i}u^-_i + (\bar{\theta}^+)^2\theta^{\mu+}\rho^i_\mu u^-_i + (\theta^+)^2(\bar{\theta}^+)^2 S^{ij}u^-_i u^-_j\,. \end{aligned} \tag{10.64}$$

The interpretation of H^{++5} as a compensator means that the complex field $M(x)$ in (10.64) has a non-vanishing constant flat space limit $M_0 = i$ (see (10.61)). Under Weyl and R transformations (see (10.53)) $\delta M = -2(a+ib)M$. Then, if $M = i + \cdots$, these local transformations are indeed compensated.*

From the transformation law (10.63) of H^{++5} together with (10.56) and (10.36) it follows that the covariant version of the terms in (10.55) involving q^+ (recall (10.59)) is

$$\begin{aligned} S^{q^+} &= \int du\, d\zeta^{(-4)}\, \frac{1}{2}\hat{q}^{+r}\mathcal{D}^{++}\hat{q}^+_r \\ &= \int du\, d\zeta^{(-4)} \left[\frac{1}{2}q^{+r}\mathcal{D}^{++}q^+_r + \frac{i\xi}{2\kappa}\tilde{q}^+ H^{++5} q^+\right]. \end{aligned} \tag{10.65}$$

So, from here on we shall concentrate our efforts on the covariantization of the pure Maxwell term.

10.3.3 Covariant harmonic derivative $\mathcal{D}^{--}$

Now we recall that the pure Maxwell term in (10.55) also involves the superfield H^{--5}. In rigid superspace it has the meaning of the gauge connection for the harmonic derivative D^{--}. In the Kaluza–Klein picture it should appear as a vielbein term in $\mathcal{D}^{--}$.

The covariant harmonic derivative $\mathcal{D}^{--}$ can be defined by analogy with $\mathcal{D}^{++}$:

$$\mathcal{D}^{--} = \partial^{--} + H^{--m,5}\partial_{m,5} + H^{--\hat{\mu}+}\partial_{\hat{\mu}+} + H^{--\hat{\mu}-}\partial_{\hat{\mu}-}\,. \tag{10.66}$$

Its transformation law generalizes that of the flat space derivative (9.33):

$$\delta\mathcal{D}^{--} = -(\mathcal{D}^{--}\lambda^{++})\mathcal{D}^{--}\,. \tag{10.67}$$

Note that (10.67) is consistent with the group composition law (10.35). For the vielbeins H^{--} this implies

$$\delta H^{--m,5,\hat{\mu}\pm} = -(\mathcal{D}^{--}\lambda^{++})H^{--m,5,\hat{\mu}\pm} + \mathcal{D}^{--}\lambda^{m,5,\hat{\mu}\pm}\,. \tag{10.68}$$

The absence of an analog of the $\mathcal{D}^{++}$ vielbein H^{++++} in $\mathcal{D}^{--}$ is explained by the asymmetric transformation laws of the harmonic variables.

We have seen earlier that the vielbeins H^{++++}, $H^{++m,\hat{\mu}\pm}$ are the prepotentials of conformal supergravity, and H^{++5} is the Maxwell prepotential. Therefore, one should try to express the vielbeins H^{--} in terms of H^{++}. Recall

* Note that if one uses a q^+ as the second compensating multiplet, one of the scalars in it is also shifted by Weyl transformations. So, in this case the true Weyl compensating field is a combination of the two scalars.

the similar situation in $N = 2$ SYM where the gauge connection for $\mathcal{D}^{--}$ is related to that for $\mathcal{D}^{++}$ by the commutation relation (7.77)

$$[D^{++} + iV^{++}, D^{--} + iV^{--}] = D^0 . \tag{10.69}$$

This equation can be regarded as the Yang–Mills covariant version of the flat space relation

$$[D^{++}, D^{--}] = D^0 . \tag{10.70}$$

The idea now is to covariantize (10.70) with respect to the group (10.30)–(10.34), (10.58). It is not hard to see that the correct covariant version of (10.70) is

$$[\mathcal{D}^{++} - H^{++++}\mathcal{D}^{--}, \mathcal{D}^{--}] = D^0 \tag{10.71}$$

(use (10.37), (10.38), (10.43), (10.67)). Note that in principle the right-hand side of (10.71) might contain some non-trivial torsion or curvature tensors. However, they would correspond to certain component fields which are not present in the Wess–Zumino gauge for H^{++M}, H^{++5}.

The main point of this subsection is to convince ourselves that the vielbeins H^{--} are completely determined by H^{++} as solutions of eq. (10.71). In more detail, eq. (10.71) reads

$$\begin{aligned} &\mathcal{D}^{++}H^{--m,5} - \mathcal{D}^{--}H^{++m,5} + (\mathcal{D}^{--}H^{++++})H^{--m,5} = 0 , \\ &\mathcal{D}^{++}H^{--\hat{\mu}\pm} - \mathcal{D}^{--}H^{++\hat{\mu}\pm} + (\mathcal{D}^{--}H^{++++})H^{--\hat{\mu}\pm} = \pm\theta^{\hat{\mu}\pm} . \end{aligned} \tag{10.72}$$

From the general arguments of Chapter 4 it follows that eqs. (10.72) always have a unique perturbative solution (details can be found in [Z3]). The situation here is rather similar to the $N = 2$ SYM one (see (7.79)).

Concluding this subsection we would like to point out that in what follows we shall not need the explicit solution of (10.72).

10.3.4 Building blocks and superspace densities

Our main problem now is the covariantization of the pure Maxwell term in (10.55):

$$S_{\text{Maxwell}} = -\frac{1}{4\kappa^2}\int du\, d^{12}X_A\, H^{++5}H^{--5} . \tag{10.73}$$

In the curved case the supervolume transforms under the diffeomorphisms (10.30)–(10.34) as follows:

$$\delta(du\, d^{12}X_A) = (\partial^{--}\lambda^{++} + \partial_m\lambda^m - \partial_{\hat{\mu}+}\lambda^{\hat{\mu}+} - \partial_{\hat{\mu}-}\lambda^{\hat{\mu}-})du\, d^{12}X_A . \tag{10.74}$$

Further, the integrand in (10.73) transforms under (10.30)–(10.34) with a weight factor

$$\delta(H^{++5}H^{--5}) = -(\mathcal{D}^{--}\lambda^{++})H^{++5}H^{--5} \tag{10.75}$$

(see (10.68); we postpone the discussion of the central charge gauge transformations (10.63) until the end of this subsection). Considering (10.74) and (10.75), one sees that in order to make the action term (10.73) invariant one needs a density E:

$$S_{\text{Maxwell}} = -\frac{1}{4\kappa^2}\int du\, d^{12}X_A\; E\; H^{++5}H^{--5} \tag{10.76}$$

with the following transformation law:

$$\delta E = -(\partial^{--}\lambda^{++} + \partial_m\lambda^m - \partial_{\hat\mu\pm}\lambda^{\hat\mu\pm} - \mathcal{D}^{--}\lambda^{++})E\,. \tag{10.77}$$

The problem now is how to construct the density E. The idea is to first find suitable objects with relatively simple transformation laws ('building blocks'; cf. the construction of the density for $N = 1$ supergravity in Section 10.2) and then to build the desired density out of these blocks. A useful trick to find the building blocks is a temporary extension of the dimension of space-time to six [G23]. Here we skip the details and give the result directly.

The first step is to construct two matrices, one 4×4

$$e_{\hat\alpha}{}^{\hat\mu} = \partial_{\hat\alpha-}H^{--\hat\mu+} \tag{10.78}$$

and one 6×5

$$e^{\hat m}_{[\hat\alpha\hat\beta]} = \partial_{\hat\alpha-}\partial_{\hat\beta-}H^{--\hat m} - \partial_{\hat\alpha-}e_{\hat\beta}{}^{\hat\mu}(e^{-1})^{\hat\nu}_{\hat\mu}\partial_{\hat\nu-}H^{--\hat m}\,. \tag{10.79}$$

Here $\hat m = m, 5$ is a five-component index and the antisymmetrization $[\hat\alpha\hat\beta]$ corresponds to a six-component index. It can be shown that the diffeomorphism group (10.30)–(10.34), (10.58) acts on these object as follows:

$$\delta e^{\hat m}_{[\hat\alpha\hat\beta]} = -(\mathcal{D}^{--}\lambda^{++})e^{\hat m}_{[\hat\alpha\hat\beta]} + e^{\hat n}_{[\hat\alpha\hat\beta]}\lambda^{\hat m}_{\hat n} - \partial_{\hat\alpha-}\lambda^{\hat\gamma-}e^{\hat m}_{[\hat\gamma\hat\beta]} - \partial_{\hat\beta-}\lambda^{\hat\gamma-}e^{\hat m}_{[\hat\alpha\hat\gamma]}, \tag{10.80}$$

$$\begin{aligned}\delta e_{\hat\alpha}{}^{\hat\mu} &= -(\mathcal{D}^{--}\lambda^{++})e_{\hat\alpha}{}^{\hat\mu} - \partial_{\hat\mu-}\lambda^{\hat\rho-}e_{\hat\rho}{}^{\hat\mu} + e_{\hat\mu}{}^{\hat\rho}\partial_{\hat\rho+}\lambda^{\hat\mu+} + \partial_{\hat\alpha-}H^{--\hat m}\partial_{\hat m}\lambda^{\hat\mu+}\\ &\quad -(\partial_{\hat\alpha-}H^{--\hat m}\partial_{\hat m}\lambda^{++} + e_{\hat\alpha}{}^{\hat\rho}\partial_{\hat\rho+}\lambda^{++})H^{--\hat\mu+}\,,\end{aligned} \tag{10.81}$$

where

$$\begin{aligned}\lambda^{\hat m}_{\hat n} &= \partial_{\hat n}\lambda^{\hat m} - \partial_{\hat n}\lambda^{\hat\mu+}(e^{-1})^{\hat\nu}_{\hat\mu}\partial_{\hat\nu-}H^{--\hat m}\\ &\quad -\partial_{\hat n}\lambda^{++}(H^{--\hat m} - H^{--\hat\mu+}(e^{-1})^{\hat\nu}_{\hat\mu}\partial_{\hat\nu-}H^{--\hat m})\,.\end{aligned} \tag{10.82}$$

Further, from the matrix $e^{\hat m}_{[\hat\alpha\hat\beta]}$ one can construct a 'metric'

$$g^{\hat m\hat n} = \frac{1}{32}\epsilon^{\hat\alpha\hat\beta\hat\gamma\hat\delta}e^{\hat m}_{[\hat\alpha\hat\beta]}e^{\hat n}_{[\hat\gamma\hat\delta]}\,, \tag{10.83}$$

which has a very simple transformation law:

$$\delta g^{\hat m\hat n} = -(2\mathcal{D}^{--}\lambda^{++} + \partial_{\hat\alpha-}\lambda^{\hat\alpha-})g^{\hat m\hat n} + g^{\hat k\hat n}\lambda^{\hat m}_{\hat k} + g^{\hat m\hat k}\lambda^{\hat n}_{\hat k} \tag{10.84}$$

(the numerical coefficient in (10.83) was chosen so as to ensure the correct flat limit, $g^{mn} \to \eta^{mn}$).

In fact, since none of the superfields depend on x^5, the first four components of the vector indices $\hat{m}, \hat{n}$ in (10.84) do not mix up with the fifth. Therefore, the submatrix g^{mn}, $m, n = (0, 1, 2, 3)$ transforms in the same way as $g^{\hat{m}\hat{n}}$ in (10.84). Then we consider the transformation laws of the determinants of g^{mn} and $g^{\hat{m}\hat{n}}$:

$$\delta(\det g^{mn}) = [2\lambda_k^k - 4(2\mathcal{D}^{--}\lambda^{++} + \partial_{\hat{\alpha}-}\lambda^{\hat{\alpha}-})] \det g^{mn} , \quad (10.85)$$

$$\delta(\det g^{\hat{m}\hat{n}}) = [2\lambda_k^k - 5(2\mathcal{D}^{--}\lambda^{++} + \partial_{\hat{\alpha}-}\lambda^{\hat{\alpha}-})] \det g^{\hat{m}\hat{n}} . \quad (10.86)$$

The only difference between (10.85) and (10.86) is in the weights -4 and -5 related to the number of components in m and $\hat{m}$. From (10.82) one obtains

$$\lambda_{\hat{k}}^{\hat{k}} = \lambda_k^k = \partial_k \lambda^k - \partial_k \lambda^{\hat{\mu}+} (e^{-1})_{\hat{\mu}}^{\hat{\nu}} \partial_{\hat{\nu}-} H^{--k} - \partial_k \lambda^{++} (H^{--k} - H^{--\hat{\mu}+} (e^{-1})_{\hat{\mu}}^{\hat{\nu}} \partial_{\hat{\nu}-} H^{--k}) . \quad (10.87)$$

Our third and last building block with a homogeneous transformation law is the determinant of the matrix $e_{\hat{\alpha}}{}^{\hat{\mu}}$ which transforms as follows (see (10.81)):

$$\delta(\det e_{\hat{\alpha}}{}^{\hat{\mu}}) = \big[- 4(\mathcal{D}^{--}\lambda^{++}) - \partial_{\hat{\alpha}-}\lambda^{\hat{\alpha}-} + \partial_{\hat{\alpha}+}\lambda^{\hat{\alpha}+} + \partial_{\hat{\alpha}-} H^{--m} \partial_m \lambda^{\hat{\nu}+} (e^{-1})_{\hat{\nu}}^{\hat{\alpha}} + H^{--\hat{\nu}+} (e^{-1})_{\hat{\nu}}^{\hat{\alpha}} \partial_{\hat{\alpha}-} H^{--m} \partial_m \lambda^{++} + H^{--\hat{\nu}+} \partial_{\hat{\nu}+} \lambda^{++} \big] \det e_{\hat{\alpha}}{}^{\hat{\mu}} . \quad (10.88)$$

Finally, combining (10.88) with (10.85) and (10.86) and using (10.87), (10.66), we find the following two harmonic superspace densities:

$$E = |\det g^{mn}|^{-1/2} \det e_{\hat{\alpha}}{}^{\hat{\mu}} , \quad (10.89)$$

$$E_A = |\det g^{\hat{m}\hat{n}}|\,|\det g^{mn}|^{-3/2} \det e_{\hat{\alpha}}{}^{\hat{\mu}} \quad (10.90)$$

with transformation laws:

$$\delta E = -(\partial^{--}\lambda^{++} + \partial_m \lambda^m - \partial_{\hat{\mu}+}\lambda^{\hat{\mu}+} - \partial_{\hat{\mu}-}\lambda^{\hat{\mu}-} - \mathcal{D}^{--}\lambda^{++}) E , \quad (10.91)$$

$$\delta E_A = -(\partial^{--}\lambda^{++} + \partial_m \lambda^m - \partial_{\hat{\mu}+}\lambda^{\hat{\mu}+}) E_A . \quad (10.92)$$

The first of them is precisely the density needed for the covariantization of the Maxwell action (see (10.76), (10.77)). It is important to realize that it is built out of conformal supergravity prepotentials only (i.e., the Maxwell compensator H^{--5} does not appear in (10.89)).

The second density E_A (10.90) seems to have the right transformation law for defining the invariant supervolume of the analytic superspace $\int du\, d\zeta^{(-4)}\, E_A$. However, this cannot be done because E_A is not analytic itself. This important problem will be discussed in Section 10.4.

10.3.5 Abelian gauge invariance of the Maxwell action

Having found the density E, we succeeded in covariantizing the Maxwell action (10.76) with respect to the superconformal diffeomorphism group (10.30)–(10.34). This is not yet the end, however. The flat space action (10.55) is invariant under the Abelian gauge transformation (10.56), (10.57). So, we have to make sure that the covariant Maxwell action (10.76) is also invariant under the covariant version of (10.57) (see (10.63), (10.68)). Varying (10.76) and keeping in mind that E does not include $H^{\pm\pm 5}$, one obtains

$$\begin{aligned} \delta S &= -\frac{1}{4\kappa^2}\int du\, d^{12}X_A\, E\, [(\mathcal{D}^{++}\lambda^5)H^{--5} + H^{++5}\mathcal{D}^{--}\lambda^5] \\ &= -\frac{1}{4\kappa^2}\int du\, d^{12}X_A\, E\, [\mathcal{D}^{++}(\lambda^5 H^{--5}) + \mathcal{D}^{--}(\lambda^5 H^{++5}) \\ &\quad -\lambda^5(\mathcal{D}^{++}H^{--5} + \mathcal{D}^{--}H^{++5})]\,. \end{aligned} \tag{10.93}$$

Using the relationship (10.72) between H^{++5} and H^{--5}, one gets

$$\begin{aligned} \delta S &= -\frac{1}{4\kappa^2}\int du\, d^{12}X_A\, E\, [\mathcal{D}^{++}(\lambda^5 H^{--5}) + \lambda^5(\mathcal{D}^{--}H^{++++})H^{--5}] \\ &\quad -\frac{1}{4\kappa^2}\int du\, d^{12}X_A\, E\, \mathcal{D}^{--}(\lambda^5 H^{++5}) \\ &\quad +\frac{1}{2\kappa^2}\int du\, d^{12}X_A\, E\, \lambda^5\mathcal{D}^{--}H^{++5}\,. \end{aligned} \tag{10.94}$$

Each of the three terms in (10.94) vanishes separately. In the first integral one can integrate by parts the partial derivatives in $\mathcal{D}^{++}$ (see (10.41)), after which it is reduced to

$$\int du\, d^{12}X_A\, E\, [-\mathcal{D}^{++}\ln E - (-1)^M\partial_M H^{++M} + \mathcal{D}^{--}H^{++++}]\, \lambda^5 H^{--5}\,. \tag{10.95}$$

Using the transformation laws of the coordinates, of the vielbeins H^{++}, of $\mathcal{D}^{--}$ and of E, one can convince oneself that the expression in the square brackets in (10.95) is an invariant scalar of dimension 0 and charge +2. Since such scalars are not found in the Wess–Zumino gauge (10.52), one concludes that this expression must vanish. The second integral in (10.94) is treated in the same way.

In the third integral in (10.94) we split the Grassmann measure as $du\, d^{12}X_A = du\, d\zeta^{(-4)}(\partial_-)^4$. Then one uses the analyticity of λ^5, H^{++5} and the explicit form of $\mathcal{D}^{--}$ (10.66) to obtain

$$\int du\, d\zeta^{(-4)}\, \lambda^5(\partial_-)^4(E\mathcal{D}^{--}H^{++5}) = \int du\, d\zeta^{(-4)}\, \lambda^5[(\partial_-)^4(E\mathcal{H}^{--M})]\partial_M H^{++5}\,, \tag{10.96}$$

where $M = (++, m, \hat{\mu}+)$ and $\mathcal{H}^{--M} = (1, H^{--m}, H^{--\hat{\mu}+})$. Now we show that

$$W^{++M} \equiv (\partial_-)^4(E\mathcal{H}^{--M}) = 0\,. \qquad (10.97)$$

To do this let us consider the transformation of W^{++M}. Using (10.91), (10.68) and the analyticity of the parameters λ^M, we find

$$\begin{aligned}
\delta^* W^{++M} &\simeq W^{++M'}(\zeta, u) - W^{++M}(\zeta, u) \\
&= (\partial_-)^4\big[-((-1)^N\partial_N\lambda^N + \lambda^N\partial_N)(E\mathcal{H}^{--M}) \\
&\quad + (\partial_{\hat{\alpha}-}\lambda^{\hat{\alpha}-} - \lambda^{\hat{\alpha}-}\partial_{\hat{\alpha}-})(E\mathcal{H}^{--M}) + E\mathcal{D}^{--}\lambda^M\big] \\
&= -\big[(-1)^N\partial_N\lambda^N + \lambda^N\partial_N\big]W^{++M} + W^{++N}\partial_N\lambda^M \\
&\quad +(\partial_-)^4\big[\partial_{\hat{\alpha}-}(\lambda^{\hat{\alpha}-}E\mathcal{H}^{--M})\big]\,. \qquad (10.98)
\end{aligned}$$

The last term in (10.98) vanishes ($(\partial_-)^5 = 0$), so W^{++M} transforms homogeneously. Then it should give rise to component fields in the Wess–Zumino gauge (10.52) having dimensions 2,1,3/2 and $U(1)$ charges +4,+2,+3 (for $M = ++, m, \hat{\mu}+$). The latter do not exist, so one concludes that (10.97) holds.

This completes the proof of the λ^5 invariance of the Maxwell action in the superconformal background.

10.4 Different versions of $N = 2$ supergravity and matter couplings

10.4.1 Principal version of $N = 2$ supergravity and general matter couplings

In the preceding section we found the action of one of the possible versions of $N = 2$ Einstein supergravity. It consists of the conformally covariantized actions of two compensating multiplets, a hypermultiplet q^+ (10.65) and a Maxwell multiplet (10.76):

$$\begin{aligned}
S &= \int du\, d\zeta^{(-4)}\left[\frac{1}{2}q^{+r}\mathcal{D}^{++}q^+_r + \frac{i\xi}{2\kappa}\tilde{q}^+H^{++5}q^+\right] \\
&\quad - \frac{1}{4\kappa^2}\int du\, d^{12}X_A\, E\, H^{++5}H^{--5}\,. \qquad (10.99)
\end{aligned}$$

The pure Maxwell part of this action involves a finite set of 32+32 off-shell fields (24+24 fields from the conformal supergravity prepotentials (10.52) and 8+8 from the Maxwell compensator H^{++5} (10.64)). This is what is called 'minimal off-shell representation' [B10, D16].

Now we can explain why neither the Maxwell term nor the hypermultiplet alone are sufficient to obtain a meaningful component action. To this end we may concentrate on just a few component fields, namely, on all the scalars (except for the auxiliary field S^{ij} of the Maxwell multiplet). So, we take the following simplified form of the Wess–Zumino gauge (10.52) and (10.64) for the prepotentials:

$$\begin{aligned} H^{++m} &= -2i\theta^+\sigma^m\bar\theta^+ , \\ H^{++\mu+} &= H^{++\dot\mu+} = 0 , \\ H^{++++} &= (\theta^+)^2(\bar\theta^+)^2 D(x) , \\ H^{++5} &= (\theta^+)^2 M(x) + (\bar\theta^+)^2 \bar M(x) . \end{aligned} \tag{10.100}$$

It is then straightforward to work out the solution of the equations (10.72) for the vielbeins H^{--}:

$$\begin{aligned} H^{--m} &= -2i\theta^-\sigma^m\bar\theta^- , \\ H^{--\mu+} &= \theta^{-\mu} + [\theta^{+\mu}(\bar\theta^+\bar\theta^-)(\theta^-)^2 - \frac{1}{2}\theta^{-\mu}(\theta^+)^2(\bar\theta^-)^2]D(x) , \\ H^{--5} &= (\theta^-)^2[M(x) + 2i\theta^+\sigma^m\bar\theta^-\partial_m M(x) - (\theta^+)^2(\bar\theta^-)^2\Box M(x)] + \text{c.c.} \end{aligned} \tag{10.101}$$

(the vielbein $H^{--\mu-}$ is not relevant to this calculation). Now we recall the discussion after eq. (10.64) from which it follows that the real part of the scalar field $M(x)$ is a compensator for the local R symmetry which is part of the residual gauge freedom in the Wess–Zumino gauge. This allows us to choose the R gauge $M(x) = -\bar M = i\varphi(x)$. Then we substitute these results into the Maxwell part of the action (10.99) and obtain the following component action term:

$$S_1 = -\frac{1}{2\kappa^2}\int d^4x\ (\partial_m\varphi\,\partial^m\varphi + D\varphi^2) . \tag{10.102}$$

Looking at this action, one understands why the Maxwell compensating multiplet alone is not sufficient: Varying (10.102) with respect to D, one obtains the equation $\varphi^2 = 0$ which contradicts the flat limit (10.61). So, we need a second compensating multiplet, which we have chosen as a q^+ hypermultiplet.

Let us now turn to the q^+ part of the action (10.99). In the expansion of the compensating superfield q^+_r one finds four scalars of physical dimension: $q^+_r = f^i_r(x)u^+_i + \cdots$. Three of them are compensators for the local $SU(2)$ remaining in the Wess–Zumino gauge. We can profit from this freedom to gauge them away, after which $f^i_r(x) = \sqrt{2}\delta^i_r\phi(x)$. Finally, eliminating all the auxiliary fields in q^+ (here this is an easy exercise, due to the simple structure of the vielbeins H^{++}), we find the following component action term:

$$S_2 = \int d^4x\ \left(-2\partial_m\phi\,\partial^m\phi + D\phi^2 + \frac{\xi^2}{2\kappa^2}\varphi^2\phi^2\right) . \tag{10.103}$$

Once again, if q^+ were the only compensating multiplet, then the scalar ϕ would be the Weyl compensator, which contradicts the field equation $\phi^2 = 0$ following from the variation with respect to D in (10.103). We can obtain a meaningful action by combining the two compensating multiplets. Considering the action term $S_1 + S_2$, we see that now the D field equation simply identifies the fields φ and ϕ:

$$\varphi^2 = 2\kappa^2\phi^2 \,. \tag{10.104}$$

Finally, substituting eq. (10.104) into the action $S_1 + S_2$, we obtain

$$S_1 + S_2 = \int d^4x \; (3\phi\Box\phi + \xi^2\phi^4) \,. \tag{10.105}$$

This is precisely the flat-space limit (10.7) of the $N = 0$ compensating action (10.5). If we now restore the gravitational field, we are sure to obtain the right sign of the Einstein term R, as in (10.5). This mechanism explains, firstly, why one needs two compensating multiplets and, secondly, why both of their actions must have the wrong sign.

Let us now come back to the discussion of the $N = 2$ Einstein supergravity action (10.99). It is important to realize that the compensator q^+ brings in infinitely many auxiliary fields and thus gives rise to a very unusual version of $N = 2$ supergravity. As we shall show later on, all the other versions can be reduced to this one by duality transformations. Moreover, this version allows for the most general matter couplings. For these reasons we call it the 'principal version' of $N = 2$ Einstein supergravity.

It should be pointed out that the principal version cannot be discovered in the ordinary component approach where one deals with finite sets of fields only. The reason is that one has already used almost the whole gauge freedom (10.30)–(10.34), (10.58) in order to gauge away the infinite towers of components in the prepotentials $H^{++|++,m,5,\mu+}$. The remaining gauge parameters that still need to be compensated are the local $SU(2)$ parameters in λ^{++} (see (10.53)). They are compensated by just three of the components of the hypermultiplet q^{+a}, leaving an infinite tower of auxiliary fields. To see how q^{+a} compensates the local $SU(2)$ part of the λ^{++} transformations, one assigns a non-vanishing flat space limit to it,

$$(q_i^+)_{\text{flat}} = \kappa^{-1} u_i^+ \,. \tag{10.106}$$

Then from (10.32), (10.33), (10.36) one finds that the parameter λ^{++} is compensated by the combination

$$\delta \left(\frac{u_i^+ q^{+i}}{u_j^- q^{+j}} \right) = \lambda^{++} \,. \tag{10.107}$$

The most remarkable feature of the principal version is the existence of a dimensionless analytic density. It is given by the projection

$$u_i^- q^{+i} \,, \qquad \delta(u_i^- q^{+i}) = -\frac{1}{2}(-1)^M \partial_M \lambda^M (u_i^- q^{+i}) \,. \tag{10.108}$$

Since it is analytic, one may use it to construct an invariant volume element for the analytic superspace $\mathbb{HA}^{4+2|4}$:

$$du\,d\zeta^{(-4)}\,\kappa^2(u_i^- q^{+i})^2\,, \qquad \delta[du\,d\zeta^{(-4)}\,\kappa^2(u_i^- q^{+i})^2] = 0 \tag{10.109}$$

(the gravitational constant κ is needed to compensate for the dimension of the superfield q^+). Note that this supervolume is not invariant under the Abelian central charge transformations (10.56), therefore the density (10.108) can be used in the absence of the cosmological term ($\xi = 0$) only. In this case the superfield q^+, and hence the density (10.108) become inert under the central charge transformations.

Let us now consider $N = 2$ matter in the background of the principal version of $N = 2$ supergravity. From Chapter 5 we know that the most general $N = 2$ matter self-interactions in flat space correspond to the following action of n hypermultiplets $Q^{+r}, r = 1, \ldots, 2n$:

$$S_Q = \frac{1}{2\gamma^2}\int du\,d\zeta^{(-4)}\,[L_r^+(Q,u)D^{++}Q^{+r} + L^{+4}(Q,u)]\,. \tag{10.110}$$

Here we have factored out the sigma model coupling constant γ of dimension $[\gamma] = -1$, so that the matter superfields Q^+ can be considered dimensionless. The sigma model potentials L_r^+ and L^{+4} are arbitrary functions of Q^+ and of the harmonic variables.

This action can be coupled to the principal version of $N = 2$ supergravity in three steps.

Firstly, one replaces D^{++} by the following covariant derivative:

$$\nabla^{++} = \mathcal{D}^{++} + \left(\frac{u_i^+ q^{+i}}{u_j^- q^{+j}}\right) D^0\,. \tag{10.111}$$

Using (10.37), (10.38) and (10.107), and assuming that Q^+ transforms as a weightless scalar, one can check that $\nabla^{++}Q^+$ is a scalar as well.

Further, one should replace the harmonic variables $u^\pm$ appearing in (10.110) explicitly (but not the coordinates $u^\pm$ of superspace) by the following composite variables:*

$$\begin{aligned} v_i^+ &= u_i^+ - \frac{u_i^+ q^{+i}}{u_j^- q^{+j}}\, u_i^-\,, \\ v_i^- &= u_i^-\,, \end{aligned} \tag{10.112}$$

which are inert under the supergravity group.

* This change of variables is analogous to the relation between the different choices of harmonic variables discussed in Chapter 9.

Finally, covariantizing the analytic supervolume as shown in (10.109), one obtains the general $N = 2$ matter self-couplings in the background of $N = 2$ supergravity:

$$S_Q^{\text{curved}} = \frac{\kappa^2}{2\gamma^2} \int du\, d\zeta^{(-4)}\, (u^- q^+)^2 \left[L_r^+(Q, v)\nabla^{++} Q^{+r} + L^{+4}(Q, v)\right]. \tag{10.113}$$

Clearly, this procedure does not restrict the matter Lagrangian in any way,* owing to the existence of the analytic density (10.108). As we shall see, this is an exclusive feature of the principal version of $N = 2$ supergravity.

10.4.2 Other versions of $N = 2$ supergravity

The other versions of $N = 2$ supergravity are obtained by replacing the q^+ compensator by different compensating matter multiplets. There are three alternatives, the non-linear, the tensor and the central-charged multiplets [D17, D19, D20, F15]. In Chapter 6 we saw that in flat superspace the conformally invariant actions for the non-linear and tensor multiplets could be obtained from the free q^+ action by means of duality transformations. The same can be done in curved superspace as well. Take, for instance, the non-linear multiplet version of $N = 2$ supergravity. It can be obtained by the following change of variables (see (6.79)):

$$q_i^+ = (u_i^+ - N^{++} u_i^-)\, \omega\,, \tag{10.114}$$

or vice versa:

$$\omega = u_i^- q^{+i}\,, \qquad N^{++} = \frac{u_i^+ q^{+i}}{u_j^- q^{+j}}\,. \tag{10.115}$$

Putting this into the q^+ action (10.39) one finds

$$S_{\omega,N} = \frac{1}{2} \int du\, d\zeta_A^{(-4)}\, \omega^2 \left[H^{++++} - \mathcal{D}^{++} N^{++} - (N^{++})^2\right]. \tag{10.116}$$

Varying (10.116) with respect to ω produces the covariant version of the non-linear multiplet constraint (6.43):

$$\mathcal{D}^{++} N^{++} + (N^{++})^2 - H^{++++} = 0\,. \tag{10.117}$$

In other words, in this case H^{++++} is no longer an independent prepotential, it is expressed in terms of the arbitrary analytic superfield N^{++} (and the other vielbeins $H^{++|m,\mu+}$).

---

* As we pointed out earlier, in the presence of a cosmological term the supervolume and the harmonics $v^{\pm}$ are not invariant under the central charge Abelian transformations (10.56). This severely restricts the possible matter Lagrangians with a cosmological term.

From the expression (10.115) for N^{++} and from (10.107) it follows that

$$\delta N^{++} = \lambda^{++} . \tag{10.118}$$

The parameter λ^{++} is an arbitrary analytic superfunction, so it can be used to completely gauge away the analytic superfield N^{++}. In the gauge $N^{++} = 0$ the parameter is completely fixed, $\lambda^{++} = 0$. Moreover, in this gauge we have $H^{++++} = 0$ in virtue of the constraint (10.117). The absence of λ^{++} and H^{++++} in this version considerably simplifies the formalism. In particular, the equations for the vielbeins H^{--} (10.72) become linear and one can find an explicit perturbative solution. Another simplification, in contrast with the principal version, is that the non-linear one involves only a finite number $(40 + 40)$ of components. Therefore, it has been discovered by traditional component field methods [D19, D20, F15].

At the same time, the non-linear multiplet version suffers from a serious drawback. There it is impossible to construct an invariant analytic supervolume, and hence its couplings to matter are severely restricted. In fact, in all versions of $N = 2$ supergravity there exists the density E_A (10.90), which has the right transformation law (10.92). However, E_A is not analytic itself. To see this it is sufficient to consider E_A in the linearized approximation. There we keep only the vielbein $H^{--5} = (\theta^-)^2 + (\bar\theta^-)^2 + h^{--5}$ (h^{--5} is small) and put all the rest equal to their flat limits. Then, with the help of (10.79), (10.78), (10.83), (10.90), it is not hard to obtain

$$\begin{aligned} \partial^+_\alpha E_A &\sim \partial^+_\alpha [1 + (\partial^+)^2 h^{--5} + (\bar\partial^+)^2 h^{--5}] \\ &= \partial^+_\alpha (\bar\partial^+)^2 h^{--5} . \end{aligned} \tag{10.119}$$

Further, using the Wess–Zumino gauge (10.52) we find from (10.72)

$$\partial^+_\alpha (\bar\partial^+)^2 h^{--5} = \rho^i_\alpha u^+_i + \cdots \neq 0 , \tag{10.120}$$

i.e., E_A is indeed not analytic. Then, supposing that there exists another (analytic) density with the same transformation law (10.92), one sees that its ratio with E_A should be an invariant dimensionless scalar. However, one does not find such a field among the $40 + 40$ components of the non-linear multiplet version. This proves our statement that there is no analytic density in this version.

The tensor multiplet version can be treated similarly. One transforms q^+ into a tensor multiplet L^{++}, as in the flat case (see Section 6.6.1). The covariant form of the constraint (9.55) is

$$(\mathcal{D}^{++} + \Gamma^{++}) L^{++} = 0 . \tag{10.121}$$

Here Γ^{++} is defined in (10.46) and the transformation law of L^{++} is the same as in the rigid case (9.56),

$$\delta L^{++} = 2\Lambda L^{++} \tag{10.122}$$

with Λ from (10.36). The invariant action generalizing the rigid one (9.58) is

$$S_{\text{impr}} = \frac{1}{\gamma^2}\int du\, d\zeta^{(-4)}\ \left[(g^{++})^2 - \Gamma^{++}g^{++}c^{+-} - H^{++++}(1+2g^{++}c^{--})\right], \tag{10.123}$$

where g^{++} is defined in (9.58), (9.59). The resulting version of $N = 2$ Einstein supergravity has $40 + 40$ fields. Once again, one does not find a dimensionless scalar among the components, so this version does not possess an analytic density either.

Finally, the version employing a hypermultiplet with a non-trivial central charge* is described as follows. We replace the analytic superfield $q^{+a}(\zeta, u)$ by an analytic superfield $\phi^{+a}(\zeta, x^5, u)$ which explicitly depends on x^5 (like in (10.59)) and is subject to the covariant constraint

$$\left(\mathcal{D}^{++} + \frac{1}{2}\Gamma^{++}\right)\phi^{+a} = 0\,. \tag{10.124}$$

Here $\mathcal{D}^{++}$ involves the vielbein H^{++5}; ϕ^{+a} transforms as a density with respect to the supergravity group:

$$\delta\phi^{+a} = \Lambda(\zeta, u)\phi^{+a}(\zeta, x^5, u)\,. \tag{10.125}$$

The action for this multiplet is rather unusual:

$$S_{\phi^+} = -\frac{1}{2}\int du\, d\zeta^{(-4)}\ H^{++5}\phi^{+r}\partial_5\phi^+_r\ . \tag{10.126}$$

Using the constraint (10.124), one can show that the action is invariant. Moreover, $\partial_5[H^{++5}\phi^{+r}\partial_5\phi^+_r]$ is a total derivative, therefore there is no need to integrate over x^5 in (10.126). This version of $N = 2$ Einstein supergravity also has $40 + 40$ fields off shell.

Like in the principal version above, in this one there exists an analytic density, $u^-_r\phi^{+r}$ (cf. (10.108)). However, it is not suitable for the construction of an invariant analytic volume, since it explicitly depends on x^5. Therefore, this version is very restrictive for matter couplings, like the other two versions with finite sets of components. We should mention that we do not know if there exists a duality transformation from this version of $N = 2$ supergravity to the principal one.

In conclusion, we may say that the principal version of $N = 2$ supergravity is a close analog of the minimal version of $N = 1$ supergravity (see Section 10.2). In both cases the matter compensators are unconstrained analytic superfields ($q^+(\zeta, u)$ for $N = 2$, and a chiral superfield $\phi(\zeta_L)$ for $N = 1$). Both compensators can be used as densities for the corresponding analytic superspace

* This off-shell $N = 2$ multiplet was proposed in [S12]. Its harmonic superspace formulation is discussed in [D24, D26].

integrals. This allows one to couple supergravity to matter in the most general way. All the other off-shell versions of these theories* are classically equivalent to these two by duality transformations, but only in the absence of matter.

10.5 Geometry of $N = 2$ matter in $N = 2$ supergravity background

Here we demonstrate that the equations of motion following from the general harmonic superspace off-shell action of $N = 2$ matter coupled to $N = 2$ supergravity have a simple interpretation in terms of quaternionic geometry in the unconstrained harmonic space formulation [G7]. Surprisingly, the components

$$\omega = u_i^- q^{+i}, \qquad N^{++} = \frac{u_i^+ q^{+i}}{u_j^- q^{+j}} \tag{10.127}$$

of the supergravity hypermultiplet compensator q^{+i} acquire a clear geometric meaning as 'central charge' coordinates z_A^0, z_A^{++} of the analytic space $\{z_A^0, z_A^{++}, w^{\pm i}, x^{\mu+}\}$ of the quaternionic geometry, while the $N = 2$ matter hypermultiplet superfields Q^{+r} can be identified with the coordinates $x^{\mu+}$ of this target subspace, like in the hyper-Kähler case (see Chapter 11).

Let us start with the $N = 2$ matter–supergravity action obtained by combining the compensator term (10.116) and the matter term (10.113):†

$$\begin{aligned} S^{N=2}_{\text{SG+matter}} &= \frac{1}{2}\int d\zeta^{(-4)} du\, \omega^2 \Big\{ H^{++++} - \mathcal{D}^{++}N^{++} - (N^{++})^2 \\ &\quad + \frac{\kappa^2}{\gamma^2}\left[L_r^+(Q,v)\nabla^{++}Q^{+r} + L^{+4}(Q,v)\right]\Big\}, \end{aligned} \tag{10.128}$$

where the shifted harmonics $v^{\pm}$ are defined in (10.112) and ∇^{++} in (10.111).

The action (10.128) is invariant under two kinds of gauge transformations with analytic parameters, diffeomorphisms (parameters $\lambda^{+r}(Q,v)$) and the quaternionic analog of the hyper-Kähler transformations (parameter $\Lambda^{++}(Q,v)$):

$$\delta Q^{+r} = \lambda^{+r}(Q,v), \tag{10.129}$$

$$\delta \mathcal{D}_v^{++} = \frac{\kappa^2}{\gamma^2}\Lambda^{++}\,\mathcal{D}^0,$$

$$\delta N^{++} = \frac{\kappa^2}{\gamma^2}\,\Lambda^{++},$$

$$\delta\omega = \frac{\kappa^2}{2\gamma^2}\,(\partial_v^{--}\Lambda^{++} - L_r^+\partial_v^{--}\lambda^{+r})\,\omega, \tag{10.130}$$

---

* With the possible exception of the version with a Fayet–Sohnius compensator.

† The full matter–supergravity action also includes the Maxwell compensator term, which is irrelevant for our purposes here.

$$\begin{aligned}
\delta L^+_r &= \partial_{r+}\Lambda^{++} - L^+_s\,\partial_{r+}\lambda^{+s} + \frac{\kappa^2}{\gamma^2}L^+_r(-\partial^{--}_v\Lambda^{++} + L^+_s\partial^{--}_v\lambda^{+s})\,,\\
\delta L^{+4} &= \partial^{++}_v\Lambda^{++} - L^+_r\partial^{++}_v\lambda^{+r}\\
&\quad -\frac{\kappa^2}{\gamma^2}\,(\,\partial^{--}_v\Lambda^{++}\,L^{+4} - \partial^{--}_v\lambda^{+r}\,L^+_rL^{+4} + \Lambda^{++}\,L^+_rQ^{+r}\,)\,.
\end{aligned}$$

One sees that in the flat limit $\kappa \to 0$ the transformation laws (10.130) are reduced to those from hyper-Kähler geometry (see (11.156), (11.163), (11.226), (11.241)), if one identifies the compensators N^{++} and ω with the central charge coordinates from Section 11.3.7:

$$N^{++} \equiv \frac{\kappa^2}{\gamma^2}z^{++}_A\,, \qquad \ln(\kappa\omega) \equiv \frac{\kappa^2}{\gamma^2}z^0_A\,. \tag{10.131}$$

Using the diffeomorphism transformations (10.129) one can fix the gauge (cf. (11.169) in the hyper-Kähler case):

$$L^+_r = \Omega_{rs}Q^{+s} \equiv Q^+_r\,. \tag{10.132}$$

In this gauge, with the help of the equations of motion for N^{++} and ω, one obtains the following equation for Q^+_r:

$$(\mathcal{D}^{++}_u + N^{++})Q^+_r = \frac{1}{2}\partial_{r+}L^{+4} - \frac{\kappa^2}{2\gamma^2}Q^+_r\partial^{--}_vL^{+4}\,. \tag{10.133}$$

In the flat limit $\kappa \to 0$, $N^{++} = 0$ (see (10.131)), so one recovers the main differential equation of hyper-Kähler geometry (11.267).

In conclusion, the coupling of hypermultiplet matter to $N = 2$ supergravity gives rise to the most general quaternionic manifold [B4]. The main difference between the quaternionic and the hyper-Kähler cases is the presence of an extra constant scalar curvature in the former,

$$R = -8n(n+2)\,\frac{\kappa^2}{\gamma^2}\,. \tag{10.134}$$

The detailed correspondence between quaternionic geometry and $N = 2$ supergravity-matter systems in the harmonic superspace approach has been studied in [G7]. In particular, it has been shown that the sigma model potential L^{+4} from the action (10.128) is the basic unconstrained potential of quaternionic geometry, in close analogy with the hyper-Kähler case from Chapter 11. It is worth mentioning that even with $L^{+4} = 0$ in (10.128) the physical bosonic fields of the hypermultiplets Q^{+r} parametrize a curved manifold, the homogeneous quaternionic manifold $Sp(1,n)/Sp(1) \times Sp(n)$, i.e., for them, a non-trivial sigma model action arises. Its precise form can be found [I12] by extending to this case the simple computation given in the beginning of Section 10.4.1. In order to obtain the correct action, one should take into account the contribution of the non-propagating $SU(2)$ gauge field $V_m^{(ik)}$ defined in eq. (10.52).

Note that in [G1, G2] were considered some particular quaternionic sigma models corresponding to matter couplings in the off-shell versions of $N = 2$ Einstein supergravity with finite sets of auxiliary fields (i.e., obtained by using conventional compensators). These sigma models necessarily possess certain isometries, in contrast to general matter couplings in the principal version which can have no isometries at all.

11
Hyper-Kähler geometry in harmonic space

The aim of this chapter is to show that the concept of harmonic analyticity* has deep implications not only in $N = 2$ (and $N > 2$) supersymmetry, but also in purely bosonic ($N = 0$) gauge theories. Namely, this concept allows one to obtain unconstrained geometric formulations of self-dual Yang–Mills theory and hyper-Kähler geometry. These formulations closely parallel those of $N = 2$ Yang–Mills theory and $N = 2$ supergravity in harmonic superspace. The basic objects are unconstrained potentials defined on an analytic subspace of the harmonic extension of the original space (in the general case it is $\mathbb{R}^{4n} \times S^2$). They encode all the information about the quantities present in the conventional formulations, self-dual Yang–Mills connections in the first case and hyper-Kähler metrics in the second one. We show that the harmonic analytic potential of the most general hyper-Kähler manifold serves as the Lagrangian of the most general $N = 2$ sigma model (after identifying the coordinates of the analytic subspace of this manifold with the analytic superfields q^+ describing $N = 2$ matter). This establishes a direct, one-to-one correspondence between $N = 2$ sigma models and hyper-Kähler manifolds.

11.1 Introduction

In the previous chapters we demonstrated the relevance of the harmonic variables for obtaining an adequate formulation of $N = 2$ supersymmetric theories. The idea was to combine the principle of Grassmann analyticity [G8] with manifest invariance under the automorphism group $SU(2)$ of the superalgebra. The main consequence of this is the existence of Grassmann analytic superspaces closed under both supersymmetry and the automorphism group. These analytic superspaces play a fundamental rôle in all the $N = 2$ theories (as well as in

* The term 'harmonic analyticity' used throughout this chapter should not be confused with, e.g., the interpretation of the free-field hypermultiplet equation $D^{++}q^+ = 0$ as an analyticity condition on S^2.

$N = 3$ SYM theory, see Chapter 12). Their $N = 1$ analog is the well-known chiral superspace, a manifestation of $N = 1$ Grassmann analyticity (chirality). The preservation of $N = 1, 2, 3$ analyticities is the geometric principle of the unconstrained, manifestly invariant geometric formulations of the $N = 1$ and $N = 2$ SYM and supergravity theories, as well as of $N = 3$ SYM theory. $N = 1$ and $N = 2$ matter is also naturally described by analytic superfields.

The principle of analyticity plays a crucial rôle not only in supersymmetric theories: It has deep implications in ordinary bosonic ($N = 0$) theories as well. The preservation of analytic representations in a gauge field background governs such classical problems as the theory of YM instantons, Kähler and hyper-Kähler geometry. Moreover, it turns out that the different kinds of Grassmann analyticity have direct bosonic analogs. For instance, $N = 1$ Grassmann analyticity is very similar to the ordinary Cauchy–Riemann analyticity. Correspondingly, $N = 1$ SYM theory has as a prototype the so-called Yang complex theory [Y2]. $N = 2$ SYM theory is the supersymmetric analog of the self-dual $N = 0$ YM theory (the theory of instantons). As we shall see later, the latter is based on a harmonic generalization of Cauchy–Riemann analyticity. The same kind of analyticity is the corner stone of the hyper-Kähler and quaternionic geometries, which in many aspects resemble $N = 2$ supergravity in harmonic superspace.

The interplay between Grassmann analyticity and its $N = 0$ analogs becomes even more explicit in the context of supersymmetric matter. Both in the $N = 1$ and $N = 2$ supersymmetric sigma models the geometry of the superspace where the unconstrained sigma model superfields are defined, and the geometry of the target space are determined by the preservation of similar kinds of analyticity. In the case of $N = 1$ sigma models the target space is a Kähler manifold, and the base superspace is chiral. Both spaces are directly related to complex analyticity (the ordinary Cauchy–Riemann analyticity and its supersymmetric version – chirality). A similar phenomenon in $N = 2$ sigma models can only be revealed in the framework of harmonic superspace. It is well known that the geometry of the bosonic target space of an $N = 2$ sigma model is hyper-Kähler in the case of rigid $N = 2$ supersymmetry [A2, A3] and quaternionic in the local case [B4]. In [G7, G19, G20, G21] we have shown that the deep reason for this correspondence is the fact that the geometry of hyper-Kähler and quaternionic manifolds is determined by the preservation of a harmonic analyticity of the same kind* as the Grassmann $N = 2$ analyticity associated with $N = 2$ supersymmetric matter.

The present chapter is devoted to the applications of harmonic analyticity to the self-dual YM theory and hyper-Kähler geometry. The main result is that the constraints determining these geometries can be solved in terms of unconstrained potentials, which are defined in appropriate harmonic-analytic

* Precisely the same harmonic analyticity is relevant to the self-dual YM theory.

subspaces of the $SU(2)$-harmonic extensions of the initial manifolds. These potentials encode all the information about the given theory in the sense that all the geometric objects – connections, vielbeins, metric – can be expressed in terms of the potentials. This resembles very much, e.g., the rôle of the prepotential V^{++} in $N = 2$ SYM theory (see Chapter 7). It should be pointed out that the explicit solution to the above constraints involves an object similar to the $N = 2$ SYM bridge v. Like in the supersymmetric case, the bridge is obtained from the potential as a solution to a differential equation on S^2. This, in general, is not an easy task, but we give examples where the explicit solution can be found.

The results of this chapter confirm the one-to-one correspondence between $N = 2$ sigma models and hyper-Kähler manifolds anticipated in Chapter 5.* The harmonic-analytic potential of the most general hyper-Kähler manifold can be identified with the Lagrangian of the most general $N = 2$ sigma model (after replacing the coordinates of the manifold by analytic superfields q^+ describing $N = 2$ matter). The metric obtained from the supersymmetric Lagrangian upon elimination of the infinite set of auxiliary fields coincides with that derived from the potential of hyper-Kähler geometry.

Not surprisingly, our treatment of $N = 2$ self-dual YM theory and of hyper-Kähler geometry has much in common with the well-known twistor approach [P2, P3, P4, W1, W2, W3, W4] to the same problems. From our point of view, the main merits of the harmonic approach are, firstly, that it best suits the purpose of understanding the relationship between geometry and $N = 2$ supersymmetry and, secondly, that it exploits the standard differential geometry technique familiar to many physicists with a field-theoretical background.

11.2 Preliminaries: Self-dual Yang–Mills equations and Kähler geometry

In this section we illustrate the main points of our approach to $N = 0$ gauge theories by two well-known examples. The first is self-dual Yang–Mills (SDYM) theory. This problem can be formulated in a form which is remarkably similar to the constrained formulation of $N = 2$ SYM theory. Consequently, one can employ the same technique of introducing $SU(2)/U(1)$ harmonic variables, interpreting the SDYM equations as integrability conditions for a certain type of harmonic analyticity, and finally solving them in terms of an unconstrained analytic potential. The second example is that of Kähler geometry. Once again, the corresponding constraints are related to the preservation of (ordinary) analytic representations, and can be solved by introducing bridges to an analytic basis and subsequently expressing all the geometric objects in terms of a single

* The correspondence between $N = 2$ sigma models coupled to supergravity and quaternionic manifolds is briefly discussed in Chapter 10.

potential (the Kähler potential). The latter has a nice interpretation as a vielbein associated with an extra 'central charge' coordinate.

11.2.1 Harmonic analyticity and SDYM theory

The simplest example of how harmonic analyticity works in the $N = 0$ case is the SDYM theory in the four-dimensional* Euclidean space $\mathbb{R}^4$.

The Lorentz group of $\mathbb{R}^4$ is $SO(4) \sim SU(2)_L \times SU(2)_R$, so one can parametrize $\mathbb{R}^4$ as follows:

$$\mathbb{R}^4 = \frac{ISO(4)}{SU(2)_L \times SU(2)_R} = \{x^{\mu i}\}, \qquad \overline{x^{\mu i}} = \epsilon_{\mu\nu}\epsilon_{ij}x^{\nu j} . \tag{11.1}$$

Here the indices $\mu = 1, 2$ and $i = 1, 2$ transform under $SU(2)_L$ and $SU(2)_R$, respectively. Introducing Yang–Mills covariant derivatives $\mathcal{D}_{\mu i} = \partial_{\mu i} + iA_{\mu i}(x)$, one has the following bi-spinor definition of the field strength:

$$[\mathcal{D}_{\mu i}, \mathcal{D}_{\nu j}] = i\epsilon_{\mu\nu}F_{(ij)} + i\epsilon_{ij}F_{(\mu\nu)} . \tag{11.2}$$

The condition of self-duality means that one half of this tensor vanishes, e.g.,

$$F_{(ij)} = 0 , \quad \Leftrightarrow \quad [\mathcal{D}_{\mu(i}, \mathcal{D}_{\nu j)}] = 0 \tag{11.3}$$

(anti-self-duality would mean vanishing of the other half $F_{(\mu\nu)}$).

The similarity between (11.3) and the $N = 2$ SYM defining constraints (7.57)–(7.59) is obvious. This immediately suggests using the same strategy for solving the SDYM equations (11.3). Firstly, one introduces harmonic variables associated with $SU(2)_R$:

$$\{u^{\pm i}\} = \frac{SU(2)_R}{U(1)_R} , \qquad u^{+i}u_i^- = 1 , \qquad \overline{u^{+i}} = u_i^- \tag{11.4}$$

and extends $\mathbb{R}^4$ to the harmonic space $\mathbb{R}^{4+2}$:

$$\mathbb{R}^4 \quad \Rightarrow \quad \mathbb{R}^{4+2} = \frac{ISO(4)}{SU(2)_L \times U(1)_R} = \{x^{\mu i}, u^{\pm i}\} . \tag{11.5}$$

The harmonics $u^{\pm i}$ have all the properties of those used in $N = 2$ supersymmetry, the only difference being that the former are related to the $SU(2)_R$ part of the Lorentz group, while the latter have to do with the automorphism group $SU(2)_A$ of the $N = 2$ superspace $\mathbb{R}^{4|8}$.

In complete analogy with the supersymmetric case, we introduce a new parametrization of $\mathbb{R}^{4+2}$ ($x^{\mu i}$ are now the analogs of $\theta^{\alpha i}, \bar{\theta}^{\dot{\alpha} i}$):

$$\{x^{\mu i}, u^{\pm i}\} \rightarrow \{x^{\mu\pm}, u^{\pm i}\} , \qquad x^{\mu\pm} \equiv x^{\mu i}u_i^{\pm} ,$$

---

* In [G20] we considered a generalization to the space $\mathbb{R}^{4n}$ with Lorentz group $Sp(1) \times Sp(n)$.

$$\partial_{\mu i} \quad \to \quad \partial_{\mu\pm} = \partial/\partial x^{\mu\pm} \equiv \mp u^{\mp i}\partial_{\mu i}\,, \tag{11.6}$$

and observe that the subspaces $\{x^{\mu+}, u^{\pm i}\}$ or $\{x^{\mu-}, u^{\pm i}\}$ are closed under the action of the whole inhomogeneous Euclidean group $ISO(4)$. These subspaces are the analogs of the analytic and anti-analytic $N = 2$ superspaces. We can define analytic fields:

$$\partial_{\mu-}\phi(x,u) = 0 \quad \Leftrightarrow \quad \phi = \phi(x^{\mu+}, u^{\pm i})\,. \tag{11.7}$$

Then it becomes clear that the SDYM equations (11.3) are nothing but the integrability conditions for the existence of such fields (just like the $N = 2$ SYM constraints guarantee the existence of analytic superfields):

$$[\mathcal{D}^+_\mu, \mathcal{D}^+_\nu] = 0\,, \tag{11.8}$$

$$[\partial^{++}, \mathcal{D}^+_\mu] = 0\,. \tag{11.9}$$

As in the $N = 2$ case, (11.9) implies $\mathcal{D}^+_\mu = \partial_{\mu-} + iu^{+i}A_{\mu i}(x) = u^{+i}\mathcal{D}_{\mu i}$, then (11.8) is equivalent to (11.3). The constraint (11.8) is the integrability condition for the covariant version of (11.7):

$$\mathcal{D}^+_\mu\phi(x,u) = 0\,, \qquad \phi' = e^{i\tau(x)}\phi\,, \tag{11.10}$$

where $\tau(x)$ is an ordinary, u-independent gauge parameter. The general solution of (11.8) is given in terms of a gauge 'bridge' $v(x,u)$ from the u-independent τ gauge frame to the analytic λ gauge frame:

$$A^+_\mu = -ie^{-iv(x,u)}\partial_{\mu-}e^{iv(x,u)}\,; \tag{11.11}$$

$$e^{iv'} = e^{i\lambda}e^{iv}e^{-i\tau}\,, \qquad \partial_{\mu-}\lambda = 0 \to \lambda = \lambda(x^+, u)\,. \tag{11.12}$$

The λ frame in which analyticity is manifest is defined quite analogously to the case of $N = 2$ SYM. The covariantly analytic field (11.10) becomes manifestly analytic:

$$\Phi = e^{iv}\phi\,, \quad \Phi' = e^{i\lambda}\Phi \quad \Rightarrow \quad \partial_{\mu-}\Phi = 0 \quad \Leftrightarrow \quad \Phi = \Phi(x^+, u)\,. \tag{11.13}$$

The covariant derivatives $\mathcal{D}^+_\mu$ and $\mathcal{D}^{++}$ in the λ frame are

$$(\mathcal{D}^+_\mu)_\lambda = e^{iv}\mathcal{D}^+_\mu e^{-iv} = \partial_{\mu-}\,, \tag{11.14}$$

$$(\mathcal{D}^{++})_\lambda = e^{iv}\partial^{++}e^{-iv} \equiv \partial^{++} + iV^{++}\,. \tag{11.15}$$

Substituting (11.14) and (11.15) into (11.9) one concludes that the harmonic connection V^{++} must be analytic:

$$[\mathcal{D}^{++}, \mathcal{D}^+_\mu] = 0 \quad \Rightarrow \quad \partial_{\mu-}V^{++} = 0 \quad \Rightarrow \quad V^{++} = V^{++}(x^+, u)\,;$$

$$V^{++\prime} = e^{i\lambda}V^{++}e^{-i\lambda} - ie^{i\lambda}\partial^{++}e^{-i\lambda}\,. \tag{11.16}$$

The conclusion is that all the information about the solution of the SDYM equations is encoded in the arbitrary analytic field $V^{++}(x^+, u)$. Given a V^{++} one can, in principle, solve eq. (11.15) for the bridge $v(x, u)$ and then construct the τ-frame connection A^+_μ (11.11); the latter is guaranteed by (11.9) to depend linearly on u^{+i}, so one can extract the self-dual connection $A_{\mu i}(x)$ satisfying (11.3).

The non-trivial step in this program is to find the bridge v (i.e., to solve the non-linear differential equation (11.15) on S^2). As an alternative (recall the case of $N = 2$ SYM), one can stay in the λ frame where one needs the harmonic connection V^{--} to complete the differential geometry formalism. The latter can be found as the (unique) solution of the linear differential equation

$$\partial^{++}V^{--} - \partial^{--}V^{++} + i[V^{++}, V^{--}] = 0\,, \tag{11.17}$$

following from the conventional constraint

$$[\mathcal{D}^{++}, \mathcal{D}^{--}] = D^0\,. \tag{11.18}$$

Then the connection A^-_μ is defined as follows:

$$\mathcal{D}^-_\mu = [\mathcal{D}^{--}, \mathcal{D}^+_\mu] \quad \Rightarrow \quad A^-_\mu = -\partial_{\mu-}V^{--}\,, \tag{11.19}$$

and the non-vanishing part of the field strength is simply

$$F_{\mu\nu} = -i[\mathcal{D}^+_\mu, \mathcal{D}^-_\nu] = -\partial_{\mu-}\partial_{\nu-}V^{--}\,. \tag{11.20}$$

It is covariantly u independent ($\mathcal{D}^{++}F_{\mu\nu} = \mathcal{D}^{--}F_{\mu\nu} = 0$) in the λ frame. Since the global characteristics of the self-dual solutions, such as the topological charge and the action, are gauge-frame independent, one may calculate them in the λ frame as well.

In conclusion, we should point out that, given an analytic prepotential $V^{++}(x^+, u)$, one does not know if the corresponding self-dual solution will be physically meaningful, i.e., if it will have finite energy. Finding those V^{++} which yield the instanton solutions is a non-trivial problem. The simplest example [G16] is the V^{++} which generates the 1-instanton solution of [B5]:

$$(V^{++})_i^j = -\frac{1}{\rho^2}x_i^+x^{+j}\,. \tag{11.21}$$

The multi-instanton V^{++}'s have been found in [K3]. The harmonic space version of the general ADHM construction [A7, A9] was discussed in [O2]. Note also the similarity between the harmonic space approach to SDYM and the inverse scattering problem approach of [B6]. In particular, the complex spectral parameter introduced in [B6] can be interpreted as an $SU(2)$ harmonic in a special parametrization.

11.2.2 Comparison with the twistor space approach

Above we gave a treatment of the SDYM problem which followed closely that of the $N = 2$ SYM constrained system. On the other hand, the reader may be familiar with the well-known twistor approach to the same problem. Here we briefly discuss the relationship between the harmonic space and the twistor space approaches (more specifically, the twistor construction of Ward [W1]).

Common for both approaches is the interpretation of the SDYM equations as integrability conditions obtained with the help of additional variables related to the sphere S^2. We describe S^2 *globally* by considering functions defined on $SU(2) = \{u_i^{\pm}\}$ and requiring them to possess a definite $U(1)$ charge. Thus we avoid using an explicit parametrization of S^2. Another feature is the manifest covariance with respect to $SU(2)$ acting on the indices $i, j, \ldots$. It allows one to control the pattern of $SU(2)$ breaking.

In the twistor approach one considers functions of two complex variables π^i $(i = 1, 2)$ which are holomorphic (i.e., do not depend on $\bar{\pi}^i$) and homogeneous, so they effectively depend on, e.g., $\xi = \pi^1/\pi^2$. This clearly provides a parametrization of only one part of S^2 (leaving out the point $\pi^2 = 0$). In our language this would correspond to using only the single complex variable $\xi = u^{+1}/u^{+2}$ (but not u^-) and replacing the $U(1)$ charge by a degree of homogeneity in u^{+i}. Consequently, we would not be able to use the derivative $D^{++} = u^{+i}\partial/\partial u^{-i} \sim \partial/\partial\bar{\xi}$. Therefore, the way in which we introduce the prepotential $V^{++}(x^+, u^{\pm})$ as the connection for $\mathcal{D}^{++}$ and the basic equation (11.15) determining the bridge $v(x^{\pm}, u^{\pm})$ in terms of V^{++} are not directly applicable in the twistor approach. Instead, one does the following. The self-duality equations imply that the connection $A_\mu^+ = A_{\mu i}\pi^i$ can be represented in the 'pure gauge' form

$$A_\mu^+ = H^{-1}\partial_{\mu-}H\,. \tag{11.22}$$

The condition that A_μ^+ depends on π^i only linearly (which in our language has the form of a differential constraint, $D^{++}A_\mu^+ = 0$) is now formulated as the condition that A_μ^+ is regular in π (taking into account the homogeneity in π) on the whole sphere S^2. Such a restriction imposed on H would be too strong (H would become a gauge transformation and A_μ^+ would be empty). Therefore, it is weakened by requiring that H be analytic (regular) only in one part Ω_1 of S^2 (the approximate equivalent in our language is that we do not require $D^{++}v(x, u) = 0$). However, the presentation (11.22) is not unique, since one can consider another function $\hat{H}$ giving the same A_μ^+:

$$A_\mu^+ = \hat{H}^{-1}\partial_{\mu-}\hat{H}\,, \tag{11.23}$$

but analytic in another region $\Omega_2 \subset S^2$, $\Omega_1 \cup \Omega_2 = S^2$. Comparing (11.22) and (11.23) one concludes that

$$G = H\hat{H}^{-1}\,, \qquad \partial_{\mu-}G = 0\,, \tag{11.24}$$

i.e., G is analytic in π on $\Omega_1 \cap \Omega_2$ and depends on $x^{\mu+} = x^{\mu i}\pi_i$ only, $G = G(x^+, \pi)$. The argument can now be reversed: given a potential $G(x^+, \pi)$ one may try to perform the 'splitting' (11.24) into H and $\tilde{H}$ with overlapping regions of analyticity (a variant of the well-known Riemann–Hilbert problem), and subsequently construct the connection A^+_μ (11.22) which will automatically satisfy the SDYM equations. We conclude that this procedure is to a certain extent analogous to ours, with H replacing the bridge v, G replacing the $\mathcal{D}^{++}$ connection V^{++}, and the splitting (11.24) replacing the non-linear differential relation (11.15) between V^{++} and v.

Closer to our approach seems to be that developed by Newman *et al.* (see [K8, N1] and references therein) in parallel with the twistor one. There the central problem is to solve a certain differential equation (referred to as the Sparling equation in [K8]). The latter looks almost identical with our equation (11.15). Furthermore, an essential point in that approach is the use of harmonic expansions on the sphere S^2 (with the name 'spin-weight' standing for the $U(1)$ charge and with vector harmonics instead of the spinor ones in our formalism).

11.2.3 Complex analyticity and Kähler geometry

Above we have demonstrated how the principle of preservation of analyticity leads to an adequate geometric formulation of a constrained gauge theory (the SDYM theory). The same principle is the key to understanding the intrinsic geometry of certain constrained theories of gravitational type (i.e., gauge theories of the diffeomorphism group of some Riemannian manifolds). The simplest example is the so-called Kähler geometry. We use it as an introduction to the much more complicated hyper-Kähler geometry discussed in the rest of this chapter.

Standard Kähler geometry One way to define Kähler geometry is to introduce a covariantly constant complex structure in a Riemannian manifold. This means the following. Consider a $2n$-dimensional real manifold

$$\mathbb{R}^{2n} = \{X^M\}, \qquad M = 1, \ldots, 2n, \qquad \overline{X^M} = X^M \tag{11.25}$$

with a Riemannian metric $g_{MN}(x)$ and diffeomorphism group

$$\delta X^M = \tau^M(X). \tag{11.26}$$

In order to develop the differential geometry formalism, we introduce the tangent space group $O(2n)$ (the Lorentz group of $\mathbb{R}^{2n}$) and define $O(2n)$ covariant derivatives:

$$\mathcal{D}_A = e^M_A \frac{\partial}{\partial X^M} + \omega_{ABC} L^{BC}. \tag{11.27}$$

Here $e_A^M(X)$ is the vielbein transforming as an $O(2n)$ vector (index $A = 1, \ldots, 2n$) with respect to the tangent space group and as a contravariant vector (index M) with respect to the diffeomorphism group (11.26).* The connection $\omega_{ABC} = -\omega_{ACB}$ is Lie-algebra valued (L^{BC} are the $O(2n)$ generators). The invariant tensor in the theory is the Riemann tensor (curvature) defined by the commutator

$$[\mathcal{D}_A, \mathcal{D}_B] = R_{ABCD} L^{CD} . \qquad (11.28)$$

The absence of a torsion term in (11.28) (the manifold is Riemannian)† implies that ω_A can be expressed in terms of e_A^M.

Now, define a covariantly constant complex structure as a tensor I_A^B with the following properties:

$$I_A^B I_B^C = -\delta_A^C , \qquad (11.29)$$

$$I_{AB} \equiv I_A^C \eta_{BC} = -I_{BA} , \qquad (11.30)$$

$$\mathcal{D}_C I_A^B = 0 . \qquad (11.31)$$

The existence of such an object is one of the definitions of Kähler geometry.

Equations (11.29)–(11.31) impose certain restrictions on the curvature. One way to find them is to study the integrability conditions for eq. (11.31). We prefer a simpler approach in which the properties (11.29)–(11.31) are made manifest. It can be shown that using the full $O(2n)$ tangent space gauge freedom one can always choose a frame in which the complex structure becomes a constant matrix. In the process the tangent space group $O(2n)$ is reduced to $U(n)$. This amounts to replacing the index A by a pair $(\alpha, \bar{\alpha})$ $(\alpha, \bar{\alpha} = 1, \ldots, n)$ and choosing I in the form

$$I_\alpha^\beta = i\delta_\alpha^\beta , \qquad I_{\bar{\alpha}}^{\bar{\beta}} = -i\delta_{\bar{\alpha}}^{\bar{\beta}} , \qquad I_\alpha^{\bar{\beta}} = I_{\bar{\alpha}}^\beta = 0 . \qquad (11.32)$$

Substituting (11.32) into (11.31) one finds that the off-diagonal connection components should vanish,

$$\omega_{A\beta}{}^{\bar{\gamma}} \equiv \omega_{A\beta\gamma} = 0 , \qquad \omega_{A\bar{\beta}}{}^{\gamma} \equiv \omega_{A\bar{\beta}\bar{\gamma}} = 0 . \qquad (11.33)$$

This in turn yields the vanishing of the curvature components

$$R_{AB\gamma\delta} = R_{AB\bar{\gamma}\bar{\delta}} = 0 . \qquad (11.34)$$

Then, using the Bianchi identity $R_{ABCD} + R_{BCAD} + R_{CABD} = 0$ one sees that

$$R_{\alpha\beta\gamma\bar{\delta}} = R_{\bar{\alpha}\bar{\beta}\bar{\gamma}\delta} = 0 . \qquad (11.35)$$

Thus, the only remaining non-vanishing component of the curvature is $R_{\alpha\bar{\beta}\gamma\bar{\delta}}$. Equations (11.34) as well as their corollaries (11.35) constitute an equivalent

---

* The inverse metric g^{MN} is the 'square' of the vielbeins, $g^{MN} = e_A^M \eta^{AB} e_B^N$.

† The case of manifolds with torsion is discussed in [D7].

definition of a Kähler manifold as a Riemannian manifold whose holonomy group (the group generated by the curvature tensor) lies in $U(n)$ [E6, K9].

We now proceed to solving the Kähler geometry constraints. Firstly, it is convenient to introduce a new, complex parametrization of the manifold:

$$x^\mu = X^\mu + iX^{\mu+n}, \qquad x^{\bar\mu} \equiv \overline{x^\mu} = X^\mu - iX^{\mu+n}, \qquad \mu = 1, \ldots, n. \quad (11.36)$$

In this setup the diffeomorphism group (11.26) has the form

$$\delta x^\mu = \tau^\mu(x, \bar x), \qquad \delta x^{\bar\mu} = \tau^{\bar\mu}(x, \bar x), \qquad \tau^{\bar\mu} = \overline{\tau^\mu} \quad (11.37)$$

and the covariant derivatives (11.27) become

$$\begin{aligned} \mathcal{D}_\alpha &= e^\mu_\alpha \partial_\mu + e^{\bar\mu}_\alpha \partial_{\bar\mu} + \omega_\alpha, \\ \mathcal{D}_{\bar\alpha} &= \overline{\mathcal{D}_\alpha} = e^{\bar\mu}_{\bar\alpha} \partial_{\bar\mu} + e^\mu_{\bar\alpha} \partial_\mu + \omega_{\bar\alpha}. \end{aligned} \quad (11.38)$$

We recall that the tangent space group $U(n)$ acts as follows:

$$\delta \mathcal{D}_\alpha = \Lambda_{\alpha\bar\beta}(x, \bar x) \mathcal{D}_\beta, \qquad \Lambda_{\alpha\bar\beta} = -\Lambda_{\bar\beta\alpha} = -\overline{\Lambda_{\beta\bar\alpha}}. \quad (11.39)$$

The algebra of the covariant derivatives corresponding to the above constraints has the following form

$$[\mathcal{D}_\alpha, \mathcal{D}_\beta] = [\mathcal{D}_{\bar\alpha}, \mathcal{D}_{\bar\beta}] = 0 \quad (11.40)$$

and

$$[\mathcal{D}_\alpha, \mathcal{D}_{\bar\beta}] = R_{\alpha\bar\beta}. \quad (11.41)$$

These algebraic relations provide yet another equivalent definition of Kähler geometry. Note the absence of torsion terms in these commutators. The vanishing of torsion is the standard postulate of Riemannian geometry, which normally allows one to express the connection ω in terms of the vielbeins e without restricting the latter in any way. However, in our case some of the connections have been set to zero (see (11.33)), therefore some of the torsion constraints become non-trivial and imply differential conditions on the vielbein.

So, we need to solve the following torsion constraints:

$$\begin{aligned} &(a) \quad T_{\bar\alpha\bar\beta\bar\gamma} = 0, \\ &(b) \quad T_{\alpha\bar\beta\gamma} = 0, \\ &(c) \quad T_{\alpha\beta\bar\gamma} = 0 \end{aligned} \quad (11.42)$$

(and their complex conjugates). Note that the curvature constraint $R_{\alpha\beta} = 0$ contained in (11.40) is in fact a corollary of (11.42) via the Bianchi identities. Once again, we are in the familiar situation where the constraints defining the theory have the form of integrability conditions for the existence of covariantly analytic fields satisfying the Cauchy–Riemann condition

$$\mathcal{D}_{\bar\alpha} \phi(x, \bar x) = 0. \quad (11.43)$$

The crucial observation concerning the constraint (11.42a) is that it implies the existence of an analytic basis where the analyticity (11.43) becomes manifest. Indeed, let us look at the detailed form of eq. (11.42a):

$$T_{\bar{\alpha}\bar{\beta}\bar{\gamma}} = e_{\bar{\alpha}}^{M} \partial_M e_{\bar{\beta}}^{N} e_{N\bar{\gamma}} - (\bar{\alpha} \leftrightarrow \bar{\beta}) = 0 \,, \tag{11.44}$$

where $M = (\mu, \bar{\mu})$ and $e_{N\bar{\gamma}}$ is the inverse vielbein. In order to understand its meaning, let us linearize it by setting $e_{\bar{\alpha}}^{\bar{\mu}} = \delta_{\bar{\alpha}}^{\bar{\mu}} + \varepsilon_{\bar{\alpha}}^{\bar{\mu}}$, $e_{\bar{\alpha}}^{\mu} = \varepsilon_{\bar{\alpha}}^{\mu}$, etc., where all $\varepsilon \ll 1$. The result is

$$\partial_{\bar{\alpha}} \varepsilon_{\bar{\beta}}^{\mu} - \partial_{\bar{\beta}} \varepsilon_{\bar{\alpha}}^{\mu} = 0 \,,$$

which has the obvious solution

$$\varepsilon_{\bar{\alpha}}^{\mu} = \partial_{\bar{\alpha}} v^{\mu} \tag{11.45}$$

with an arbitrary *complex* vector $v^{\mu}(x, \bar{x})$. The full non-linear version of the solution (11.45) is given by the equation

$$e_{\bar{\alpha}}^{\bar{\nu}} \partial_{\bar{\nu}} v^{\mu} + e_{\bar{\alpha}}^{\nu} (\delta_{\nu}^{\mu} + \partial_{\nu} v^{\mu}) = 0 \tag{11.46}$$

which allows us to express the vielbein $e_{\bar{\alpha}}^{\nu}$ in terms of $e_{\bar{\alpha}}^{\bar{\nu}}$ and of the *arbitrary vector* $v^{\mu}(x, \bar{x})$ (the matrix $\delta_{\nu}^{\mu} + \partial_{\nu} v^{\mu}$ is supposed invertible).

Now, this result has very important implications. Let us use the vector $v^{\mu}(x, \bar{x})$ to perform the change of variables:

$$x_A^{\mu} = x^{\mu} + v^{\mu}(x, \bar{x}) \,, \qquad x_A^{\bar{\mu}} \equiv \overline{(x_A^{\mu})} = x^{\bar{\mu}} + v^{\bar{\mu}}(x, \bar{x}) \,. \tag{11.47}$$

The vielbeins $e_{\bar{\alpha}}^{\bar{\mu}}$ and $e_{\bar{\alpha}}^{\mu}$ are transformed and we get, in particular,

$$E_{\bar{\alpha}}^{\mu} = \mathcal{D}_{\bar{\alpha}} x_A^{\mu} = e_{\bar{\alpha}}^{\bar{\nu}} \partial_{\bar{\nu}} v^{\mu} + e_{\bar{\alpha}}^{\nu} (\delta_{\nu}^{\mu} + \partial_{\nu} v^{\mu}) = 0 \tag{11.48}$$

as a consequence of (11.46). In other words, the *bridge* $v^{\mu}(x, \bar{x})$ defines the new basis $x_A, \bar{x}_A$ (11.47) in which the covariant derivative $\mathcal{D}_{\bar{\alpha}}$ becomes simpler,

$$\mathcal{D}_{\bar{\alpha}} = E_{\bar{\alpha}}^{\bar{\mu}} \frac{\partial}{\partial x_A^{\bar{\mu}}} + \omega_{\bar{\alpha}} \,, \tag{11.49}$$

so that the Cauchy–Riemann condition (11.43) can be solved explicitly:

$$\mathcal{D}_{\bar{\alpha}} \phi = 0 \quad \Rightarrow \quad \phi = \phi(x_A^{\mu}) \,. \tag{11.50}$$

We can say that the field ϕ is manifestly analytic and can call (11.47) an *analytic basis*.*

* Note that if the field ϕ in (11.43) has a $U(n)$ tangent space index, the presence of a connection in (11.49) appears as an obstruction to obtaining the manifestly analytic solution (11.50). In fact, the vanishing of the curvature $R_{\bar{\alpha}\bar{\beta}}$ allows us to make a $U(n)$ frame rotation such that the connection $\omega_{\bar{\alpha}}$ is gauged away. However, such a gauge will be essentially complex, i.e., the connection ω_{α} will be non-vanishing.

It is clear that the bridge-defining equation (11.46) allows x^μ_A to transform with an arbitrary *analytic* parameter:

$$\delta x^\mu_A = \lambda^\mu(x_A)\,, \qquad \delta x^{\bar\mu}_A = \lambda^{\bar\mu}(\bar x_A)\,. \tag{11.51}$$

This means that v^μ transforms as follows:

$$\delta v^\mu(x,\bar x) = \lambda^\mu(x_A) - \tau^\mu(x,\bar x)\,, \tag{11.52}$$

so it serves as a bridge from the central basis $\{x^\mu, x^{\bar\mu}\}$ with a τ diffeomorphism group to the analytic basis $\{x^\mu_A, x^{\bar\mu}_A\}$ with an analytic λ diffeomorphism group. The latter is compatible with the notion of manifestly analytic fields (11.50).

The reader may notice that the bridge v^μ is in fact a pure gauge, since the parameter τ^μ in (11.52) is of exactly the same type as v^μ. For instance, one can fix the gauge

$$v^\mu = 0 \quad \Leftrightarrow \quad x^\mu_A = x^\mu \quad \Rightarrow \quad \tau^\mu(x,\bar x) = \lambda^\mu(x_A)\,. \tag{11.53}$$

In this gauge there is no difference between the old basis and the new, analytic one, which allows the derivative $\mathcal{D}_{\bar\alpha}$ to keep its short form in both bases. This is a peculiarity of the x-type bridges in the Kähler case. As we shall see in the next subsection where we introduce an extra ('central charge') coordinate, the bridge for this additional coordinate will carry gauge-independent degrees of freedom (actually, its imaginary part will be the Kähler potential, the basic object of Kähler geometry). Later on we shall show that in the hyper-Kähler case even the x-type bridges are not pure gauges. Another remark concerning the gauge (11.53) is that the initial τ diffeomorphism group with arbitrary complex parameters is replaced by a λ group with analytic parameters. This is in accord with the notion of manifest analyticity exhibited in eq. (11.50). In what follows we shall always work in the gauge (11.53).

Let us now proceed to solving the rest of the constraints (11.42). Equation (11.42b) is of the usual type in Riemannian geometry, its solution being just an expression of the connection in terms of the vielbeins:

$$T_{\alpha\bar\beta\gamma} = 0 \quad \Rightarrow \quad \omega_{\alpha\bar\beta\gamma} = -E^\nu_\alpha \partial_\nu E^{\bar\mu}_{\bar\beta} E_{\bar\mu\gamma}\,. \tag{11.54}$$

Then we can insert (11.54) into (11.42c) to find a constraint on the Kähler *metric* constructed from the inverse vielbeins:

$$g_{\bar\mu\nu} = E_{\bar\mu\alpha} E_{\nu\bar\alpha} \tag{11.55}$$

(note that in the gauge (11.53) there are no metric components of the type $g_{\mu\nu}$ or $g_{\bar\mu\bar\nu}$). The result is the well-known Kähler condition on the metric:

$$T_{\bar\alpha\bar\beta\gamma} = 0 \quad \Rightarrow \quad \partial_{[\bar\mu} g_{\bar\nu]\lambda} = 0\,. \tag{11.56}$$

It has the general local solution

$$g_{\bar{\mu}\nu} = \partial_{\bar{\mu}}\partial_{\nu}K\,, \tag{11.57}$$

where the arbitrary real function $K(x, \bar{x}) = \bar{K}$ is the Kähler potential,* the primary object of Kähler geometry. Note that K is defined up to the pregauge freedom

$$\delta K = \frac{1}{2i}(\lambda(x_A) - \bar{\lambda}(\bar{x}_A))\,, \tag{11.58}$$

where $\lambda(x_A)$ is an arbitrary analytic function. This 'pregauge' invariance is the so-called Kähler invariance.

This completes the procedure of solving the constraints (11.40) defining Kähler geometry. Our aim here was to underline the importance of the concept of analyticity, both for formulating the constraints and for solving them.

It should be pointed out that both the Kähler potential and its pregauge invariance do not have a natural geometric meaning in the present framework. Their rôle becomes clear in an extended space involving a central charge coordinate.

11.2.4 Central charge as the origin of the Kähler potential

In order to incorporate K as a geometric object, we extend $\mathbb{R}^{2n}$ by adding a new real coordinate $z = \bar{z}$ with the dimension of [length]^2:

$$\mathbb{R}^{2n+1} = \{x^{\mu}, x^{\bar{\mu}}, z\}\,. \tag{11.59}$$

In this space one can realize the following 'central charge' extension of the Poincaré algebra:

$$\begin{aligned} &[P_{\mu}, P_{\bar{\nu}}] = 2\delta_{\mu\bar{\nu}}Z\,, \\ &[P_{\mu}, Z] = [P_{\bar{\mu}}, Z] = [P_{\mu}, P_{\nu}] = [P_{\bar{\mu}}, P_{\bar{\nu}}] = 0\,. \end{aligned} \tag{11.60}$$

This algebra still has $U(n)$ as its automorphism group, the real central charge being a singlet. The transformations realizing (11.60) in $\mathbb{R}^{2n+1}$ are

$$\delta x^{\mu} = a^{\mu}\,, \qquad \delta x^{\bar{\mu}} = a^{\bar{\mu}}\,, \qquad \delta z = a + i(a^{\bar{\mu}}x^{\mu} - a^{\mu}x^{\bar{\mu}})\,. \tag{11.61}$$

The covariant derivatives in $\mathbb{R}^{2n+1}$ (commuting with the generators $P, \bar{P}, Z$) have the form

$$D_{\mu} = \partial_{\mu} + ix^{\bar{\mu}}\partial_z\,, \qquad D_{\bar{\mu}} = \partial_{\bar{\mu}} - x^{\mu}\partial_z\,, \qquad D_z = \partial_z \tag{11.62}$$

* The name 'prepotential' would be more adequate if one calls the metric the 'potential' and the curvature the 'field strength tensor' of the theory, in analogy with Yang–Mills theory.

and satisfy the algebra

$$[D_\mu, D_{\bar\nu}] = -2i\delta_{\mu\bar\nu} D_z\,, \qquad (11.63)$$
$$[D_\mu, D_z] = [D_{\bar\mu}, D_z] = [D_\mu, D_\nu] = [D_{\bar\mu}, D_{\bar\nu}] = 0\,.$$

The crucial observation is that the algebra (11.63) is still consistent with the analyticity (11.50). This becomes obvious in a special *analytic basis* in $\mathbb{R}^{2n+1}$:

$$x_A^{\mu,\bar\mu} = x^{\mu,\bar\mu}\,, \qquad z_A = z + ix^\mu x^{\bar\mu}\,; \qquad (11.64)$$
$$\delta x_A^{\mu,\bar\mu} = a^{\mu,\bar\mu}\,, \qquad \delta z_A = a + 2ia^{\bar\mu} x_A^\mu\,. \qquad (11.65)$$

Note that the new coordinate z_A is complex, unlike the old one z. We see that the subspace x_A^μ, z_A is invariant under the transformations (11.65). Further, the covariant derivative $\bar D$ is 'short' in the new basis

$$D_\mu = \partial_{A\mu} + 2ix_A^{\bar\mu}\partial_{Az}\,, \qquad D_{\bar\mu} = \partial_{A\bar\mu}\,, \qquad D_z = \partial_{Az}\,, \qquad (11.66)$$

which allows us to solve the analyticity condition (11.50) in the central-charge extended case by simply writing down $\phi = \phi(x_A, z_A)$. Of course, we can still have ordinary z-independent analytic functions $\phi(x_A)$ as well.

The idea now is to generalize the above flat framework to the curved case. One of the basic assumptions is that no geometric object should depend on the auxiliary coordinate z. In particular, this means that the diffeomorphism parameters in $\mathbb{R}^{2n+1}$ should be z-independent. This includes the transformations of z itself, $\delta z = \tau(x, \bar x)$, $\tau = \bar\tau$. Next, we need covariant derivatives. They are obtained from the old ones $\mathcal{D}_{\alpha,\bar\alpha}$ in the framework without central charge by adding a new vielbein term:

$$\nabla_{\alpha,\bar\alpha} = \mathcal{D}_{\alpha,\bar\alpha} + E^z_{\alpha,\bar\alpha}\partial_z\,. \qquad (11.67)$$

As for the derivative ∇_z, the z-independence of the diffeomorphism and, by analogy, of the tangent space groups allows it to keep its flat form

$$\nabla_z = \frac{\partial}{\partial z}\,. \qquad (11.68)$$

Then we have to impose constraints on the above covariant derivatives. Obviously, the introduction of the auxiliary variable z should not give rise to new tensors, other than the curvature component $R_{\alpha\bar\beta}$ of Kähler geometry in (11.41). At the same time, we should incorporate the flat torsion $T^z_{\alpha\bar\beta} = -2i\delta_{\alpha\bar\beta}$ from eq. (11.63). Thus, we are led to choose

$$\text{(a)} \qquad [\nabla_\alpha, \nabla_\beta] = [\nabla_{\bar\alpha}, \nabla_{\bar\beta}] = 0\,,$$
$$\text{(b)} \qquad [\nabla_\alpha, \nabla_{\bar\beta}] = -2i\delta_{\alpha\bar\beta}\nabla_z + R_{\alpha\bar\beta}\,, \qquad (11.69)$$
$$\text{(c)} \qquad [\nabla_\alpha, \nabla_z] = [\nabla_{\bar\alpha}, \nabla_z] = 0\,.$$

Clearly, the constraint (11.69c) simply expresses the fact that the vielbeins and connections do not depend on the central charge. As before, the constraint (11.69a) is the integrability condition for the existence of analytic fields. It also guarantees the existence of an analytic basis. To show this, let us examine its implications on the newly introduced central charge vielbein $E^z_{\bar\alpha}$. Assuming that the old algebra (11.40), (11.41) of the derivatives $\mathcal{D}_{\alpha,\bar\alpha}$ holds, from (11.69a) and (11.40) we obtain the constraint

$$[\nabla_{\bar\alpha}, \nabla_{\bar\beta}] = [\mathcal{D}_{\bar\alpha}, \mathcal{D}_{\bar\beta}] + \mathcal{D}_{[\bar\alpha}E^z_{\bar\beta]}\partial_z = 0 \quad \Rightarrow \quad \mathcal{D}_{[\bar\alpha}E^z_{\bar\beta]} = 0\,, \tag{11.70}$$

which has the obvious general solution

$$E^z_{\bar\alpha} = -\mathcal{D}_{\bar\alpha} w \tag{11.71}$$

with an arbitrary *complex* scalar function $w(x,\bar x)$ (the minus sign is introduced for convenience). Finally, the constraint (11.69b) together with (11.41) imply

$$[\nabla_\alpha, \nabla_{\bar\beta}] = [\mathcal{D}_\alpha, \mathcal{D}_{\bar\beta}] + (\mathcal{D}_\alpha E^z_{\bar\beta} - \mathcal{D}_{\bar\beta}E^z_\alpha)\partial_z = -2i\delta_{\alpha\bar\beta}\nabla_z + R_{\alpha\bar\beta} \quad \Rightarrow$$

$$\mathcal{D}_\alpha E^z_{\bar\beta} - \mathcal{D}_{\bar\beta}E^z_\alpha = -2i\delta_{\alpha\bar\beta}\,. \tag{11.72}$$

Inserting the solution (11.71) into (11.72) and using (11.41) once again, we find

$$\mathcal{D}_\alpha\mathcal{D}_{\bar\beta}w - \mathcal{D}_{\bar\beta}\mathcal{D}_\alpha\bar w = [\mathcal{D}_\alpha, \mathcal{D}_{\bar\beta}]\,\mathrm{Re}\,w + i\{\mathcal{D}_\alpha, \mathcal{D}_{\bar\beta}\}\,\mathrm{Im}\,w = 2i\mathcal{D}_\alpha\mathcal{D}_{\bar\beta}\,\mathrm{Im}\,w = 2i\delta_{\alpha\bar\beta}\,.$$

We see that the real part of w drops out whereas the imaginary part

$$K \equiv \mathrm{Im}\,w \tag{11.73}$$

satisfies the equation (obtained by using the explicit form of the connection (11.54))

$$\mathcal{D}_\alpha\mathcal{D}_{\bar\beta}K = E^\mu_\alpha E^{\bar\nu}_{\bar\beta}\partial_\mu\partial_{\bar\nu}K = \delta_{\alpha\bar\beta} \quad \Rightarrow \quad g_{\mu\bar\nu} \equiv E_{\mu\bar\gamma}E_{\bar\nu\gamma} = \partial_\mu\partial_{\bar\nu}K\,. \tag{11.74}$$

So, we have arrived at the familiar expression of the Kähler metric in terms of the Kähler potential once again. The new point now is that it came out as an algebraic consequence of the central charge constraints, and not as a solution to the differential equation (11.56) in the framework without central charge.

Another achievement is that we have found the geometric origin of the Kähler potential K. Indeed, as mentioned before, eq. (11.69a) is the integrability condition for the existence of analytic fields. To make analyticity manifest we need a new basis. In addition to the coordinate changes (11.47), we define a new complex coordinate z_A:

$$z_A = z + w(x,\bar x)\,, \qquad \delta z_A = \lambda(x_A)\,, \qquad \delta w = \lambda(x_A) - \tau(x,\bar x)\,. \tag{11.75}$$

The bridge $w(x,\bar x)$ is just the object we found when solving the constraint on $E^z_{\bar\alpha}$ above, eq. (11.71). After this additional change of variables the derivative

$\nabla_{\tilde{\alpha}}$ (11.67) takes the short form (11.49) which makes analyticity manifest. The important point now is that, unlike $v^{\mu}(x,\bar{x})$ (11.52), the bridge w cannot be completely gauged away because it is complex and the parameter τ is real. The best we can do is to gauge away the real part of w:

$$\mathrm{Re}\, w = 0 \quad \to \quad \tau(x,\bar{x}) = \frac{1}{2}(\lambda(x_A) + \bar{\lambda}(\bar{x}_A))\,. \tag{11.76}$$

The remaining part of w is

$$K = \mathrm{Im}\, w = \mathrm{Im}\, z_A\,, \qquad \delta K = \frac{1}{2i}(\lambda(x_A) - \bar{\lambda}(\bar{x}_A))\,. \tag{11.77}$$

Thus, we have identified the Kähler potential with the imaginary part of the central charge bridge or, equivalently, of the complex central charge coordinate z_A in the analytic basis. In addition, we have found a natural explanation for the Kähler pregauge freedom (11.58).

The picture we have described in this subsection has close analogs both in the $N = 1$ SYM (Chapter 7) and supergravity (Chapter 10) theories. In these supersymmetric theories one identifies the gauge prepotentials with the imaginary part of the coordinates of a complexified space (the group space in SYM [I3] or space-time in supergravity [O6]).

11.3 Harmonic analyticity and hyper-Kähler potentials

Like in the Kähler case, the constraints on hyper-Kähler geometry follow from the existence of complex structures on the manifold. This time one deals with three such structures, not just one.*

Consider a $4n$-dimensional Riemannian real manifold $\mathbb{R}^{4n} = \{X^M\}$, $M = 1, \ldots, 4n$ with tangent space group $O(4n)$. The covariant derivatives $\mathcal{D}_A$ have the usual form (see (11.27)). Further, suppose that there exist *three* complex structures $(I_a)_A^B$, $a = 1, 2, 3$. All of them have the properties (11.29)–(11.31). In addition, they satisfy the relation

$$I_a I_b + I_b I_a = 0\,, \qquad a \neq b\,. \tag{11.78}$$

The existence of three such anticommuting covariantly constant complex structures is one of the possible definitions of a hyper-Kähler manifold.

One can show that using part of the $O(4n)$ tangent space freedom one can gauge I_a into three constant matrices:

$$(I_a)_{\alpha i}^{\beta j} = i\delta_{\alpha}^{\beta}(\sigma_a)_i^j\,. \tag{11.79}$$

* It is easy to see that having two complex structures automatically yields a third one.

Here we have split the $4n$-vector index A into a pair αi ($\alpha = 1, \ldots, 2n, i = 1, 2$) and its complex conjugate; σ_a are the 2×2 Pauli matrices. The reality condition on a vector V^A is

$$\overline{V^A} = V^A \quad \Rightarrow \quad \overline{V^{\alpha i}} = \Omega_{\alpha\beta}\epsilon_{ij}V^{\beta j}\,, \tag{11.80}$$

where $\Omega_{\alpha\beta} = -\Omega_{\beta\alpha}$ is the $Sp(n)$ invariant tensor. The complex structures (11.79) are required to be covariantly constant (cf. (11.29)),

$$\mathcal{D}_{\gamma k}(I_a)_{\alpha i}^{\beta j} = 0\,, \tag{11.81}$$

where

$$\mathcal{D}_{\alpha i} = e_{\alpha i}^M \frac{\partial}{\partial X^M} + \omega_{\alpha i\, \beta j\, \gamma k} L^{\beta j\, \gamma k}\,.$$

Substituting (11.79) into the condition (11.81) we find

$$\omega_{\alpha i\, \beta j\, \gamma k} = \omega_{\alpha i \beta\gamma}\epsilon_{jk}\,, \qquad \omega_{\alpha i\beta\gamma} = \omega_{\alpha i\gamma\beta}\,. \tag{11.82}$$

Thus, the tangent space group $O(4n)$ is reduced to a local $Sp(n)$ (indices $\alpha, \beta, \ldots$; generators $L^{(\alpha\beta)}$) times a rigid $Sp(1) \sim SU(2)$ group (indices $i, j, \ldots$).

This restriction on the connection yields constraints on the curvature:

$$[\mathcal{D}_{\alpha i}, \mathcal{D}_{\beta j}] = \epsilon_{ij} R_{(\alpha\beta)(\gamma\delta)} L^{(\gamma\delta)}\,. \tag{11.83}$$

The non-vanishing curvature $R_{(\alpha\beta)}$ in (11.83) generates a holonomy group which lies in $Sp(n)$, rather than the original holonomy group $O(4n)$ of $\mathbb{R}^{4n}$. This, together with the zero-torsion condition (the postulate of Riemannian geometry), is an equivalent definition of a hyper-Kähler manifold [E6, K9].

Note that the absence of torsion in (11.83) is not just a conventional constraint allowing one to express the connection in terms of the vielbeins. Like in the Kähler case, some of these torsion constraints impose differential restrictions on the vielbeins, as we shall see shortly.

From (11.83) and the Bianchi identities it follows that the curvature is totally symmetric, $R_{(\alpha\beta)(\gamma\delta)} = R_{(\alpha\beta\gamma\delta)}$. In the case of $\mathbb{R}^4$ (i.e., $n = 1$) this is just the self-dual part of the Weyl tensor, so in four dimensions the hyper-Kähler constraints are equivalent to the condition of self-duality for the Weyl tensor (i.e., the vanishing of its anti-self-dual part).*

* In four dimensions this requirement is the only possible definition of hyper-Kähler manifolds since the definition via the restriction on the holonomy group ceases to be meaningful (the holonomy group of the *general* four-dimensional Riemannian manifold is $O(4) \sim Sp(1) \times Sp(1)$).

11.3.1 Constraints in harmonic space

As our previous experience has shown, one should try to interpret the constraints (11.83) as the integrability conditions for some kind of analyticity. Equation (11.83) can be rewritten in the following equivalent form

$$[\mathcal{D}_{\alpha(i}, \mathcal{D}_{\beta j)}] = 0\,, \qquad (11.84)$$

after which it looks exactly like the SDYM constraint (11.3). This suggests that the appropriate analyticity should be of the harmonic kind.

To begin with, it is convenient to use a parametrization of $\mathbb{R}^{4n}$ by $x^{\mu i}$, $\overline{x^{\mu i}} = \Omega_{\mu\nu}\epsilon_{ij}x^{\nu j}$, $\mu = 1, \ldots, n$, $i = 1, 2$. Next, as usual, we introduce harmonics $u^{\pm i}$ for the rigid $Sp(1) \sim SU(2)$ part of the tangent space group $Sp(n) \times Sp(1)$. They extend $\mathbb{R}^{4n}$ to the harmonic space $\mathbb{R}^{4n+2} = \{x^{\mu i}, u^{\pm i}\}$. The diffeomorphism group in $\mathbb{R}^{4n+2}$ is given by

$$\delta x^{\mu i} = \tau^{\mu i}(x)\,, \qquad \delta u_i^{\pm} = 0\,. \qquad (11.85)$$

Then, proceeding as in the case of SDYM, we rewrite (11.84) as a set of two constraints in harmonic space (cf. (11.8), (11.9)):

$$\left[\mathcal{D}_\alpha^+, \mathcal{D}_\beta^+\right] = 0\,, \qquad (11.86)$$
$$\left[\mathcal{D}^{++}, \mathcal{D}_\alpha^+\right] = 0\,. \qquad (11.87)$$

The second one, (11.87), guarantees that $\mathcal{D}_\alpha^+$ depends on u^{+i} only linearly, $\mathcal{D}_\alpha^+ = u^{+i}\mathcal{D}_{\alpha i}$ (so far $\mathcal{D}^{++}$ coincides with $\partial^{++} = u^{+i}\partial/\partial u^{-i}$). Then one derives from (11.86):

$$[\mathcal{D}_{\alpha i}, \mathcal{D}_{\beta j}] = \epsilon_{ij}\left(T_{(\alpha\beta)}{}^{\gamma k}\mathcal{D}_{\gamma k} + R_{(\alpha\beta)}\right). \qquad (11.88)$$

The difference between (11.88) and (11.83) is the presence of a torsion term in (11.88). As we wish to have a Riemannian geometry, we require it to vanish. This is equivalent to demanding

$$[\mathcal{D}_\alpha^+, \mathcal{D}_\beta^-] = R^{+-}_{(\alpha\beta)}\,, \qquad (11.89)$$

in addition to (11.86), (11.87). This set of constraints should be completed by the standard conventional ones:

$$\left[\mathcal{D}^{++}, \mathcal{D}^{--}\right] = D^0\,, \qquad (11.90)$$
$$\left[\mathcal{D}^{--}, \mathcal{D}_\alpha^+\right] = \mathcal{D}_\alpha^-\,. \qquad (11.91)$$

The set of constraints (11.86)–(11.89) and (11.90), (11.91) gives an equivalent definition of the standard hyper-Kähler geometry in the extended framework of harmonic space $\mathbb{R}^{4n+2} = \{x^{\mu i}, u^{\pm i}\}$.

We stress that so far the harmonic variables are just auxiliary, neither the diffeomorphism nor the tangent space group parameters depend on them and they do not transform under any of the gauge groups. This picture will change when we switch to the harmonic analytic basis.

From the commutators (11.86)–(11.89) follow a number of constraints on the harmonic $U(1)$ projections of the torsion and curvature occurring, e.g., in

$$[\mathcal{D}^+_\alpha, \mathcal{D}^+_\beta] = T^{++\gamma+}_{\alpha\,\beta}\mathcal{D}^-_\gamma + T^{++\gamma-}_{\alpha\,\beta}\mathcal{D}^+_\gamma + R^{++}_{\alpha\,\beta}\,. \tag{11.92}$$

It should be realized that not all of these constraints are independent. A possible choice of a set of independent constraints is

- one of the torsion constraints contained in eq. (11.86):

$$T^{++\gamma+}_{\alpha\,\beta} = 0\,; \tag{11.93}$$

- all of the torsion constraints contained in eq. (11.87):

$$\text{(a)} \qquad T^{+++\beta+}_{\quad\alpha} = 0\,, \qquad \text{(b)} \qquad T^{+++\beta-}_{\quad\alpha} = 0\,; \tag{11.94}$$

- one of the torsion constraints contained in eq. (11.89):

$$T^{+-\gamma+}_{\alpha\,\beta} = 0\,. \tag{11.95}$$

It can be shown that the rest of the torsion and curvature constraints follow from the Bianchi identities. Take, for example, the Bianchi identity

$$R^{++\ \ \delta}_{\alpha\,\beta\,\gamma} = \mathcal{D}^+_\alpha T^{+-\delta+}_{\beta\,\gamma} - TT + \text{cycle}\,. \tag{11.96}$$

Assuming that all the torsions vanish, we derive the curvature constraint

$$R^{++}_{\alpha\,\beta} = 0\,. \tag{11.97}$$

Another, less trivial example is the constraint $T^{+-\gamma-}_{\alpha\,\beta} = 0$ contained in eq. (11.89). From (11.87), (11.90), (11.91) one finds the Bianchi identity

$$[\mathcal{D}^{++}, \mathcal{D}^-_\alpha] = \mathcal{D}^+_\alpha\,. \tag{11.98}$$

Further, from (11.98), (11.86), (11.87) one obtains another Bianchi identity

$$[\mathcal{D}^{++}, [\mathcal{D}^+_\alpha, \mathcal{D}^-_\beta]] = 0\,, \tag{11.99}$$

which, in particular, contains

$$\mathcal{D}^{++}T^{+-\gamma-}_{\alpha\,\beta} + T^{+-\gamma+}_{\alpha\,\beta} = 0\,. \tag{11.100}$$

As a consequence of the constraint (11.95), eq. (11.100) is reduced to $\mathcal{D}^{++}T^{+-\gamma-}_{\alpha\,\beta} = 0$. Since in the present frame $\mathcal{D}^{++} = \partial^{++}$, one concludes that $T^{+-\gamma-}_{\alpha\,\beta} = 0$.

Note as well that the remaining commutators of covariant derivatives can also be obtained from the Bianchi identities

$$[\mathcal{D}^{++}, [\mathcal{D}^{--}, \mathcal{D}^-_\alpha]] = 0 \quad \Rightarrow \quad [\mathcal{D}^{--}, \mathcal{D}^-_\alpha] = 0\,, \tag{11.101}$$

$$[\mathcal{D}^{++}, [\mathcal{D}^-_\alpha, \mathcal{D}^-_\beta]] = 0 \quad \Rightarrow \quad [\mathcal{D}^-_\alpha, \mathcal{D}^-_\beta] = 0\,. \tag{11.102}$$

Finally, we quote two important relations for the curvature following from the Bianchi identities. Commuting both sides of eq. (11.89) with $\mathcal{D}^+_\gamma$ and $\mathcal{D}^{++}$ and using (11.86), (11.87) and (11.98), we get, in particular,

$$R^{+-}_{(\alpha\beta)(\gamma\rho)} - R^{+-}_{(\gamma\beta)(\alpha\rho)} = 0 \quad \Rightarrow \quad R^{+-}_{(\alpha\beta)(\gamma\rho)} = R^{+-}_{(\alpha\beta\gamma\rho)} \tag{11.103}$$

$$\mathcal{D}^{++} R^{+-}_{(\alpha\beta)} = 0\,. \tag{11.104}$$

These relations mean that the curvature is totally symmetric in its $Sp(n)$ indices (as noticed earlier) and does not depend on the harmonics in the τ basis and frame.

We now proceed to solving the constraints (11.93)–(11.95).

11.3.2 *Harmonic analyticity*

The key to solving the above constraints is the observation of the underlying analytic structure. The constraint (11.86) has the meaning of the integrability condition for the existence of harmonic-analytic fields defined by

$$\mathcal{D}^+_\alpha \phi_{\beta j\ldots}(x, u) = 0\,. \tag{11.105}$$

In the flat case (11.105) means that $\phi_{\beta j\ldots} = \phi_{\beta j\ldots}(x^{\mu k} u^+_k, u)$. In the curved case this analyticity is not manifest because the diffeomorphism group as well as the local $Sp(n)$ tangent space group mix up the dependence on $x^{\mu +} = x^{\mu i} u^+_i$ and $x^{\mu -} = x^{\mu i} u^-_i$. Therefore, we should try to define a new basis and a new tangent frame with analytic gauge groups.

The existence of such an analytic basis follows from the torsion constraint (11.93). Indeed, using $x^{\mu\pm}$ as independent variables, the covariant derivative $\mathcal{D}^+_\alpha$ can be written down as follows:

$$\mathcal{D}^+_\alpha = e^{+\mu -}_\alpha \partial_{\mu -} + e^{+\mu +}_\alpha \partial_{\mu +} + \omega^+_\alpha\,, \tag{11.106}$$

where

$$\partial_{\mu\pm} \equiv \frac{\partial}{\partial x^{\mu\pm}}\,.$$

Then, the linearized torsion constraint (11.93) implies

$$T^{++\gamma +}_{\alpha\;\beta} = 0 \quad \Rightarrow \quad \partial_{[\alpha -} e^{+\gamma +}_{\beta]} = 0 \quad \Rightarrow \quad e^{+\mu +}_\alpha = -\partial_{\alpha -} v^{\mu +}\,, \tag{11.107}$$

where $v^{\mu+}(x,u)$ is the *harmonic-dependent bridge to the analytic basis*. The full non-linear version of (11.107) is

$$e_\alpha^{+\mu+}(\delta^\nu_\mu + \partial_{\mu+}v^{\nu+}) + e_\alpha^{+\mu-}\partial_{\mu-}v^{\nu+} = 0\,. \tag{11.108}$$

The analytic basis is defined by the change of variables:

$$\underline{\text{Analytic basis:}} \qquad x_A^{\mu\pm} = x^{\mu i}u_i^\pm + v^{\mu\pm}(x,u)\,, \tag{11.109}$$

where the second bridge $v^{\mu-}(x,u)$ is introduced for future convenience. The effect of this change of variables is such that the covariant derivative $\mathcal{D}_\alpha^+$ becomes

$$\mathcal{D}_\alpha^+ = \mathcal{E}_\alpha^{+\mu-}\partial_{A\mu-} + \omega_\alpha^+\,. \tag{11.110}$$

The term with $\partial/\partial x_A^{\mu+}$ in (11.110)) has been eliminated with the help of the bridge introduced in eq. (11.108). This becomes clear after rewriting eq. (11.108) in the form

$$\mathcal{D}_\alpha^+ x_A^{\mu+}(x,u) = 0\,. \tag{11.111}$$

The situation here is similar to that in the Kähler case (cf. eq. (11.48)). However, the principal difference is that in the Kähler case the bridges have the form of a general τ transformation (and can thus be gauged away), whereas in the hyper-Kähler case they essentially depend on the harmonics unlike the harmonic-independent τ transformations.

The condition (11.111) allows $x_A^{\mu+}$ to transform with an analytic diffeomorphism parameter:

$$\delta x_A^{\mu+} = \lambda^{\mu+}(x_A^{\nu+},u)\,, \tag{11.112}$$

whereas $x_A^{\mu-}$ may transform in a general way:

$$\delta x_A^{\mu-} = \lambda^{\mu-}(x_A^{\nu+},x_A^{\nu-},u)\,. \tag{11.113}$$

Correspondingly, the bridges $v^{\mu\pm}$ transform as follows:

$$\delta v^{\mu+} = \lambda^{\mu+}(x_A^+,u) - \tau^{\mu i}(x)u_i^+\,, \tag{11.114}$$

$$\delta v^{\mu-} = \lambda^{\mu-}(x_A^+,x_A^-,u) - \tau^{\mu i}(x)u_i^-\,. \tag{11.115}$$

Note that the bridge $v^{\mu-}$ can be entirely gauged away by using the parameter $\lambda^{\mu-}$ (although in what follows we fix a more convenient gauge). We emphasize that the harmonics $u^\pm$ do not transform under either the τ or the λ diffeomorphism groups.

The second step needed to make the analyticity (11.105) manifest is to choose an analytic tangent frame. This is possible because the curvature component $R^{++}_{\alpha\,\beta}$ vanishes (see (11.97)), so the projection ω_α^+ of the connection is 'pure gauge':

$$R^{++}_{\alpha\,\beta} = 0 \quad\Rightarrow\quad \omega^+_{\alpha\,\beta}{}^\gamma = (M^{-1})_\beta^\delta\,\mathcal{E}_\alpha^{+\mu-}\partial_{A\mu-}\,M_\delta^\gamma\,. \tag{11.116}$$

Here $M_\alpha^\beta(x,u)$ is an appropriate $Sp(n)$ matrix 'bridge'. It is the analog of the bridge $e^{iv(x,u)}$ in the SDYM case (see (11.11)). Like the latter, it transforms under two $Sp(n)$ groups: A new, analytic one* and the old (u-independent) tangent space one:

$$\delta M_\alpha^\beta(x,u) = \lambda_\alpha^\gamma(x_A^+,u)M_\gamma^\beta - M_\alpha^\gamma \tau_\gamma^\beta(x) . \tag{11.117}$$

Going to the new frame means that all fields are rotated, e.g.,

$$\Phi_\alpha(x,u) = M_\alpha^\beta(x,u)\phi_\beta(x,u) , \tag{11.118}$$

after which they start transforming under the analytic $Sp(n)$ tangent space group:

$$\delta\Phi_\alpha = \lambda_\alpha^\beta \Phi_\beta . \tag{11.119}$$

In this tangent frame the covariant derivative $\mathcal{D}_\alpha^+$ has no connection:

$$\mathcal{D}_\alpha^+ = E_\alpha^{+\mu-}\partial_{A\mu-} , \qquad E_\alpha^{+\mu-} = M_\alpha^\beta \, \mathcal{E}_\beta^{+\mu-} . \tag{11.120}$$

So, in the analytic basis and frame the analyticity condition (11.105) can be solved by a manifestly analytic field:

$$\mathcal{D}_\alpha^+\Phi_{\beta j\ldots}(x,u) = 0 \quad \Rightarrow \quad \Phi_{\beta j\ldots} = \Phi_{\beta j\ldots}(x_A^+,u) . \tag{11.121}$$

In what follows we shall mostly work in the analytic, or 'λ world' (basis and frame), so we shall often drop the index A. The new vielbeins $E_\alpha^{+\mu-}$ transform under the λ groups as follows:

$$\delta E_\alpha^{+\mu-} = \lambda_\alpha^\beta E_\beta^{+\mu-} + E_\alpha^{+\nu-}\partial_{\nu-}\lambda^{\mu-} . \tag{11.122}$$

They are still subject to constraints following from (11.93)–(11.95), but we postpone the discussion until we have studied the covariant harmonic derivatives in the analytic basis.

11.3.3 Harmonic derivatives in the λ world

The introduction of the analytic basis (11.109) and frame (11.116) allowed us to solve the torsion constraint (11.93) (as well the curvature constraint (11.97)). As a result, we have made the covariant derivative $\mathcal{D}_\alpha^+$ 'short' (see (11.120)). The harmonic derivatives $\mathcal{D}^{++}$, $\mathcal{D}^{--}$ have acquired vielbeins and connections:

$$\mathcal{D}^{\pm\pm} = \partial^{\pm\pm} + H^{\pm\pm\mu+}\partial_{\mu+} + H^{\pm\pm\mu-}\partial_{\mu-} + \omega^{\pm\pm} \equiv \Delta^{\pm\pm} + \omega^{\pm\pm} . \tag{11.123}$$

Here

$$\begin{aligned} H^{\pm\pm\mu+} &= \Delta^{\pm\pm}x_A^{\mu+} = \Delta^{\pm\pm}(x^{\mu i}u_i^+ + v^{\mu+}) , & (11.124)\\ H^{\pm\pm\mu-} &= \Delta^{\pm\pm}x_A^{\mu-} = \Delta^{\pm\pm}(x^{\mu i}u_i^- + v^{\mu-}) , & (11.125)\\ \omega^{\pm\pm}{}_\alpha^\beta &= M_\alpha^\gamma \, \Delta^{\pm\pm}(M^{-1})_\gamma^\beta . & (11.126) \end{aligned}$$

* The $Sp(n)$ parameters are real in the sense of $\tilde{}$ conjugation.

They transform under the λ diffeomorphism and tangent space groups as follows:

$$\delta H^{\pm\pm\mu+} = \Delta^{\pm\pm}\lambda^{\mu+}\,, \qquad \delta H^{\pm\pm\mu-} = \Delta^{\pm\pm}\lambda^{\mu-}\,, \qquad \delta\omega^{\pm\pm\beta}_{\ \ \ \alpha} = -\Delta^{\pm\pm}\lambda^{\beta}_{\alpha}\,. \tag{11.127}$$

The vielbein $H^{++\mu-}(x^{\pm}, u)$ is a general harmonic function of charge $+1$. It transforms by the parameter $\lambda^{\mu-}(x^{\pm}, u)$ which is a general harmonic function of charge -1. This allows us to gauge-fix $H^{++\mu-}$ to its flat space limit $x^{\mu+}_A$:

$$\underline{\lambda^{\mu-}\text{ gauge:}} \qquad H^{++\mu-} = x^{\mu+}_A \;\Rightarrow\; \Delta^{++}v^{\mu-} = v^{\mu+} \;\Rightarrow\; \Delta^{++}\lambda^{\mu-} = \lambda^{\mu+}\,. \tag{11.128}$$

This very convenient gauge will always be used from now on.

The newly introduced vielbeins satisfy constraints following from the relations (11.94a) and (11.90).

The constraint (11.94a) implies that the vielbein $H^{++\mu+}$ must be analytic. Indeed, inserting (11.123) into (11.94a), we find

$$T^{+++\mu+}_{\ \ \alpha} = 0 \;\Rightarrow\; \partial_{\nu-}H^{++\mu+} = 0 \;\Rightarrow\; H^{++\mu+} = H^{++\mu+}(x^{+}_A, u)\,. \tag{11.129}$$

Previous experience with $N = 2$ SYM and SDYM suggests that such an object may be the main potential of the theory. This is almost true here: We shall see that all the objects of differential geometry derive from $H^{++\mu+}$; however, it is subject to further constraints and can be expressed in terms of some other unconstrained analytic field.

The way from the analytic vielbein $H^{++\mu+}$ back to the ordinary, τ-world harmonic-independent vielbeins and connections goes through finding the bridges $v^{\mu\pm}$ (up to gauge freedom). The bridge $v^{\mu+}$ is determined as a solution to the differential equation following from the definition (11.124) rewritten in the τ basis:

$$H^{++\mu+}(x^{\mu i}u^{+}_{i} + v^{\mu+}, u) = \partial^{++}v^{\mu+}(x^{\lambda j}, u)\,. \tag{11.130}$$

This is in full analogy with the rôle of the prepotential and the bridge in the SDYM case. Equation (11.130) is a non-linear differential equation on S^2. Solving it is a highly non-trivial task, in general. Section 11.3.8 contains a class of examples where the explicit solution can be found.

The other bridge $v^{\mu-}$ is determined from the gauge-fixing condition (11.128), which has the following form in the τ basis:

$$\partial^{++}v^{\mu-}(x^{\nu i}, u) = v^{\mu+}(x^{\nu i}, u)\,. \tag{11.131}$$

The solution of eq. (11.131) is easy to find:

$$v^{\mu-}(x^{\nu i}, u) = \int dw \frac{1}{(u^{+}w^{+})} v^{\mu+}(x^{\nu i}, w) \tag{11.132}$$

(see Section 11.3.8).

Coming back to the discussion of the constraints, we see that the conventional constraint (11.90) relates the vielbeins $H^{--\mu\pm}$ in the covariant derivative $\mathcal{D}^{--}$ to the analytic vielbein $H^{++\mu+}$. Using the gauge (11.128) and taking into account that in the analytic basis $D^0 = \partial^0 + x_A^{\mu+}\partial_{\mu+} - x_A^{\mu-}\partial_{\mu-}$, we obtain from (11.90):

$$\Delta^{++}H^{--\mu+} - \Delta^{--}H^{++\mu+} = x_A^{\mu+}\,, \tag{11.133}$$

$$\Delta^{++}H^{--\mu-} - H^{--\mu+} = -x_A^{\mu-}\,. \tag{11.134}$$

These are differential equations for $H^{--\mu\pm}$ on the harmonic sphere S^2. In principle, they have a unique (local) solution for any given $H^{++\mu+}(x_A^+, u)$. However, in practice it is not easy to find the explicit solution.* In what follows we shall use these two constraints without explicitly solving them. In any case, if one has managed to solve the bridge equations (11.130) and (11.131), the vielbeins $H^{--\mu\pm}$ can be computed from their definitions (11.124), (11.125).

Note that the conventional constraint (11.90) also relates the harmonic connections ω^{++} and ω^{--}:

$$R^{++--} = 0 \quad \Rightarrow \quad \Delta^{++}\omega^{--} - \Delta^{--}\omega^{++} + [\omega^{++}, \omega^{--}] = 0\,. \tag{11.135}$$

In addition, the vanishing of the curvature $R^{+++}_{\ \ \alpha}$ in the commutation relation (11.87) implies that the connection ω^{++} is analytic:

$$R^{+++}_{\ \ \alpha} = 0 \quad \Rightarrow \quad \partial_{\mu-}\omega^{++} = 0 \quad \Rightarrow \quad \omega^{++} = \omega^{++}(x_A^+, u)\,. \tag{11.136}$$

In fact, both constraints (11.135) and (11.136) are secondary, they follow from the Bianchi identities.

We stress once again that solving the differential equations for the bridges or for $H^{--\mu\pm}$ is the essential and most non-trivial part of our treatment of the constraints of hyper-Kähler geometry. Once this has been achieved, all the remaining constraints can be solved explicitly, as we explain below.

11.3.4 Hyper-Kähler potentials

So far we have solved two of the defining torsion constraints of hyper-Kähler geometry, eqs. (11.93) and (11.94a). The first of them leads to the introduction of the analytic basis (11.109) and the second one yields the analyticity of $H^{++\mu+}$, eq. (11.129). In this subsection we solve the two remaining constraints, first eq. (11.95) and then eq. (11.94b). We see that the latter leads to the unconstrained potentials of hyper-Kähler geometry.

Our first step is to use the conventional constraint (11.91) to express the vielbeins and connection in the covariant derivative $\mathcal{D}^-_\alpha$:

$$[\mathcal{D}^{--}, \mathcal{D}^+_\alpha] = \mathcal{D}^-_\alpha \equiv E_\alpha^{-\mu-}\partial_{\mu-} - e^\mu_\alpha\partial_{\mu+} + \omega^-_\alpha \quad \Rightarrow \tag{11.137}$$

* A perturbative solution can be constructed by analogy with the $N = 2$ SYM case (Chapter 7).

$$E_\alpha^{-\mu-} = \Delta^{--}E_\alpha^{+\mu-} + \omega^{--\beta}{}_\alpha E_\beta^{+\mu-} - E_\alpha^{+\nu-}\partial_{\nu-}H^{--\mu-}, \quad (11.138)$$
$$e_\alpha^\mu = E_\alpha^{+\nu-}\partial_{\nu-}H^{--\mu+}, \quad (11.139)$$
$$\omega_\alpha^- = -E_\alpha^{+\mu-}\partial_{\mu-}\omega^{--}. \quad (11.140)$$

Next we turn to the constraint (11.95). It implies that the vielbein e_α^μ must be analytic:

$$T^{+-\gamma+}_{\alpha\,\beta} = 0 \quad \Rightarrow \quad \partial_{\nu-}e_\alpha^\mu = 0. \quad (11.141)$$

This is the second analytic object in the theory (besides $H^{++\mu+}$) and it will play an important rôle in constructing the full set of vielbeins and connections. In particular, we can turn eq. (11.138) around and express the vielbein $E_\alpha^{+\mu-}$ in terms of e_α^μ and the matrix $(\partial H)^\mu_\nu \equiv \partial_{\nu-}H^{--\mu+}$:

$$E_\alpha^{+\mu-} = e_\alpha^\nu(\partial H)^{-1\mu}{}_\nu. \quad (11.142)$$

The remaining unsolved constraint is eq. (11.94b):

$$T^{+++\beta-}_{\alpha} = 0 \quad \Rightarrow \quad \Delta^{++}E_\alpha^{+\mu-} + \omega^{++\beta}{}_\alpha E_\beta^{+\mu-} = 0. \quad (11.143)$$

Inserting in it the expression (11.142) for $E_\alpha^{+\mu-}$ and using the corollary

$$\Delta^{++}\partial_{\nu-}H^{--\rho+} = \partial_{\nu-}(x_A^{\rho+} + \Delta^{--}H^{++\rho+}) = (\partial H)^\sigma_{\nu-}\partial_{\sigma+}H^{++\rho+} \quad (11.144)$$

of the harmonic constraint (11.133), we can rewrite eq. (11.143) as follows:

$$\Delta^{++}e_\alpha^\mu + \omega^{++\beta}{}_\alpha e_\beta^\mu - e_\alpha^\nu\partial_{\nu+}H^{++\mu+} = 0. \quad (11.145)$$

After multiplication of this equation by the inverse vielbein $e_{\mu\beta}$ ($e_\alpha^\mu e_{\mu\beta} = -\Omega_{\alpha\beta}$), its part symmetric in α, β just gives an expression for the harmonic connection:

$$\omega^{++}{}_{\alpha\beta} = e^\mu_{(\alpha}\Delta^{++}e_{\mu\beta)} + e^\mu_{(\alpha}\partial_{\mu+}H^{++\nu+}e_{\nu\beta)}, \quad (11.146)$$

while the antisymmetric part amounts to a differential constraint on the vielbein e_α^μ:

$$e^\mu_{[\alpha}\Delta^{++}e_{\mu\beta]} + e^\mu_{[\alpha}\partial_{\mu+}H^{++\nu+}e_{\nu\beta]} = 0. \quad (11.147)$$

To find out the meaning of this constraint, note that part of e_α^μ is arbitrary (it corresponds to the tangent $Sp(n)$ freedom). Its gauge invariant part is represented by the analytic 'symplectic metric'

$$H_{\mu\nu} = -e_\mu^\alpha\,\Omega_{\alpha\beta}\,e_\nu^\beta = -H_{\nu\mu}, \qquad \partial_{\lambda-}H_{\mu\nu} = 0. \quad (11.148)$$

In terms of this new object eq. (11.147) becomes

$$\Delta^{++}H_{\mu\nu} - 2\partial_{[\mu+}H^{++\rho+}H_{\nu]\rho} = 0. \quad (11.149)$$

This equation has two important consequences. Firstly, it implies

$$\partial_{[\mu+}H_{\nu\lambda]} = 0. \quad (11.150)$$

To see this, one differentiates (11.149) with $\partial_{\lambda+}$ and uses the relation

$$\partial_{\lambda+}\Delta^{++}H_{\mu\nu} = \Delta^{++}\partial_{\lambda+}H_{\mu\nu} + \partial_{\lambda+}H^{++\rho+}\partial_{\rho+}H_{\mu\nu} \tag{11.151}$$

(following from the definition of Δ^{++} (11.123) and the analyticity of $H_{\mu\nu}$ and $H^{++\mu+}$). The result can be written in terms of the totally antisymmetric tensor $H^{-}_{\mu\nu\lambda} \equiv \partial_{[\mu+}H_{\nu\lambda]}$:

$$\Delta^{++}H^{-}_{\mu\nu\lambda} + 3\partial_{[\mu+}H^{++\rho+}H^{-}_{\nu\lambda]\rho} = 0\,. \tag{11.152}$$

In terms of the tangent space tensor

$$H^{-}_{\alpha\beta\gamma} = e^{\mu}_{\alpha}e^{\nu}_{\beta}e^{\lambda}_{\gamma}H^{-}_{\mu\nu\lambda}\,, \tag{11.153}$$

eq. (11.152) reads (recall (11.146))

$$\mathcal{D}^{++}H^{-}_{\alpha\beta\gamma} = 0 \quad \Rightarrow \quad H^{-}_{\alpha\beta\gamma} = 0 \quad \Rightarrow \quad H^{-}_{\mu\nu\lambda} = 0\,. \tag{11.154}$$

Note that the same result (11.150) can be directly obtained from the secondary torsion constraint $T^{--+}_{[\alpha\beta\gamma]} = 0$.

The constraint (11.150) has the following general (local) solution:

$$H_{\mu\nu} = \partial_{[\mu+}L^{+}_{\nu]}\,. \tag{11.155}$$

Here L^{+}_{μ} is an *arbitrary* world vector under the λ diffeomorphism group. Note that it is defined up to terms of the type $\partial_{\mu+}\lambda^{++}$ (the pregauge freedom in (11.155)). So, the transformation law of L^{+}_{μ} is

$$\delta L^{+}_{\mu} = -\partial_{\mu+}\lambda^{\nu+}L^{+}_{\nu} + \partial_{\mu+}\lambda^{++}\,. \tag{11.156}$$

Without loss of generality both L^{+}_{μ} and λ^{++} can be assumed analytic,

$$\partial_{\nu-}L^{+}_{\mu} = \partial_{\nu-}\lambda^{++} = 0 \tag{11.157}$$

(the non-analytic, i.e., $x^{\mu-}_{A}$-dependent part of L^{+}_{μ} must be of pregauge form, because $H_{\mu\nu}$ is analytic; then it can be gauged away by the corresponding part of λ^{++}).

The second corollary of eq. (11.149) is obtained by inserting eq. (11.155) into it:

$$\partial_{[\mu+}(\Delta^{++}L^{+}_{\nu]} + \partial_{\nu]+}H^{++\rho+}L^{+}_{\rho}) = 0\,. \tag{11.158}$$

This equation has the following general local solution:

$$\Delta^{++}L^{+}_{\mu} + \partial_{\mu+}H^{++\nu+}L^{+}_{\nu} = \partial_{\mu+}H^{+4}\,. \tag{11.159}$$

Once again, one can choose the new unconstrained object H^{+4} to be analytic,

$$\partial_{\mu-}H^{+4} = 0\,. \tag{11.160}$$

Taking into account (11.112), (11.156), (11.127), it can be checked that

$$\delta H^{+4} = \Delta^{++}\lambda^{++} . \tag{11.161}$$

Finally, it is convenient to introduce yet another analytic object:

$$L^{+4} = H^{+4} - H^{++\mu+}L^{+}_{\mu}, \qquad \partial_{\mu-}L^{+4} = 0\,, \tag{11.162}$$

$$\delta L^{+4} = \partial^{++}\lambda^{++} - \partial^{++}\lambda^{\mu+}L^{+}_{\mu} . \tag{11.163}$$

This allows us to solve (11.159) for $H^{++\mu+}$:

$$H^{++\mu+} = \frac{1}{2}H^{\mu\nu}(\partial^{++}L^{+}_{\nu} - \partial_{\nu+}L^{+4})\,, \qquad H^{\mu\nu}H_{\nu\lambda} = \delta^{\mu}_{\lambda} . \tag{11.164}$$

Note that L^{+4} (11.163) is real in the sense of $\widetilde{\ }$ conjugation. In fact, it is precisely this last object L^{+4} which coincides with the most general self-interaction term for $N = 2$ hypermultiplets (see Section 11.4).

Summarizing the rather lengthy derivation above, we can say that all the constraints of hyper-Kähler geometry have been solved in terms of *two unconstrained analytic potentials* L^{+}_{μ} and L^{+4}. Like the Kähler potential K (11.58), they have their own pregauge transformations (11.156) and (11.163) with the analytic parameter λ^{++}. The dimension of these potentials and parameter are peculiar ($[L^{+}_{\mu}] = -1$, $[L^{+4}] = [\lambda^{++}] = -2$), their geometric meaning is obscure in the present scheme. The origin of these objects is clarified in an extended framework involving central charge coordinates (see Section 11.3.7).

Concluding this subsection, we note that the remaining object of differential geometry, the harmonic connection ω^{--}, can be determined from eq. (11.143). Indeed, (11.143) can be written down as $\mathcal{D}^{++}E^{+\mu-}_{\alpha} = 0$, which means that $E^{+\mu-}_{\alpha}$ is covariantly harmonic independent (in the τ world this becomes just harmonic independence). Then the other harmonic derivative of $E^{+\mu-}_{\alpha}$ should also vanish,

$$\mathcal{D}^{--}E^{+\mu-}_{\alpha} = 0 \quad \Rightarrow \quad \Delta^{--}E^{+\mu-}_{\alpha} + \omega^{--}{}^{\beta}_{\alpha}E^{+\mu-}_{\beta} = 0\,. \tag{11.165}$$

Inserting (11.142) into (11.165), we find the following expression for ω^{--}:

$$\omega^{--}_{\alpha\beta} = e^{\mu}_{(\alpha}\Delta^{--}e_{\mu\beta)} + e^{\mu}_{(\alpha}(\partial H)^{-1\nu}{}_{\mu}\Delta^{--}(\partial H)^{\lambda}_{\nu}\, e_{\lambda\beta)}\,. \tag{11.166}$$

Another consequence of eq. (11.165) is a simplification of the expression (11.138) for the vielbein $E^{-\mu-}_{\alpha}$:

$$E^{-\mu-}_{\alpha} = -E^{+\nu-}_{\alpha}\partial_{\nu-}H^{--\mu-} = -e^{\rho}_{\alpha}(\partial H)^{-1\nu}{}_{\rho}\partial_{\nu-}H^{--\mu-}\,. \tag{11.167}$$

Finally, we note that a number of useful relations between the quantities obtained above are easier to derive directly from the Bianchi identities summarized in Section 11.3.1. For future use, we quote here one such relation which follows from the identity (11.98) applied to the λ-world derivative $\mathcal{D}^{-}_{\alpha}$, eq. (11.137),

$$\mathcal{D}^{++}\omega^{-}_{\alpha(\beta\gamma)} + e^{\mu}_{\alpha}\,\partial_{\mu+}\omega^{++}_{(\beta\gamma)} = 0\,. \tag{11.168}$$

11.3.5 Gauge choices and normal coordinates

In addition to the gauge (11.128) which fixes the parameter $\lambda^{\mu-}$ one can impose three further gauges on the parameters $\lambda^{\mu+}$, λ_{α}^{β} and λ^{++}.

From (11.156) and the fact that L^+_μ is analytic and has the flat space limit $\Omega_{\mu\nu}x_A^{\nu+}$ one concludes that the following gauge is possible:

$$\underline{\lambda^{\mu+}\text{ gauge:}} \qquad L^+_\mu = x^+_{A\mu} \quad \rightarrow \quad H_{\mu\nu} = -\Omega_{\mu\nu}\,. \tag{11.169}$$

This implies $\lambda^+_\mu + \partial_{\mu+}\lambda^{\nu+}x^+_{A\nu} - \partial_{\mu+}\lambda^{++} = 0$. Introducing the new analytic parameter

$$\hat\lambda^{++} = \lambda^{++} - \lambda^{\nu+}x^+_{A\nu}\,, \tag{11.170}$$

one can express λ^+_μ in terms of $\hat\lambda^{++}$,

$$\lambda^+_\mu = \frac{1}{2}\partial_{\mu+}\hat\lambda^{++} \quad \rightarrow \quad \partial_{[\nu+}\lambda^+_{\mu]} = 0\,. \tag{11.171}$$

After all this the remaining potential L^{+4} transforms as

$$\delta L^{+4} = \partial^{++}\hat\lambda^{++}\,. \tag{11.172}$$

The second gauge concerns the $Sp(n)$ parameter λ_{α}^{β}. Using the $Sp(n)$ transformation law of e^μ_α:

$$\delta e^\mu_\alpha = \lambda_\alpha^\beta e^\mu_\beta + e^\nu_\alpha\partial_{\nu+}\lambda^{\mu+}, \tag{11.173}$$

and the fact that e^μ_α is invertible, $e^\mu_\alpha = \delta^\mu_\alpha + \cdots$, one can fully gauge away its symmetric part $e^{(\mu\alpha)}$. This, together with (11.169) and (11.148), imply

$$\underline{Sp(n)\text{ gauge:}} \qquad e^\mu_\alpha = \delta^\mu_\alpha \quad \rightarrow \quad \lambda_{\alpha\beta} = -\partial_{(\alpha+}\lambda^+_{\beta)}\,. \tag{11.174}$$

The third gauge corresponds to finding a normal set of coordinates in the analytic basis. This means that one uses the entire remaining gauge freedom (in our case (11.172)) to gauge away as much as possible from the potential L^{+4}. The remainder is a coordinate expansion of L^{+4} at a given point, where the coefficients coincide with the values of the non-vanishing tensors at this point. To achieve this one considers the expansion of a function of charge $+q$:

$$F^{(+q)}(x^+, u) = \tag{11.175}$$
$$\sum_{\substack{n,m=0\\ m+n\geq q}}^{\infty} x^+_{\mu_1}\ldots x^+_{\mu_n}u^+_{i_1}\ldots u^+_{i_m}u^-_{i_{m+1}}\ldots u^-_{i_{n+2m-q}}\, f^{\mu_1\ldots\mu_n i_1\ldots i_{n+2m-q}}\,.$$

Comparing the expansions of L^{+4} and $\hat\lambda^{++}$, from (11.172) one derives the following normal gauge form of L^{+4}:

$$\underline{\text{Normal gauge:}} \qquad L^{+4} = \sum_{n=0}^{\infty} x^+_{\mu_1}\ldots x^+_{\mu_{n+4}}u^-_{i_1}\ldots u^-_{i_n}C^{\mu_1\ldots\mu_{n+4}i_1\ldots i_n}\,. \tag{11.176}$$

Note the absence of u_i^+ in (11.176). In the case of $\mathbb{R}^4$ the coefficient $C^{\mu_1\cdots\mu_4}$ is the value of the self-dual Weyl tensor at the point $x = 0$, and the higher-rank coefficients correspond to the totally symmetrized covariant derivatives of the Weyl tensor at that point. Still in $\mathbb{R}^4$, L^{+4} is a function of three complex variables (two from x^+ plus two from u^- minus one from the preservation of the $U(1)$ charge) (see [W2]). From the reality of the analytic space $(x^+, u^\pm)$ under the $\tilde{}$ conjugation it follows that the coefficients C are pseudo-real (since L^{+4} is real). The generalization of the above interpretation to the case of $\mathbb{R}^{4n}$ is straightforward.

In the gauge (11.176) a few constant parameters survive in the expansion of $\hat{\lambda}^{++}$:

$$\hat{\lambda}^{++} = x^+_\mu x^+_\nu \lambda^{\mu\nu} + x^+_\mu u^+_i a^{\mu i} + u^+_i u^+_j a^{ij} \,, \tag{11.177}$$

where $\lambda^{\mu\nu}$ are rigid $Sp(n)$ rotations and $a^{\mu i}$ are rigid translations. The meaning of a^{ij} ($[a] = -2$) will become clear in Section 11.3.7.

11.3.6 Summary of hyper-Kähler geometry

The procedure of solving the constraints on hyper-Kähler geometry explained above involved many steps. It may not be easy for the reader to single out the most essential points. Therefore, we present a list of all the geometric objects – vielbeins and connections – expressed in terms of the two analytic potentials L^+_μ and L^{+4} of hyper-Kähler geometry. We do this in both the analytic and ordinary (u-independent) frameworks.

Analytic basis and frame The harmonic covariant derivatives (in the gauge (11.128)) are

$$\mathcal{D}^{++} = \partial^{++} + H^{++\mu+}\partial_{\mu+} + x^{\mu+}_A \partial_{\mu-} + \omega^{++} \equiv \Delta^{++} + \omega^{++} \,, \tag{11.178}$$

$$\mathcal{D}^{--} = \partial^{--} + H^{--\mu+}\partial_{\mu+} + H^{--\mu-}\partial_{\mu-} + \omega^{--} \equiv \Delta^{--} + \omega^{--} \,. \tag{11.179}$$

The vielbein in (11.178) is expressed directly in terms of the potentials:

$$H^{++\mu+} = \frac{1}{2} H^{\mu\nu}(\partial^{++} L^+_\nu - \partial_{\nu+} L^{+4}) \,, \tag{11.180}$$

$$H^{\mu\nu} H_{\nu\lambda} = \delta^\mu_\lambda \,, \qquad H_{\mu\nu} = \partial_{[\mu+} L^+_{\nu]} \,. \tag{11.181}$$

The central problem in this framework is to find the vielbeins H^{--} in (11.179). They are the unique solutions of the linear harmonic differential equations

$$\Delta^{++} H^{--\mu+} - \Delta^{--} H^{++\mu+} = x^{\mu+}_A \,, \tag{11.182}$$

$$\Delta^{++} H^{--\mu-} - H^{--\mu+} = -x^{\mu-}_A \,. \tag{11.183}$$

The connections in (11.178), (11.179) are constructed with the help of the analytic 'square root' of the tensor $H_{\mu\nu}$ (11.181),

$$-e^{\alpha}_{\mu}\,\Omega_{\alpha\beta}\,e^{\beta}_{\nu} = H_{\mu\nu}\,, \qquad \partial_{\nu-}e^{\alpha}_{\mu} = 0\,. \qquad (11.184)$$

They are

$$\omega^{++}_{\alpha\beta} = e^{\mu}_{(\alpha}\Delta^{++}e_{\mu\beta)} + e^{\mu}_{(\alpha}\partial_{\mu+}H^{++\nu+}e_{\nu\beta)}\,, \qquad (11.185)$$

$$\omega^{--}_{\alpha\beta} = e^{\mu}_{(\alpha}\Delta^{--}e_{\mu\beta)} + e^{\mu}_{(\alpha}(\partial H)^{-1\nu}{}_{\mu}\Delta^{--}(\partial H)^{\lambda}_{\nu}e_{\lambda\beta)}\,, \qquad (11.186)$$

$$(\partial H)^{\nu}_{\mu} \equiv \partial_{\mu-}H^{--\nu+}\,, \qquad e^{\mu}_{\alpha}e_{\mu\beta} = -\Omega_{\alpha\beta}\,.$$

The covariant derivatives

$$\mathcal{D}^{\pm}_{\alpha} = E^{\pm\mu-}_{\alpha}\partial_{\mu-} + E^{\pm\mu+}_{\alpha}\partial_{\mu+} + \omega^{\pm}_{\alpha}$$

have the following vielbeins and connections:

$$E^{+\mu-}_{\alpha} = e^{\nu}_{\alpha}(\partial H)^{-1\mu}{}_{\nu}\,, \qquad E^{+\mu+}_{\alpha} = 0\,, \qquad \omega^{+}_{\alpha\beta\gamma} = 0\,; \qquad (11.187)$$

$$E^{-\mu-}_{\alpha} = -E^{+\nu-}_{\alpha}\partial_{\nu-}H^{--\mu-}\,, \quad E^{-\mu+}_{\alpha} = -e^{\mu}_{\alpha}\,, \quad \omega^{-}_{\alpha\beta\gamma} = -E^{+\nu-}_{\alpha}\partial_{\nu-}\omega^{--}_{\beta\gamma}\,. \qquad (11.188)$$

From (11.187) and (11.188) one can derive the hyper-Kähler (contravariant) metric in the λ basis:

$$\begin{aligned} g^{\mu+\nu+} &= 0\,, \\ g^{\mu+\nu-} &= g^{\nu-\mu+} = \Omega^{\alpha\beta}E^{-\mu+}_{\alpha}E^{+\nu-}_{\beta} = H^{\mu\lambda}(\partial H)^{-1\nu}{}_{\lambda}\,, \\ g^{\mu-\nu-} &= -\,\Omega^{\alpha\beta}(E^{+\mu-}_{\alpha}E^{-\nu-}_{\beta} + E^{+\nu-}_{\alpha}E^{-\mu-}_{\beta}) \\ &= -\,2H^{\rho\sigma}(\partial H)^{-1\lambda}{}_{\sigma}(\partial H)^{-1(\mu}{}_{\rho}\partial_{\lambda-}H^{--\nu)-}\,. \end{aligned} \qquad (11.189)$$

The transformation law of a world vector ($A^{\mu+}$, $A^{\mu-}$) in the λ basis is asymmetric ($\lambda^{\mu+}$ is analytic, but $\lambda^{\mu-}$ is not):

$$\delta A^{\mu+} = A^{\nu+}\partial_{\nu+}\lambda^{\mu+}\,, \qquad \delta A^{\mu-} = A^{\nu-}\partial_{\nu-}\lambda^{\mu-} + A^{\nu+}\partial_{\nu+}\lambda^{\mu-}\,. \qquad (11.190)$$

A peculiarity of the metric (11.189) is its dependence on the harmonic variables, $g = g(x^{\pm}, u^{\pm})$. This is quite natural in the λ basis. However, the metric is covariantly independent of $u^{\pm}$:

$$\begin{aligned} \mathcal{D}^{++}g^{\mu+\nu-} &= \mathcal{D}^{++}g^{\mu-\nu+} = 0\,, \\ \mathcal{D}^{++}g^{\mu-\nu-} &= g^{\mu+\nu-} + g^{\mu-\nu+}\,, \end{aligned} \qquad (11.191)$$

where $\mathcal{D}^{++}$ contains suitable Christoffel terms.

Many of the above relations are radically simplified in the gauges (11.169), (11.174). In particular, eq. (11.184) becomes an identity while (11.180), (11.185), (11.186) take the form

$$H^{++\mu+} = \frac{1}{2}\,\Omega^{\mu\nu}\partial_{\nu+}L^{+4}\,, \tag{11.192}$$

$$\omega^{++}_{\alpha\beta} = \frac{1}{2}\,\partial_{\alpha+}\partial_{\beta+}L^{+4}\,, \tag{11.193}$$

$$\omega^{--}_{\alpha\beta} = (\partial H)^{-1\nu}{}_{(\alpha}\Delta^{--}(\partial H)_{\nu\beta)}\,. \tag{11.194}$$

The relation (11.168) becomes

$$\mathcal{D}^{++}\omega^{-}_{\alpha(\beta\gamma)} + \frac{1}{2}\,\partial_{\alpha+}\partial_{\beta+}\partial_{\gamma+}L^{+4} = 0\,. \tag{11.195}$$

This implies, e.g.,

$$\omega^{-}_{\alpha(\beta\gamma)} = \omega^{-}_{(\alpha\beta\gamma)}\,. \tag{11.196}$$

Harmonic-independent hyper-Kähler metrics Here we outline the procedure of constructing hyper-Kähler metrics from the analytic potentials in the more familiar u-independent framework (τ frame and basis). For convenience we use the gauges (11.169), (11.174). The relation between the λ and τ bases is given by the bridges $v^{\mu\pm}$:

$$x^{\mu\pm}_{A} = x^{\mu i}u^{+}_{i} + v^{\mu\pm}(x^{\nu i}, u^{\pm})\,. \tag{11.197}$$

These bridges can be found in three steps. Firstly, one obtains the λ-basis analytic vielbein $H^{++\mu+}$ from the relation (11.192). It is then substituted in the equation (11.124) relating $H^{++\mu+}$ and the bridge $v^{\mu+}$:

$$H^{++\mu+}(x^{\mu i}u^{+}_{i} + v^{\mu+}(x,u), u^{\pm}) = \partial^{++}v^{\mu+}(x^{\lambda i}, u^{\pm})\,. \tag{11.198}$$

Finally, the other bridge $v^{\mu-}$ is found from the equation

$$\partial^{++}v^{\mu-} = v^{\mu+} \quad\Rightarrow\quad v^{\mu-} = \int dw\,\frac{1}{(u^{+}w^{+})}\,v^{\mu+}(x^{\nu i}, w^{\pm i})\,. \tag{11.199}$$

Equation (11.198) is a non-linear differential equations on S^2. Finding its solution (up to gauge freedom) is the non-trivial part of this program.

Once the relation between the two bases is known, it is not hard to rewrite the expressions (11.187), (11.188) for the vielbeins of the covariant derivatives $\mathcal{D}^{\pm}_{\alpha}$ in the τ basis:

$$E^{+\rho k}_{\alpha} = E^{+\mu-}_{\alpha}\partial_{A\mu-}x^{\rho k} = (\partial H)^{-1\mu}{}_{\alpha}\partial_{A\mu-}x^{\rho k}, \tag{11.200}$$

$$\begin{aligned} E^{-\rho k}_{\alpha} &= E^{-\mu+}_{\alpha}\partial_{A\mu+}x^{\rho k} + E^{-\mu-}_{\alpha}\partial_{A\mu-}x^{\rho k} \\ &= -\partial_{A\alpha+}x^{\rho k} - (\partial H)^{-1\mu}{}_{\alpha}\,\partial_{A\mu-}x^{\nu i}\,\partial_{\nu i}\partial^{--}x^{\nu-}_{A}\,\partial_{A\nu-}x^{\rho k} \\ &= (\partial H)^{-1\mu}{}_{\alpha}\partial^{--}\partial_{A\mu-}x^{\rho k}\,. \end{aligned} \tag{11.201}$$

Here

$$(\partial H)^{\nu}_{\mu} = \partial_{A\mu-} H^{--\nu+} = \partial_{A\mu-} x^{\rho i}\, \partial_{\rho i} \partial^{--} x^{\nu+}_{A}\,. \tag{11.202}$$

Then the τ-basis metric is

$$g^{\mu i,\nu j} = \Omega^{\alpha\beta} \left(E^{-\mu i}_{\alpha} E^{+\nu j}_{\beta} - E^{+\mu i}_{\alpha} E^{-\nu j}_{\beta} \right). \tag{11.203}$$

It does not depend on $u^{\pm}$:

$$\partial^{++} g^{\mu i,\nu j} = 0\,, \tag{11.204}$$

which follows from the relations

$$\begin{aligned} \partial^{++} E^{+\mu i}_{\alpha} + \omega^{++}{}_{\alpha}^{\beta} E^{+\mu i}_{\beta} &= 0\,, \\ \partial^{--} E^{+\mu i}_{\alpha} + \omega^{--}{}_{\alpha}^{\beta} E^{+\mu i}_{\beta} &= E^{-\mu i}_{\alpha}\,. \end{aligned} \tag{11.205}$$

Note that the vielbeins (11.200), (11.201) are τ basis, but still λ frame. They can be rotated to the τ frame with the help of the $Sp(n)$ bridge $M^{\beta}_{\alpha}(x,u)$ (11.120). Obviously, this does not affect the expression for the metric (11.203).

The three covariantly constant complex structures of the hyper-Kähler manifold (11.78) are obtained from

$$I^{++[\mu i,\nu j]} = I^{[\mu i,\nu j]}_{(kl)} u^{+k} u^{+l} = E^{+\mu i}_{\alpha} E^{+\nu j}_{\beta} \Omega^{\alpha\beta} \tag{11.206}$$

by removing the harmonic variables.

It is useful to have the expressions for the inverse vielbein and the ordinary covariant metric (with subscript world indices):

$$\begin{aligned} E^{\alpha+}_{\rho k} &= -\partial_{\rho k} x^{\alpha+}\,, & (11.207)\\ E^{\alpha-}_{\rho k} &= \partial_{\rho k} x^{\gamma-} (\partial H)^{\alpha}_{\gamma} - \partial_{\rho k} x^{\delta+} (\partial H)^{-1\phi}_{\delta}\, \partial_{\phi-} x^{\beta i}\, \partial_{\beta i} \partial^{--} x^{\gamma-} (\partial H)^{\alpha}_{\gamma}\,, & (11.208)\\ g_{\mu i,\nu j} &= \Omega_{\alpha\beta} \left(E^{\alpha-}_{\mu i} E^{\beta+}_{\nu j} - E^{\alpha+}_{\mu i} E^{\beta-}_{\nu j} \right). & (11.209) \end{aligned}$$

It is straightforward to check the corresponding orthonormalization conditions and to show, e.g.,

$$\partial^{++} E^{\alpha-}_{\rho k} - \omega^{++\alpha}{}_{\beta} E^{\beta-}_{\rho k} = E^{\alpha+}_{\rho k} = -\partial_{\rho k} x^{\alpha+}\,, \tag{11.210}$$

whence the harmonic independence of the metric (11.209) follows.

11.3.7 Central charges as the origin of the hyper-Kähler potentials

Above we presented a solution of the hyper-Kähler constraints. We saw that all the objects of differential geometry (vielbeins, connections, etc.) could be expressed in terms of two analytic potentials L^{+}_{μ} and L^{+4} which have their own pregauge freedom with parameter λ^{++}. However, neither the potentials nor the pregauge invariance naturally fit in the standard geometric framework. As we saw in Section 11.2.4, in the context of Kähler geometry the same problem is solved by introducing an auxiliary central charge coordinate. Here we develop an analogous treatment of hyper-Kähler geometry in an extended harmonic space with an $Sp(1) \sim SU(2)$ triplet of central charge coordinates:

$$\mathbb{R}^{4n+3} = \{x^{\mu i}, z^{ij}\}\,. \tag{11.211}$$

In it one can realize the following central charge extension of the Poincaré group:

$$\begin{aligned} &[P_{\mu i}, P_{\nu j}] = 2i\Omega_{\mu\nu} Z_{ij}\,, \\ &[P_{\mu i}, Z_{kl}] = [Z_{ij}, Z_{kl}] = 0\,, \qquad Z_{ik} = Z_{ki}\,. \end{aligned} \tag{11.212}$$

The corresponding coordinate transformations are

$$\delta x^{\mu i} = a^{\mu i}\,, \qquad \delta z^{ij} = a^{ij} - a^{\mu(i} x^{j)}_{\mu}\,. \tag{11.213}$$

Further, one can introduce covariant derivatives which commute with the generators:

$$D_{\mu i} = \partial_{\mu i} + \Omega_{\mu\nu} x^{\nu j} \partial^z_{ij}\,, \qquad \left(\partial^z_{ij} \equiv \frac{\partial}{\partial z^{ij}}\right) \tag{11.214}$$

and satisfy the following algebra with torsion:

$$[D_{\mu i}, D_{\nu j}] = -2\Omega_{\mu\nu}\partial^z_{ij}\,. \tag{11.215}$$

The harmonic extension of $\mathbb{R}^{4n+3}$ is defined in the standard way:

$$\mathbb{R}^{4n+5} = \{x^{\mu i}, z^{ij}, u^{\pm i}\} = \{x^{\mu\pm}, z^{++}, z^{--}, z^0, u^{\pm i}\}\,, \tag{11.216}$$

where

$$x^{\mu\pm} = x^{\mu i} u^{\pm}_i\,, \qquad z^{\pm\pm} = z^{ij} u^{\pm}_i u^{\pm}_j\,, \qquad z^0 = z^{ij} u^+_i u^-_j\,.$$

In this harmonic space we can find three analytic subspaces closed under the transformations (11.213):

$$\{x^{\mu+}\} \subset \{x^{\mu+}, z^{++}\} \subset \{x^{\mu+}, z^{++}, z^0_A\} \subset \mathbb{R}^{4n+5}\,, \tag{11.217}$$

where

$$z^0_A = z^0 + \frac{1}{2} x^{\mu+} x^-_{\mu}\,.$$

The existence of these subspaces follows from the algebra of the harmonic projections of the covariant derivatives (11.214):

$$[D^+_\mu, D^+_\nu] = -2\Omega_{\mu\nu}\partial^z_{--}\,, \quad [D^+_\mu, D^-_\nu] = \Omega_{\mu\nu}\partial^z_0\,, \quad [D^-_\mu, D^-_\nu] = -2\Omega_{\mu\nu}\partial^z_{++}\,,$$

$$[D^{\pm}_{\mu}, \partial^z] = 0\,, \qquad \left(\partial^z_{\pm\pm} \equiv \frac{\partial}{\partial z^{\pm\pm}}\,,\ \partial^z_0 \equiv \frac{\partial}{\partial z^0}\right). \tag{11.218}$$

Correspondingly, one can define analytic fields satisfying the proper integrable constraints. For instance, in the case of the biggest of the spaces (11.217) one can define an analytic field as follows:

$$D^+_\mu \phi = \partial^z_{--}\phi = 0 \quad \Rightarrow \quad \phi = \phi(x^{\mu +}, z^{++}, z^0_A)\,. \tag{11.219}$$

The manifest analyticity of $\phi(x^{\mu +}, z^{++}, z^0_A)$ results from the form of the derivative D^+_μ in the analytic basis,

$$(D^+_\mu)_\lambda = \partial_{\mu -} + x^-_\mu \partial^z_{--}\,. \tag{11.220}$$

Next we discuss the curved version of the space introduced above and locate the places where the hyper-Kähler potentials occur as geometric objects. We begin by extending the usual τ group (11.85) by adding local transformations of z^{ij}, $\delta z^{ij} = \tau^{ij}(x)$. As in the Kähler case, neither the gauge group parameters nor the geometric objects depend on z, only the matter fields are allowed to do so. The extension of the diffeomorphism group requires the introduction of new vielbein terms in the covariant derivatives, for instance,

$$\nabla^+_\alpha = \mathcal{D}^+_\alpha + E^{+3}_\alpha \partial^z_{++} + E^-_\alpha \partial^z_{--} + E^+_\alpha \partial^z_0\,. \tag{11.221}$$

The algebra of these covariant derivatives reproduces that of hyper-Kähler geometry (see Section 11.3.1) with additional flat torsion terms coming from (11.218):

$$\begin{aligned}
&\text{(a)} \qquad [\nabla^+_\alpha, \nabla^+_\beta] = -2\Omega_{\alpha\beta}\partial^z_{--}\,,\\
&\text{(b)} \qquad [\nabla^+_\alpha, \nabla^-_\beta] = \Omega_{\alpha\beta}\partial^z_0 + R^{+-}_{(\alpha\beta)}\,,\\
&\text{(c)} \qquad [\nabla^-_\alpha, \nabla^-_\beta] = -2\Omega_{\alpha\beta}\partial^z_{++}\,.
\end{aligned} \tag{11.222}$$

Of course, the derivatives $\nabla^{\pm}_\alpha$ commute with the central charge ones ∂^z, which remain flat. One should also add the proper modification of the constraint (11.87):

$$[\mathcal{D}^{++}, \nabla^+_\alpha] = 0\,. \tag{11.223}$$

Note that the constraints (11.222), (11.223) contain all the previous constraints on the x-space covariant derivatives $\mathcal{D}$ and we shall systematically make use of them in what follows.

As has by now become customary, we interpret eq. (11.222a) as the integrability condition for the existence of the curved version of an analytic basis. Indeed, using the old constraint (11.86), we find

$$\begin{aligned}
\mathcal{D}^+_{[\alpha} E^{+3}_{\beta]} = 0 \ &\Rightarrow\ E^{+3}_\alpha = -\mathcal{D}^+_\alpha v^{++}\,,\\
\mathcal{D}^+_{[\alpha} E^+_{\beta]} = 0 \ &\Rightarrow\ E^+_\alpha = -\mathcal{D}^+_\alpha v^0\,,
\end{aligned} \tag{11.224}$$

where $v^{++}(x,u)$, $v^0(x,u)$ are the bridges to the analytic basis:

$$z_A^{++} = z^{++} + v^{++}, \qquad z_A^0 = z^0 + v^0, \qquad z_A^{--} = z^{--} + v^{--} \tag{11.225}$$

(the third bridge v^{--} has been added for convenience). The analytic basis coordinates z_A transform with new parameters λ:

$$\delta z_A^{++} = \lambda^{++}(x_A^{\mu+}, u), \qquad \delta z_A^0 = \lambda^0(x_A^{\mu+}, u), \qquad \delta z_A^{--} = \lambda^{--}(x_A^{\mu+}, x_A^{\mu-}, u), \tag{11.226}$$

whereas the bridges have mixed λ and τ transformations:

$$\begin{aligned} \delta v^{++} &= \lambda^{++}(x_A^{\mu+}, u) - \tau^{ij}(x)u_i^+u_j^+, \\ \delta v^0 &= \lambda^0(x_A^{\mu+}, u) - \tau^{ij}(x)u_i^+u_j^-, \\ \delta v^{--} &= \lambda^{--}(x_A^{\mu+}, x_A^{\mu-}, u) - \tau^{ij}(x)u_i^-u_j^-. \end{aligned} \tag{11.227}$$

Hence we see that the third bridge v^{--} can be gauged away. We fix a convenient gauge later on.

The advantage of the new basis is that the covariant derivative (11.221) becomes short:

$$\nabla_\alpha^+ = \mathcal{D}_\alpha^+ + E_\alpha^- \partial_{--}^z \tag{11.228}$$

(we assume that the derivative $\mathcal{D}_\alpha^+$ has its short form (11.120) in the ordinary analytic basis and frame). This permits the existence of manifestly analytic fields (11.219). However, the harmonic derivatives acquire new, central charge vielbein terms:

$$\nabla^{++} = \mathcal{D}^{++} + H^{+4}\partial_{++}^z + (2z_A^0 + H^0)\partial_{--}^z + (z_A^{++} + H^{++})\partial_0^z, \tag{11.229}$$

$$\nabla^{--} = \mathcal{D}^{--} + (2z_A^0 + \mathcal{H}^0)\partial_{++}^z + H^{-4}\partial_{--}^z + (z_A^{--} + H^{--})\partial_0^z. \tag{11.230}$$

One of them is pure gauge (see (11.227)) and we can gauge it away:

$$\underline{\lambda^{--} \text{ gauge:}} \qquad H^{--} = 0. \tag{11.231}$$

Further, from the conventional constraint

$$[\nabla^{++}, \nabla^{--}] = \nabla^0 \tag{11.232}$$

with $\nabla^0 = D^0 + 2(z^{++}\partial_{++} - z^{--}\partial_{--})$ follow relations between the harmonic vielbeins:

$$\begin{aligned} &\Delta^{++}H^{-4} - \Delta^{--}H^0 = 0, \\ &\Delta^{++}\mathcal{H}^0 - \Delta^{--}H^{+4} + 2H^{++} = 0, \\ &H^0 - \mathcal{H}^0 = \Delta^{--}H^{++}. \end{aligned} \tag{11.233}$$

They allow us to express, in principle, H^0, H^{++} and H^{-4} in terms of H^{+4} and $\mathcal{H}^0$. In what follows we shall see that H^{+4} becomes the main potential of hyper-Kähler geometry whereas $\mathcal{H}^0$ can be eliminated in a suitable gauge.

Another conventional constraint is the definition of ∇^-_α:

$$[\nabla^{--}, \nabla^+_\alpha] = \nabla^-_\alpha = \mathcal{D}^-_\alpha - E^+_\alpha \partial^z_{++} - E^-_\alpha \partial^z_0 + (\mathcal{D}^{--} E^-_\alpha - \mathcal{D}^+_\alpha H^{-4})\partial^z_{--}\,, \tag{11.234}$$

where

$$E^+_\alpha = \mathcal{D}^+_\alpha \mathcal{H}^0\,. \tag{11.235}$$

This expression is obtained from (11.228), (11.230) and (11.91).

So far we have succeeded in reducing the set of independent vielbeins to just three, H^{+4}, $\mathcal{H}^0$ and E^-_α. Further restrictions on them are obtained from the extra torsion constraints contained in (11.222), (11.223).

Firstly,

$$T^{+++(++)}_{\alpha} = 0 \quad \Rightarrow \quad \partial_{\mu-} H^{+4} = 0\,, \tag{11.236}$$

where the notation $^{(++)}$ distinguishes the central charge index from the harmonic one (for the general definition of the torsion components see (11.92)). Equation (11.236) just states the analyticity of the main potential of hyper-Kähler geometry (recall (11.160)).

Secondly, from the constraint

$$T^{+-(++)}_{\alpha\,\beta} = 0 \quad \Rightarrow \quad \partial_{\mu-} E^+_\alpha = 0 \tag{11.237}$$

follows the analyticity of the second potential of hyper-Kähler geometry which is now introduced by (recall (11.157)):

$$E^+_\alpha \equiv e^\mu_\alpha L^+_\mu\,. \tag{11.238}$$

This condition has an important corollary. Comparing (11.235) with (11.238), we find $L^+_\mu = (\partial H)^{-1}{}^\nu_\mu \partial_{\nu-} \mathcal{H}^0$ (recall (11.120) and (11.142)). This implies

$$\partial_{\mu-} h = 0\,, \qquad h \equiv \mathcal{H}^0 - H^{--\nu+} L^+_\nu\,. \tag{11.239}$$

The analytic object h transforms as follows:

$$\delta h = -2\lambda^0 + \partial^{--}\lambda^{++} - (\partial^{--}\lambda^{\mu+}) L^+_\mu\,, \tag{11.240}$$

so it can be gauged away by means of the analytic parameter $\lambda^0(x^+, u)$:

$$\underline{\lambda^0 \text{ gauge:}} \qquad \mathcal{H}^0 - H^{--\mu+} L^+_\mu = 0\,, \qquad \lambda^0 = \frac{1}{2}[\partial^{--}\lambda^{++} - (\partial^{--}\lambda^{\mu+}) L^+_\mu]\,. \tag{11.241}$$

Thirdly, the constraint

$$T^{++-(++)}_{\alpha} = 0 \quad \Rightarrow \quad \mathcal{D}^{++} E^+_\alpha + e^\mu_\alpha \partial_{\mu+} H^{+4} = 0 \tag{11.242}$$

together with (11.238) and (11.145) imply the expression (11.164) of the harmonic derivative vielbein $H^{++\mu+}$ in terms of the potentials. Note that this constraint is contained in the central-charge extension of the relation (11.98) which follows from the basic constraints via Bianchi identities.

Fourthly, the vielbein E^-_α is determined from the constraint

$$T^{---(++)}_{\alpha} = 0 \quad \Rightarrow \quad E^-_\alpha = e^\mu_\alpha \left[\frac{1}{2}\partial^{--}L^+_\mu - (\partial_{[\mu+}L^+_{\nu]})H^{--\nu+}\right]. \quad (11.243)$$

Finally, consider the constraint

$$T^{+-(0)}_{\alpha\,\beta} = \Omega_{\alpha\beta} \quad \Rightarrow \quad \mathcal{D}^+_\alpha E^-_\beta = -\Omega_{\alpha\beta}\,. \quad (11.244)$$

Inserting (11.243) into it, we obtain the expression (11.155) of the tangent-space-gauge independent part $H_{\mu\nu}$ of the analytic vielbein e^μ_α (the same follows directly from the constraint $T^{--(++)}_{\alpha\,\beta} = -2\Omega_{\alpha\beta}$ as well).

Let us summarize this rather lengthy discussion. Although the introduction of central charges gave rise to a number of new objects, all of them as well as those in the x-space sector could finally be expressed in terms of only two independent analytic potentials, the vielbeins H^{+4} from ∇^{++} (11.229) and E^+_α from ∇^+_α (11.221). The pregauge parameter λ^{++} (11.156) also found its natural interpretation as the remaining unfixed analytic-basis central-charge diffeomorphism parameter from (11.226).

11.3.8 An explicit construction of hyper-Kähler metrics

As explained above, the only implicit (and most difficult) step in the construction of hyper-Kähler metrics starting from the analytic potentials is solving the differential equations (11.198), (11.199) for the bridges on S^2. This is a non-trivial problem and we cannot present its general solution. However, there exists a class of potentials for which the differential equations are drastically simplified and can be solved explicitly. This Ansatz is due to Ward [W2]. Here we should mention that our approach to hyper-Kähler geometry is closely related to the twistor approach to the self-dual Einstein equations in $\mathbb{R}^4$ [W2]. This relationship is similar to the one discussed in Section 11.2.2 in the context of SDYM theory.

One starts by making the following particular choice of potentials:

$$L^+_\mu = x^+_\mu\,, \qquad L^{+4} = L^{+4}(u^{+\mu}x^+_\mu, u) \quad (11.245)$$

(the choice of L^+_μ in (11.245) is just a gauge condition, recall (11.169)). For simplicity we consider a four-dimensional space, so $\mu = 1, 2$.* Note that the $Sp(1)$ symmetry is manifestly broken by the choice (11.245).

---

* This Ansatz can be generalized to $4n$ dimensions by replacing $u^{+\mu}x^+_{A\mu}$ by $P^{+\mu}x^+_\mu$ where $P^{+\mu} = P^{\mu i}u^+_i$ and $P^{\mu i}$ is a constant real vector. Note that the original Ward's Ansatz [W2] corresponds to using a third-rank tensor $P^{\mu++} = P^{\mu ij}u^+_i u^+_j$, but then $P^{\mu++}x^+_\mu$ cannot be chosen real. In the present form the Ansatz gives rise to the so-called multicenter metrics [G32], see the end of this section.

The idea of this Ansatz becomes clear after writing down the equation for the bridge $v^{\mu+}$. From (11.192) one finds

$$H^{++\mu+} = u^{+\mu} L^{(+2)} , \qquad L^{(+2)} = -\frac{1}{2}\frac{\partial L^{+4}}{\partial(u^{+\mu}x^{+}_{\mu})} . \tag{11.246}$$

Then (11.198) becomes

$$\partial^{++} v^{\mu+}(x,u) = u^{+\mu} L^{(+2)}(u^{+}x^{+}_{A}, u) . \tag{11.247}$$

It is now clear that the solution for $v^{\mu+}$ should have the form

$$v^{\mu+}(x,u) = u^{+\mu} v(x,u) . \tag{11.248}$$

Indeed, then

$$u^{+\mu}x^{+}_{A\mu} = u^{+\mu}(x^{i}_{\mu}u^{+}_{i} + u^{+}_{\mu}v) = u^{+\mu}u^{+}_{i}x^{i}_{\mu} , \tag{11.249}$$

so eq. (11.247) becomes linear:

$$\partial^{++} v(x,u) = L^{(+2)}(u^{+}x^{+}, u) . \tag{11.250}$$

The solution for $v^{\mu+}$ (up to gauge freedom) is given by the harmonic integral

$$v^{\mu+} = u^{+\mu} \int dw \frac{(u^{+}w^{-})}{(u^{+}w^{+})} L^{(+2)}(w^{+}x^{+}, w) \tag{11.251}$$

(see Chapter 4). Analogously, the solution of (11.199) is

$$v^{\mu-} = u^{+\mu} \int dw \frac{(u^{-}w^{-})}{(u^{+}w^{+})} L^{(+2)}(w^{+}x^{+}, w) . \tag{11.252}$$

Having found the bridges, one can evaluate the metric (11.203) by a straightforward calculation. The result is

$$g^{\mu i,\nu j} = \frac{1}{1+V_0}(\epsilon^{\mu\nu}\epsilon^{ij} + V^{\mu i}\epsilon^{\nu j} + V^{\nu j}\epsilon^{\mu i} + V^{2}\epsilon^{\mu i}\epsilon^{\nu j}) , \tag{11.253}$$

where

$$V_{\mu i}(x) = \int du\, u^{+}_{\mu}u^{-}_{i} L(u^{+}x^{+}, u) , \qquad V_0 = \epsilon^{\mu i}V_{\mu i} = -\int du\, L ,$$
$$L = -\frac{1}{2}\frac{\partial^{2} L^{+4}}{\partial(u^{+}x^{+})^{2}} . \tag{11.254}$$

The solution (11.253) forms a well-known class of hyper-Kähler metrics. To see this, let us look at the metric defining the invariant interval $ds^2 = g_{\mu i,\nu j}dx^{\mu i}dx^{\nu j}$ (the inverse of (11.253)):

$$g_{\mu i,\nu j} = (1+V_0)\epsilon_{\mu\nu}\epsilon_{ij} + V_{\mu i}\epsilon_{\nu j} + V_{\nu j}\epsilon_{\mu i} + \frac{2}{1+V_0}V_{\mu i}V_{\nu j} . \tag{11.255}$$

Introducing

$$\vec{V} = -i(\vec{\tau})^{\mu i} V_{\mu i}\,, \qquad \vec{x} = \frac{i}{\sqrt{2}}(\vec{\tau})^{\mu i} x_{\mu i}\,, \qquad t = -\frac{1}{\sqrt{2}}\epsilon_{\mu i} x^{\mu i}\,,$$

we can write down the interval in the form

$$ds^2 = \frac{1}{1+V_0}(dt + \vec{V}\cdot d\vec{x})^2 + (1+V_0)d\vec{x}\cdot d\vec{x}\,. \tag{11.256}$$

It is easy to see that the fields $\vec{V}$ and $1 + V_0$, by their definition, satisfy the differential equations

$$\vec{\nabla}\wedge\vec{V} = \vec{\nabla}(1+V_0)\,, \qquad \Delta(1+V_0) = 0\,, \qquad \partial_t(1+V_0) = 0\,. \tag{11.257}$$

Note that $1 + V_0$ is the general solution of the Laplace equation ('harmonic' function) written down as a 'twistor transform' [W13]. The interval (11.256) is recognized as the standard definition of the general family of multicenter hyper-Kähler metrics [G32]. In Section 5.3.2 we encountered a particular case of this family corresponding to the Taub–NUT metric (eq. (5.107)).

It is important to realize that the potential (11.245), and consequently the metric (11.255) have at least one isometry:

$$\delta x^+_\mu = C\, u^+_\mu\,, \qquad C = \text{const} \tag{11.258}$$

(or $\delta t = C$ in the form (11.256)). So, the Ansatz (11.245) clearly does not produce the most general hyper-Kähler metrics. It is known that any four-dimensional hyper-Kähler metric with at least one isometry of this type (the so-called *triholomorphic* isometry) falls into the multicenter class [G34, G35, H7]. Actually, the most general action of a single tensor $N = 2$ multiplet discussed in Section 6.2, after performing one of the possible duality transformations to the q^+ form, yields the self-interaction term $L^{+4}(u^{+i}q^+_i, u^\pm)$ (eq. (6.63)) which precisely matches the Ansatz (11.245). In view of the one-to-one correspondence between hyper-Kähler geometry and the $N = 2$ sigma models to be discussed in the next section, this proves that the q^+ self-interaction just mentioned yields the most general multicenter Ansatz for the hyper-Kähler metric in the sector of the physical bosonic fields $f^{ai}(x)$ ($q^{a+} = f^{ai}(x)u^+_i + \cdots$).

Note that there exists one more way of extending the Ansatz (11.245) to the case of $4n$-dimensional hyper-Kähler manifolds, besides the one mentioned in footnote $*$. One splits the world index μ as $\mu \to ai$, $a = 1, \ldots n$; $i = 1, 2$, and assumes that $L^{+4} = L^{+4}(u^{+i}x^+_{ai}, u)$. This potential and the corresponding metric have n commuting translational isometries

$$\delta x^+_{ai} = C_a\, u^+_i\,, \qquad C_a = \text{const}\,. \tag{11.259}$$

The hyper-Kähler manifolds of this type are called 'toric'. In the $N = 2$ sigma model language, such an Ansatz corresponds to the class of q^+ actions which is equivalent, by a duality transformation, to the most general action of n interacting tensor multiplets. The above computation can be repeated for this case and shown to give rise to the most general Ansatz for the toric hyper-Kähler manifolds metric given in [H8, L1, P1].

11.4 Geometry of $N = 2$, $d = 4$ supersymmetric sigma models

11.4.1 The geometric meaning of the general q^+ action

In [A3, Z1] a remarkable correspondence between $N = 1, 2$ sigma models in four space-time dimensions and complex (target) manifolds has been established. In the case $N = 1$ it says that any sigma model must have a metric of the Kähler type. Conversely, given a Kähler metric one can construct an *off-shell* $N = 1$ sigma model by replacing the coordinates in the corresponding Kähler potential $K(x, \bar{x})$ by chiral and antichiral superfields Φ, $\bar{\Phi}$ and then using $K(\Phi, \bar{\Phi})$ as a superspace Lagrangian.

In the case $N = 2$ the theorem of ref. [A3] states that the metric of an $N = 2$ sigma model is necessarily hyper-Kähler and vice versa, given a hyper-Kähler metric one can always construct an *on-shell* $N = 2$ sigma model. The open questions in [A3] were how to build the most general sigma models of this kind or, equivalently, how to find the most general hyper-Kähler metrics. We have answered the first form of this question in Chapter 5 where we wrote down the most general off-shell action for a set of $N = 2$ hypermultiplets q^{+a}, $a = 1, 2, \ldots, n$ (see (5.82)):*

$$S = \frac{1}{2} \int du \, d\zeta^{(-4)} \, H^{+4} , \tag{11.260}$$

where

$$H^{+4} = L^+_a(q^+, u) D^{++} q^{+a} + L^{+4}(q^+, u) \tag{11.261}$$

and $L^+_a(q^+, u)$, $L^{+4}(q^+, u)$ are *arbitrary* functions of the hypermultiplets and of the harmonics. Based on the theorem of [A3], it is natural to expect that this action also provides the answer to the second form of the same question, i.e., it yields the most general hyper-Kähler sigma model in the bosonic sector.

In the present chapter we answered the second form of the above question by solving the constraints on hyper-Kähler geometry in terms of two unconstrained analytic potentials, $L^+_a(x^+, u)$ and $L^{+4}(x^+, u)$. The similarity with eq. (11.261) is obvious, it is sufficient to replace the complex manifold coordinates $x^{\mu+}$ by hypermultiplet superfields q^{a+}. This suggests establishing a detailed one-to-one correspondence between $N = 2$ sigma models and hyper-Kähler geometry.

* In this section, in order to make contact with Chapter 5 and to avoid confusion with the Lorentz indices of the θ's, we go back to using the Latin letters $a, b, c, \ldots$ for the hypermultiplet and $Sp(n)$ indices.

Let us first examine the symmetries of the action (11.260). It is invariant under reparametrizations of q^{a+}, $\delta q^{a+} = \lambda^{a+}(q^+, u)$, provided L^+_b and L^{+4} transform as follows:

$$\delta L^+_b = -\partial_{b+}\lambda^{a+}L^+_a\,, \qquad \delta L^{+4} = -L^+_a\partial^{++}\lambda^{a+}\,. \tag{11.262}$$

Here $\partial_{a+} = \partial/\partial q^{a+}$ and ∂^{++} is the partial derivative acting on the harmonic argument u of $\lambda^{a+}(q^+, u)$ (but ignoring the u dependence of the argument q^+). In addition, the action is invariant under the following redefinitions of the self-interaction potentials:

$$\delta L^+_b = \partial_{b+}\lambda^{++}\,, \qquad \delta L^{+4} = \partial^{++}\lambda^{++} \tag{11.263}$$

with an arbitrary $\lambda^{++}(q^+, u)$. Indeed, under (11.263) the integrand in (11.260) changes by a total harmonic derivative:

$$\delta H^{+4} = \partial^{++}\lambda^{++} + D^{++}q^{a+}\partial_{a+}\lambda^{++} = D^{++}\lambda^{++}(q^+, u)\,. \tag{11.264}$$

It is now clear that eqs. (11.262) and (11.263) are the analogs of the diffeomorphism and pregauge transformations (11.156), (11.163) of the hyper-Kähler potentials.

Next, let us look at the equation of motion for $q^{\mu+}$ following from the action (11.260):

$$(\partial_{a+}L^+_b - \partial_{b+}L^+_a)D^{++}q^{b+} = -\partial_{a+}L^{+4} + \partial^{++}L^+_a\,. \tag{11.265}$$

Comparing this with eqs. (11.164), (11.155), we identify (on shell!)

$$H^{++a+}(q^+, u) = D^{++}q^{a+}\,. \tag{11.266}$$

Let us summarize the correspondence between hyper-Kähler geometry and $N = 2$ sigma models. The analytic basis coordinates $x^{\mu+}$ are replaced by Grassmann analytic superfields $q^{a+}(x, \theta^+, u)$ describing $N = 2$ hypermultiplet matter. The hyper-Kähler potentials L^+_a, L^{+4} determine the hypermultiplet self-interaction. The on-shell derivative $D^{++}q^{a+}$ expressed from the equation of motion is the vielbein H^{++a+}. Finally, the on-shell sigma model Lagrangian corresponds to the central charge vielbein H^{+4}. Thus, the pregauge invariance of the action is associated with the transformations of the central charge coordinate z^{++}_A.

We emphasize the fact that it is the *on-shell* hypermultiplet action which corresponds to a sigma model with a hyper-Kähler target manifold. The reason is that off shell the hypermultiplet superfield $q^{a+}(x, \theta^+, u)$ contains an infinite number of auxiliary fields. The rôle of the equation of motion (11.265) is to eliminate the auxiliary fields in terms of the physical ones. Only the equations of motion of the latter involve self-interactions of the familiar sigma model type. Another way to see this is to rewrite the equation of motion (11.265) as the geometric equation defining the bridge from the analytic λ basis (in which the hypermultiplets naturally live) to the τ basis in which the ordinary hyper-Kähler metric is defined. Indeed, in the gauge $L^+_a = q^+_a$ obtained with the help of the change of variables (11.262), eq. (11.265) reads

$$D^{++}q^{a+} = \frac{1}{2}\Omega^{ab}\partial_{b+}L^{+4} = H^{++a+}(q^+, u)\,. \qquad (11.267)$$

Let us make the change of variables

$$q^{a+} = q^{ai}u^+_i + v^{a+}(q^{bj}, u)\,, \qquad (11.268)$$

where $q^{ai}(x^{\alpha\dot\alpha}, \theta^{\alpha i}, \bar\theta^{\dot\alpha i})$ are ordinary (harmonic-independent) superfields. Then, rewriting eq. (11.267) in the central basis in harmonic superspace where $D^{++} = \partial^{++}$, we find

$$\partial^{++}v^{a+}(q^{bj}, u) = H^{++a+}[q^{bj}u^+_j + v^{b+}(q, u), u]\,. \qquad (11.269)$$

This is precisely the bridge-defining equation (11.130). The conclusion is that passing to the new variables q^{bi}, which correspond to the τ-basis coordinates $x^{\mu i}$ of the hyper-Kähler manifold, is equivalent to eliminating the infinite set of auxiliary fields contained in q^{b+}.* Of course, this involves solving the non-linear differential equation (11.269). A detailed analysis of the procedure of elimination of the auxiliary fields and of the way the standard on-shell sigma model emerges is given in the next subsection.

11.4.2 The component action of the general $N = 2$ sigma model

Here we explicitly demonstrate that the general off-shell q^+ action (11.260) gives rise to the standard physical component action of the general $N = 2$ sigma model [A2, A3] upon elimination of the infinite set of auxiliary fields of the q^+ hypermultiplets. The geometric objects entering this on-shell action (the bosonic target metric, the curvature tensor, etc.) are shown to precisely coincide with those derived on the purely geometric grounds in the previous sections of this chapter.

The first step is to perform the integration over the Grassmann variables in the superfield action. The θ expansion of q^{a+} reads

$$\begin{aligned} q^{a+}(\zeta, u) \;=\; & F^{a+}(x, u) + \theta^{+\alpha}\Psi^a_\alpha(x, u) + \bar\theta^+_{\dot\alpha}\bar\Psi^{\dot\alpha a}(x, u) \\ & +(\theta^+)^2 M^{a-}(x, u) + (\bar\theta^+)^2 N^{a-}(x, u) + i\theta^+\sigma^m\bar\theta^+ A^{a-}_m(x, u) \\ & +(\theta^+)^2\bar\theta^+_{\dot\alpha}\bar\chi^{\dot\alpha a(-2)}(x, u) + (\bar\theta^+)^2\theta^{+\alpha}\chi^{a(-2)}_\alpha(x, u) \\ & +(\theta^+)^2(\bar\theta^+)^2 P^{a(-3)}(x, u)\,, \end{aligned} \qquad (11.270)$$

* The equations of motion for q^{bi} in the τ-basis are a non-linear extension of the Fayet–Sohnius constraints (5.18) [S9, S10].

where, in accord with the reality condition (5.53),

$$\begin{array}{llllll} \widetilde{(F_a^+)} &= F^{a+}, & \widetilde{(\Psi_{\alpha a})} &= \bar{\Psi}^a_{\dot\alpha}, & \widetilde{(M_a^-)} &= N^{a-}, \\ \widetilde{(A^-_{am})} &= A^{a-}_m, & \widetilde{(\chi^{(-2)}_{\alpha a})} &= \bar{\chi}^{a(-2)}_{\dot\alpha}, & \widetilde{(P_a^{(-3)})} &= P^{a(-3)}. \end{array} \tag{11.271}$$

Substituting (11.270) into (11.260) in the gauge $L_a^+ = q_a^+$ (see (5.114) and (11.169)) we obtain an action written down as an integral over $\mathbb{R}^4 \times S^2$:

$$S = \int du\, d^4x\, \mathcal{L}(x, u), \tag{11.272}$$

$$\begin{aligned} \mathcal{L}(x,u) = {} & \frac{1}{4} A^{a-}_m (\mathcal{D}^{++} A^{-m}_a - 4\,\partial^m F^+_a) - M^{a-} \mathcal{D}^{++} N^-_a \\ & - P^{a(-3)} \left(\partial^{++} F^+_a - \frac{1}{2}\, \partial_{a+} L^{+4} \right) \\ & + \frac{1}{2}\, \chi^{\alpha a(-2)} \mathcal{D}^{++} \Psi_{\alpha a} + \frac{1}{2}\, \bar{\chi}^{a(-2)}_{\dot\alpha} \mathcal{D}^{++} \bar{\Psi}^{\dot\alpha}_a \\ & - \frac{1}{8} \left(M^{a-}\, \bar{\Psi}^b \bar{\Psi}^c + N^{a-}\, \Psi^b \Psi^c - i A^{a-}_m\, \Psi^b \sigma^m \bar{\Psi}^c \right) \partial_{a+} \partial_{b+} \partial_{c+} L^{+4} \\ & + \frac{i}{4} \left(\Psi^a \partial\!\!\!/ \bar{\Psi}_a - \bar{\Psi}^a \partial\!\!\!/ \Psi_a \right) + \frac{1}{32} \left(\Psi^a \Psi^b \right) \left(\bar{\Psi}^c \bar{\Psi}^d \right) \partial_{a+} \partial_{b+} \partial_{c+} \partial_{d+} L^{+4}. \end{aligned} \tag{11.273}$$

Here $L^{+4}(F^+, u) \equiv L^{+4}(q^+, u)$ at $\theta = 0$; the covariant harmonic derivative $\mathcal{D}^{++}$ contains an $Sp(n)$ connection (cf. (11.193)), e.g.,

$$\mathcal{D}^{++} \Psi_{\alpha a} = \partial^{++} \Psi_{\alpha a} - \frac{1}{2}\, \partial_{a+} \partial_{b+} L^{+4}\, \Psi^b_\alpha. \tag{11.274}$$

The kinematical equations of motion following from (11.273) are

$$\partial^{++} F^+_a - \frac{1}{2}\, \partial_{a+} L^{+4} = 0, \tag{11.275}$$

$$\mathcal{D}^{++} \Psi^\alpha_a = \mathcal{D}^{++} \bar{\Psi}^{\dot\alpha}_a = 0, \tag{11.276}$$

$$\mathcal{D}^{++} M^-_a + \frac{1}{8} (\Psi^b \Psi^c)\, \partial_{a+} \partial_{b+} \partial_{c+} L^{+4} = 0, \tag{11.277}$$

$$\mathcal{D}^{++} N^-_a + \frac{1}{8} (\bar{\Psi}^b \bar{\Psi}^c)\, \partial_{a+} \partial_{b+} \partial_{c+} L^{+4} = 0, \tag{11.278}$$

$$\mathcal{D}^{++} A^{-m}_a - 2\,\partial^m F^+_a + \frac{i}{4} (\Psi^b \sigma^m \bar{\Psi}^c)\, \partial_{a+} \partial_{b+} \partial_{c+} L^{+4} = 0. \tag{11.279}$$

The remaining two equations (coming from the variation with respect to F^+ and Ψ) are dynamical and will not be used in what follows. The kinematical equations can be substituted back into the Lagrangian in (11.273), after which it is greatly simplified:

$$\begin{aligned}\mathcal{L}(x,u) &= \frac{1}{2}A_a^{-m}\partial_m F^{a+} + \frac{i}{4}(\Psi^a \partial\!\!\!/ \bar{\Psi}_a - \bar{\Psi}^a \partial\!\!\!/ \Psi_a) \\ &+ \frac{1}{16}\left(iA_m^{a-}\Psi^b\sigma^m\bar{\Psi}^c - N^{a-}\Psi^b\Psi^c - M^{a-}\bar{\Psi}^b\bar{\Psi}^c\right)\partial_{a+}\partial_{b+}\partial_{c+}L^{+4} \\ &+ \frac{1}{32}(\Psi^a\Psi^b)(\bar{\Psi}^c\bar{\Psi}^d)\partial_{a+}\partial_{b+}\partial_{c+}\partial_{d+}L^{+4}\,. \end{aligned} \quad (11.280)$$

The harmonic fields in (11.280) are still subject to the constraints (11.275)–(11.279). Among them eqs. (11.275) and (11.276) play the leading rôles. Let us first consider eq. (11.275):

$$\partial^{++}F^{a+} = \frac{1}{2}\Omega^{ab}\,\partial_{b+}L^{+4} \equiv H^{++a+}(F^+, u)\,. \quad (11.281)$$

The new notation H^{++a+} is suggested by the analogous geometric relation (11.192). As we shall see shortly, this analogy is very deep indeed. The general solution of eq. (11.281) depends on $4n$ integration constants $f^{ai}(x)$, so we may rewrite F^{a+} as follows:

$$F^{a+}(x,u) = f^{ai}(x)u_i^+ + v^{a+}(f(x),u)\,. \quad (11.282)$$

Then (11.281) becomes

$$\partial^{++}v^{a+}(f(x),u) = H^{++a+}[f^+ + v^+(f,u),u]\,. \quad (11.283)$$

Equation (11.283) is nothing but the bridge-defining equation (11.130), (11.198) with f^{ai} playing the rôle of the coordinates x^{ai} of the hyper-Kähler manifold. This observation helps us to establish a one-to-one correspondence with the geometric construction developed earlier in this chapter. Then the remaining equations (11.276)–(11.279) can easily be solved by making use of the differential geometry techniques summarized in Section 11.3.6. Thus, it is easy to check that the solution to (11.277)–(11.279) is

$$A_m^{a-} = -2E_{bi}^{a-}\,\partial_m f^{bi} + \frac{i}{2}\Psi^b\sigma_m\bar{\Psi}^c\omega_{abc}^-\,, \quad (11.284)$$

$$M_a^- = \frac{1}{4}\Psi^b\Psi^c\omega_{abc}^-\,, \quad (11.285)$$

$$N_a^- = \frac{1}{4}\bar{\Psi}^b\bar{\Psi}^c\omega_{abc}^-\,. \quad (11.286)$$

Here E_{bi}^{a-} and ω_{abc}^- obey the equations

$$\begin{aligned}\mathcal{D}^{++}E_{bi}^{a-} &= -\partial_{bi}F^{a+}\,, \\ \mathcal{D}^{++}\omega_{abc}^- + \frac{1}{2}\,\partial_{a+}\partial_{b+}\partial_{c+}L^{+4} &= 0\,.\end{aligned} \quad (11.287)$$

Comparing them with eqs. (11.210), (11.195) of Section 11.3.6, we observe that E^{a-}_{bi} and ω^{-}_{abc} are just the λ-frame vielbein and $Sp(n)$ connection defined by (11.208), (11.188). We stress that these quantities have been explicitly constructed in terms of the bridge v^{a+} (i.e., the solution of (11.283)).

The next step is to substitute (11.284), (11.286) into (11.280), after which $\mathcal{L}(x, u)$ becomes

$$\mathcal{L} = \frac{1}{2} g_{ai,bj} \partial^m f^{ai} \partial_m f^{bj} + \frac{i}{4} \Psi^a \sigma^m \overleftrightarrow{\nabla}_m \bar{\Psi}_a + \frac{1}{32} (\Psi^a \Psi^b)(\bar{\Psi}^c \bar{\Psi}^d) \left(\partial_{a+}\partial_{b+}\partial_{c+}\partial_{d+} L^{+4} - 3\partial_{g+}\partial_{(a+}\partial_{b+} L^{+4} \omega^{g-}{}_{cd)} \right) .. \tag{11.288}$$

Here

$$g_{ai,bj} = \Omega_{cd} \left(E^{c-}_{ai} E^{d+}_{bj} - E^{c+}_{ai} E^{d-}_{bj} \right), \qquad E^{a+}_{bi} = -\partial_{bi} F^{a+}, \tag{11.289}$$

$$\nabla_m \Psi_a = \partial_m \Psi_a + \partial_m f^{bi} E^{d+}_{bi} \omega^{-}_{d,a}{}^{c} \Psi_c, \qquad \mathcal{D}^{++} \nabla_m \Psi_a = 0. \tag{11.290}$$

The metric (11.289) is exactly the one we found in Section 11.3.6 (eq. (11.209)). There it was shown that this metric is u independent, $g_{ai,bj} = g_{ai,bj}(x)$. Thus the metric arising from the $N = 2$ superspace Lagrangian is identical to the metric derived from the hyper-Kähler potentials in the purely geometric approach.

It remains to clarify the geometric meaning of the objects appearing in the fermionic terms. The spinor equation (11.276) simply means that the spinor field $\Psi^a_\alpha(x, u)$ is covariantly u independent. The reparametrization group $\delta q^{a+} = \lambda^{a+}(q, u)$ induces the following transformations for the component fields F^{a+} and Ψ^a_α in the θ expansion (11.270)

$$\delta F^{a+} = \lambda^{a+}(F, u), \qquad \delta \Psi^a_\alpha = (\partial_{b+} \lambda^{a+}) \Psi^b_\alpha . \tag{11.291}$$

In the gauge $L^{a+} = q^{a+}$ the transformation law (11.291) is reduced to an analytic $Sp(n)$ rotation (see Section 11.3.5). The quantity $\nabla_m \Psi^a_\alpha$ is the covariant derivative with respect to these induced gauge λ-frame $Sp(n)$ rotations, due to the presence of the pull-back of the $Sp(n)$ connection ω^{a-}_{bc} in it.

Finally, the coefficient of the quartic fermionic term is proportional to the λ-frame curvature tensor $R^{+-}_{(abcd)}$ defined by eq. (11.89). Indeed, using the relations (11.287), it is straightforward to find

$$\begin{aligned} R^{+-}_{abcd} &= \mathcal{D}^{+}_{a} \omega^{-}_{bcd} = (\mathcal{D}^{++} E^{-\,fi}_{a}) \partial_{fi} \omega^{-}_{bcd} \\ &= -\frac{1}{2} \partial_{a+}\partial_{b+}\partial_{c+}\partial_{d+} L^{+4} + \frac{3}{2} \partial_{g+}\partial_{(a+}\partial_{b+} L^{+4} \omega^{g-}{}_{cd)} \\ &\quad + \mathcal{D}^{++} (E^{-\,fi}_{a} \partial_{fi} \omega^{-}_{bcd}) . \end{aligned} \tag{11.292}$$

The last term effectively does not contribute to (11.272) after integrating by parts under the S^2 integral and using the constraints (11.276).

The final step in obtaining the standard $N = 2$ sigma model action is to pass to harmonic-independent fields. In Section 11.3.2 we discussed the transition from the analytic (λ) tangent frame to the u independent (τ) frame. It is achieved with the help of the tangent space bridge $M^a_b(f, u)$, satisfying the equation (see (11.116)):

$$\partial^{++} M^a_b - \frac{1}{2}\, \partial_{c+} \partial_{b+} L^{+4}\, M^{a\,c} = 0\,. \tag{11.293}$$

Then, introducing the τ-covariant spinor

$$\psi^a_\alpha = M^a_b\, \Psi^b_\alpha, \tag{11.294}$$

we can rewrite (11.276) as follows:

$$\partial^{++} \psi^a_\alpha = 0 \quad \Rightarrow \quad \psi^a_\alpha = \psi^a_\alpha(x)\,. \tag{11.295}$$

So, assuming that the two bridge equations (11.283) and (11.293) have been solved, we insert (11.294) and (11.295) into (11.288). The result is the final form of the sigma model action in terms of the physical fields $f(x)$, $\psi(x)$:

$$S = \int d^4x\, \mathcal{L}(x)\,, \tag{11.296}$$

$$\mathcal{L}(x) = \frac{1}{2} g_{ai,bj}\, \partial^m f^{ai} \partial_m f^{bj} + \frac{i}{4} \psi^a \sigma^m \overset{\leftrightarrow}{\nabla}_m \bar{\psi}_a - \frac{1}{16} (\psi^a \psi^b)(\bar{\psi}^c \bar{\psi}^d) R_{(abcd)}\,, \tag{11.297}$$

where

$$\nabla_m \psi_a = \partial_m \psi_a + \partial_m f^{bi}\, \omega_{bi,a}{}^{c}\, \psi_c\,. \tag{11.298}$$

The covariant derivative ∇_m (11.298) involves the standard τ-frame $Sp(n)$ connection (see (11.82)). The tensor $R_{(abcd)}$ is the non-vanishing part of the Riemann tensor for a hyper-Kähler manifold.

Thus we see that the sigma model Lagrangian (11.297) has exactly the form prescribed in [A2, A3]. The essential difference is that the metric in (11.297) has been *constructed* from the unconstrained potentials, whereas in [A2, A3] one uses a *given* hyper-Kähler metric. The term *constructed* in our context means that the basic bridge-defining equation (11.283) has been solved.* This is, in general, a non-trivial task, as discussed in the previous sections and in Chapter 5.

* Note that one does not really need to solve the equation (11.293) for the $Sp(n)$ tangent space bridge. Indeed, the bridge M does not enter the expression (11.289) for the metric. Moreover, the τ-frame vielbeins, from which the $Sp(n)$ connection $\omega_{ai,bc}$ and the Riemann tensor R_{abcd} are constructed, can be derived as a square root of the metric.

12

$N = 3$ supersymmetric Yang–Mills theory

The method of harmonic superspace provides the solution of one of the outstanding problems in supersymmetry. The off-shell formulation of the $N = 3$ SYM theory is only possible within this framework. The reason is that one cannot have this theory off shell with a finite number of auxiliary fields. The necessary infinite towers of auxiliary fields naturally arise in a harmonic expansion, just as in the case of the $N = 2$ complex hypermultiplets. Besides, $N = 3$ SYM theory has a number of very unusual features which are explained in this chapter.

12.1 $N = 3$ SYM on-shell constraints

The starting point in the study of $N = 3$ SYM is the differential geometry constraints in ordinary superspace which define the theory on shell. At first sight they look very similar to the $N = 2$ ones (see (7.57)–(7.59)):

$$\begin{aligned} \{\mathcal{D}^i_\alpha, \mathcal{D}^j_\beta\} &= \epsilon_{\alpha\beta}\overline{W}^{ij}\,, \\ \{\bar{\mathcal{D}}_{\dot\alpha i}, \bar{\mathcal{D}}_{\dot\beta j}\} &= \epsilon_{\dot\alpha\dot\beta}W_{ij}\,, \\ \{\mathcal{D}^i_\alpha, \bar{\mathcal{D}}_{\dot\beta j}\} &= -2i\delta^i_j\mathcal{D}_{\alpha\dot\beta}\,. \end{aligned} \tag{12.1}$$

The superfield $\overline{W}^{ij} = -\overline{W}^{ji}$ is the only independent curvature tensor in the theory. Unlike the $N = 2$ case where the constraints are kinematical, here they imply the equations of motion [S11, W14]. To see this it is sufficient to look at the linearized (Abelian) case. From (12.1) one can derive the following Bianchi identities:

$$\begin{aligned} D^i_\alpha\overline{W}^{jk} &= -D^j_\alpha\overline{W}^{ik}\,, \\ \bar{D}_{\dot\alpha i}\overline{W}^{jk} &= \frac{1}{2}(\delta^j_i\bar{D}_{\dot\alpha l}\overline{W}^{lk} - \delta^k_i\bar{D}_{\dot\alpha l}\overline{W}^{lj})\,, \end{aligned} \tag{12.2}$$

and their consequences:

$$D^i_\alpha \bar{D}_{\dot\beta k} \overline{W}^{kj} = -4i\partial_{\alpha\dot\beta}\overline{W}^{ij}\,, \qquad \bar{D}^{\dot\alpha}_i \bar{D}_{\dot\alpha j}\overline{W}^{ij} = 0\,. \tag{12.3}$$

Using all this one finds

$$\begin{aligned} -2i(\partial\bar{D}_k)_\alpha \overline{W}^{ki} &= \{D^i_\alpha, \bar{D}_{\dot\beta j}\}\bar{D}^{\dot\beta}_k \overline{W}^{kj} = -\bar{D}^{\dot\beta}_j D^i_\alpha \bar{D}_{\dot\beta k}\overline{W}^{kj} \\ &= -4i(\partial\bar{D}_k)_\alpha \overline{W}^{ki}\,, \end{aligned} \tag{12.4}$$

which means that the physical spinor field $\bar\lambda^{\dot\alpha i} = (\bar{D}^{\dot\alpha}_k \overline{W}^{ki})_{\theta=0}$ satisfies the equation of motion $\partial_{\alpha\dot\alpha}\bar\lambda^{\dot\alpha i} = 0$.

So, we have seen that the $N = 3$ constraints put the theory on shell. Therefore, we do not aim at solving these constraints. Instead, we shall try to rewrite them in an equivalent form as constraints involving covariant harmonic derivatives. These new constraints, which will still imply the equations of motion, will be obtained as variational equations from an off-shell action. This action will have the form of a Chern–Simons term for the harmonic connections.

12.2 $N = 3$ harmonic variables and interpretation of the $N = 3$ SYM constraints

From our experience with $N = 2$ supersymmetric theories we expect that the harmonic variables appropriate for the $N = 3$ case should be related to the coset space $SU(3)/H$, where $SU(3)$ is the automorphism group of the $N = 3$ SYM constraints and H is one of its subgroups. The right choice of coset space is related to the possibility of interpreting the constraints (12.1) as integrability conditions. One such interpretation was given by Witten [W14]. It is a twistor interpretation (i.e., one makes use of harmonics of the Lorentz group $SL(2,\mathbb{C})$) and is aimed at solving the equations of motion. There the space-time dependence of the corresponding analytic superfields is restricted, which is not appropriate for an off-shell unconstrained formulation.

Another interpretation of the constraints (12.1) was proposed by Rosly [R4]. The idea is to contract the $SU(3)$ indices with two non-vanishing commuting $SU(3)$ triplets ξ_i, η^j:

$$\mathcal{D}_\alpha = \xi_i \mathcal{D}^i_\alpha\,, \qquad \bar{\mathcal{D}}_{\dot\alpha} = \eta^i \bar{\mathcal{D}}_{\dot\alpha i}\,. \tag{12.5}$$

Then the original constraints (12.1) become equivalent to

$$\{\mathcal{D}_\alpha, \mathcal{D}_\beta\} = \{\bar{\mathcal{D}}_{\dot\alpha}, \bar{\mathcal{D}}_{\dot\beta}\} = \{\mathcal{D}_\alpha, \bar{\mathcal{D}}_{\dot\beta}\} = 0\,, \tag{12.6}$$

provided that the following condition holds:

$$\xi_i \eta^i = 0\,. \tag{12.7}$$

Equations (12.6) do indeed have the form of integrability conditions. They mean that the following covariant Grassmann-analytic superfields:

$$\mathcal{D}_\alpha \phi = \bar{\mathcal{D}}_{\dot\alpha} \phi = 0 \tag{12.8}$$

can be consistently defined in the $N = 3$ SYM background.

It is not hard to see that this interpretation in fact uses harmonics on $SU(3)/U(1) \times U(1)$. Indeed, the $SU(3)$ triplets ξ_i, η^i are defined modulo multiplication by non-vanishing complex numbers:

$$\xi_i \quad \to \quad \lambda \xi_i\,, \qquad \eta^i \quad \to \quad \rho \eta^i\,. \tag{12.9}$$

This allows one to fix the norm:

$$\xi_i \bar{\xi}^i = \eta^i \bar{\eta}_i = 1\,, \tag{12.10}$$

after which one is left with the freedom of $U(1) \times U(1)$ transformations:

$$\xi_i \quad \to \quad e^{i\alpha} \xi_i\,, \qquad \eta^i \quad \to \quad e^{i\beta} \eta^i\,. \tag{12.11}$$

From these objects one can construct $SU(3)/U(1) \times U(1)$ harmonics:

$$u_i^{(a,b)} = \left(\xi_i\,,\ \bar{\eta}_i\,,\ \epsilon_{ijk}\,\bar{\xi}^j \eta^k\right), \tag{12.12}$$

which form an $SU(3)$ matrix and are defined modulo the $U(1) \times U(1)$ transformations (12.11). The notation $u^{(a,b)}$ indicates the $U(1) \times U(1)$ charges of the harmonic variables (corresponding to the choice of the standard Cartan–Weyl basis for $SU(3)$):

$$u_i^{(1,0)} = \xi_i\,, \qquad u_i^{(0,-1)} = \bar{\eta}_i\,, \qquad u_i^{(-1,1)} = \epsilon_{ijk}\,\bar{\xi}^j \eta^k\,. \tag{12.13}$$

The complex conjugate harmonics have opposite charges:

$$\overline{u_i^{(a,b)}} = u^{i(-a,-b)}\,. \tag{12.14}$$

It is sometimes convenient to represent the $U(1) \times U(1)$ indices of the harmonics by a single index $I = 1, 2, 3$:

$$\begin{aligned} \left(u_i^{(1,0)}\,,\ u_i^{(0,-1)}\,,\ u_i^{(-1,1)}\right) &\equiv u_i^I = \left(u_i^1\,,\ u_i^2\,,\ u_i^3\right), \\ \left(u^{i(-1,0)}\,,\ u^{i(0,1)}\,,\ u^{i(1,-1)}\right) &\equiv u_I^i = \left(u_1^i\,,\ u_2^i\,,\ u_3^i\right). \end{aligned} \tag{12.15}$$

The fact that u_i^I, u_I^i are $SU(3)$ matrices imposes a number of algebraic restrictions on the harmonic variables (analogs of the $SU(2)$ condition $u^{+i}u_i^- = 1$ in the case $N = 2$):

$$\begin{aligned} u_i^I u_J^i &= \delta_J^I\,, \qquad u_i^I u_I^j = \delta_i^j\,, \\ \det u &= \epsilon^{ijk} u_i^1 u_j^2 u_k^3 = 1 \quad \to \quad u_1^i = \epsilon^{ijk} u_j^2 u_k^3\,, \qquad \text{etc.} \end{aligned} \tag{12.16}$$

These conditions, together with the arbitrariness of the two $U(1)$ phases, imply that the nine complex degrees of freedom in u_i^I are reduced to three independent ones (or six real). This fact will turn out to be crucial for the construction of the Chern–Simons action in Section 12.6.

The integrability and analyticity conditions (12.6) and (12.8) can be rewritten with the $U(1) \times U(1)$ charges explicitly indicated:

$$\left\{\mathcal{D}_\alpha^{(1,0)}, \mathcal{D}_\beta^{(1,0)}\right\} = \left\{\mathcal{D}_\alpha^{(1,0)}, \bar{\mathcal{D}}_{\dot\beta}^{(0,1)}\right\} = \left\{\bar{\mathcal{D}}_{\dot\alpha}^{(0,1)}, \bar{\mathcal{D}}_{\dot\beta}^{(0,1)}\right\} = 0\,, \tag{12.17}$$

$$\mathcal{D}_\alpha^{(1,0)}\phi = \bar{\mathcal{D}}_{\dot\alpha}^{(0,1)}\phi = 0\,, \tag{12.18}$$

$$\mathcal{D}_\alpha^{(1,0)} = u_i^{(1,0)}\mathcal{D}_\alpha^i\,, \qquad \bar{\mathcal{D}}_{\dot\alpha}^{(0,1)} = u^{i(0,1)}\bar{\mathcal{D}}_{\dot\alpha i}\,. \tag{12.19}$$

It is also instructive to rewrite these relations in the notation (12.15) which sometimes is technically more convenient:

$$\left\{\mathcal{D}_\alpha^1, \mathcal{D}_\beta^1\right\} = \left\{\mathcal{D}_\alpha^1, \bar{\mathcal{D}}_{\dot\beta 2}\right\} = \left\{\bar{\mathcal{D}}_{\dot\alpha 2}, \bar{\mathcal{D}}_{\dot\beta 2}\right\} = 0\,, \tag{12.20}$$

$$\mathcal{D}_\alpha^1\phi = \bar{\mathcal{D}}_{\dot\alpha 2}\phi = 0\,, \tag{12.21}$$

$$\mathcal{D}_\alpha^1 = u_i^1\mathcal{D}_\alpha^i, \qquad \bar{\mathcal{D}}_{\dot\alpha 2} = u_2^i\bar{\mathcal{D}}_{\dot\alpha i}\,. \tag{12.22}$$

We emphasize that the existence of such an interpretation of the $N = 3$ SYM constraints does not yet mean that a solution to the main problem (to find an off-shell formulation of the theory) has been found. This can be achieved through a deeper understanding of the crucial rôle played by the harmonic variables and of the properties of $N = 3$ harmonic superspace.

12.3 Elements of the harmonic analysis on $SU(3)/U(1)\times U(1)$

We begin by describing the harmonic space $SU(3)/U(1) \times U(1)$ in more detail. Firstly, let us introduce a convenient labeling for the harmonic derivatives on this space. The possible choices are either the standard Cartan–Weyl basis for $SU(3)$ corresponding to the manifest $U(1) \times U(1)$ labeling of the harmonics as in (12.13), (12.14), or the 'tensor' basis corresponding to the notation u_i^I, u_K^i. Though in the subsequent presentation we shall basically keep to the first option, one can always pass to the tensor notation using the explicit correspondence (12.15) and the analogous one for the harmonic derivatives given below. In this section we use both conventions in parallel, leaving it to the reader to make the choice.

One can divide the harmonic derivatives (as one does with the $SU(3)$ generators) into a set of three complex derivatives:

$$D^{(2,-1)} = D_3^1\,, \qquad D^{(-1,2)} = D_2^3\,, \qquad D^{(1,1)} = D_2^1\,; \tag{12.23}$$

their Hermitian conjugates $D^{(-a,-b)} = D^{(a,b)\dagger}$:

$$D^{(-2,1)} = D^3_1 , \qquad D^{(1,-2)} = D^2_3 , \qquad D^{(-1,-1)} = D^2_1 ; \tag{12.24}$$

and the two $U(1)$ charges $D^0_{(1)}$, $D^0_{(2)}$. In the $U(1) \times U(1)$ notation the commutators of the latter with the harmonic derivatives read

$$\left[D^0_{(1)}, D^{(a,b)}\right] = aD^{(a,b)} , \qquad \left[D^0_{(2)}, D^{(a,b)}\right] = bD^{(a,b)} . \tag{12.25}$$

The rest of the algebra has the following form. One of the derivatives in the set (12.23) is decomposable, i.e., it is the commutator of the other two simple derivatives:

$$D^{(1,1)} = \left[D^{(2,-1)},\ D^{(-1,2)}\right] \qquad \text{or} \qquad D^1_2 = \left[D^1_3,\ D^3_2\right], \tag{12.26}$$

the remaining commutators within the set (12.23) vanish. The algebra of the derivatives (12.24) is obtained by Hermitian conjugation. Finally, the two $U(1)$ charges appear as commutators of the simple derivatives with their conjugates:

$$\begin{aligned} \left[D^{(-2,1)},\ D^{(2,-1)}\right] &= \left[D^3_1,\ D^1_3\right] = D^0_{(1)} , \\ \left[D^{(1,-2)},\ D^{(-1,2)}\right] &= \left[D^2_3,\ D^3_2\right] = D^0_{(2)} . \end{aligned} \tag{12.27}$$

The simple harmonic derivatives $D^{(2,-1)} = D^1_3$, $D^{(-1,2)} = D^3_2$ are realized in terms of differential operators as follows:

$$\begin{aligned} D^{(2,-1)} &= D^1_3 = u^1_i \frac{\partial}{\partial u^3_i} - u^i_3 \frac{\partial}{\partial u^i_1} \equiv \partial^{(2,-1)} \equiv \partial^1_3 , \\ D^{(-1,2)} &= D^3_2 = u^3_i \frac{\partial}{\partial u^2_i} - u^i_2 \frac{\partial}{\partial u^i_3} \equiv \partial^{(-1,2)} \equiv \partial^3_2 . \end{aligned} \tag{12.28}$$

The remaining ones are then given by conjugation and by the commutators (12.26), (12.27). The two derivatives $D^0_{(1),(2)}$ just count the charges of $u^{(a,b)}$. Their explicit form, e.g., in the tensor notation, is as follows:

$$\begin{aligned} D^0_{(1)} &= u^1_i \frac{\partial}{\partial u^1_i} - u^3_i \frac{\partial}{\partial u^3_i} - u^i_1 \frac{\partial}{\partial u^i_1} + u^i_3 \frac{\partial}{\partial u^i_3} , \\ D^0_{(2)} &= u^3_i \frac{\partial}{\partial u^3_i} - u^2_i \frac{\partial}{\partial u^2_i} - u^i_3 \frac{\partial}{\partial u^i_3} + u^i_2 \frac{\partial}{\partial u^i_2} . \end{aligned} \tag{12.29}$$

Besides the usual complex conjugation (12.14) there exists a second operation $\widetilde{\ }$ (the analog of the combined complex conjugation and Weyl reflection (antipodal map) on S^2 in the case $N = 2$, (3.102)), which will help us to define a real analytic superspace. In the $N = 3$ case there are several ways to choose it because of the existence of several different Weyl reflections on $SU(3)$. The most convenient rule is

$$\begin{array}{llllll} u_i^{(1,0)} & \stackrel{\sim}{\leftrightarrow} u^{i(0,1)}, & u_i^{(0,-1)} & \stackrel{\sim}{\leftrightarrow} u^{i(-1,0)}, & u_i^{(-1,1)} & \stackrel{\sim}{\leftrightarrow} -u^{i(1,-1)} \quad \text{or} \\ u_i^1 & \stackrel{\sim}{\leftrightarrow} u_2^i, & u_i^2 & \stackrel{\sim}{\leftrightarrow} u_1^i, & u_i^3 & \stackrel{\sim}{\leftrightarrow} -u_3^i . \end{array} \tag{12.30}$$

The reader can easily check the consistency of (12.30) and (12.16). Correspondingly, the harmonic derivatives have the following properties under the $\widetilde{\ }$ conjugation:

$$\widetilde{D_3^1} = D_2^3, \qquad \widetilde{D_2^3} = D_3^1, \qquad \widetilde{D_2^1} = -D_2^1,$$

$$\widetilde{D_{(1)}^0} = D_{(2)}^0, \qquad \widetilde{D_{(2)}^0} = D_{(1)}^0 \tag{12.31}$$

(in the present case the operation $\widetilde{\ }$ has the standard complex conjugation property $\widetilde{\widetilde{A}} = A$, so it is sufficient to know these rules). One observes that the set of complex derivatives (12.23) is closed under the $\widetilde{\ }$ conjugation whereas the $U(1)$ charges are interchanged.

The final comment concerns the functions on the coset $SU(3)/U(1) \times U(1)$. They carry a pair of $U(1)$ charges, $f^{(a,b)}(u)$. The harmonic expansion of such functions goes over all the $SU(3)$ irreducible products of harmonics with overall $U(1)$ charges (a, b). These products are totally symmetric in the upper and lower case indices and are traceless. For instance,

$$\begin{aligned} f^{(1,1)}(u) &= f_i^j u_j^{(1,0)} u^{i(0,1)} + g_{(ijk)} u^{i(1,-1)} u^{j(0,1)} u^{k(0,1)} \\ &\quad + h^{(ijk)} u_i^{(1,0)} u_j^{(1,0)} u_k^{(-1,1)} + \ell_{kl}^{ij} u_i^{(1,0)} u^{k(0,1)} u_j^{(-1,1)} u^{l(1,-1)} + \cdots \end{aligned} \tag{12.32}$$

where all the coefficients are symmetric and traceless.

One can impose a reality condition on functions with equal charges (see (12.30)):

$$\widetilde{f^{(a,a)}} = f^{(a,a)} . \tag{12.33}$$

In the particular example (12.32) this implies

$$\overline{f_i^j} = f_j^i , \qquad \overline{g_{ijk}} = -h^{ijk} , \qquad \overline{\ell_{kl}^{ij}} = \ell_{ij}^{kl} , \qquad \text{etc.} \tag{12.34}$$

12.4 $N = 3$ Grassmann analyticity

In the next section we are going to solve the covariant analyticity constraints (12.18). For this purpose we need an appropriate analytic subspace which involves only part of the Grassmann variables. The harmonization of the $N = 3$ superspace $\mathbb{R}^{4|12} = \{x^{\alpha\dot\alpha}, \theta_i^\alpha, \bar\theta^{\dot\alpha i}\}$ enables one to find such a subspace. This can be done following the general Cartan's scheme of Chapter 3. Here we only list the results. Adding six real (or three complex) harmonic dimensions to $\mathbb{R}^{4|12}$ and replacing θ_i by $\theta^{(a,b)} = \theta_i u^{i(a,b)}$ one obtains the central basis of $N = 3$ harmonic superspace $\mathbb{HR}^{4+6|12}$:

$$\underline{\text{CB:}} \qquad \left\{ x^{\alpha\dot\alpha},\ \theta^{(-1,0)}_\alpha,\ \theta^{(0,1)}_\alpha,\ \theta^{(1,-1)}_\alpha,\ \bar\theta^{(1,0)}_{\dot\alpha},\ \bar\theta^{(0,-1)}_{\dot\alpha},\ \bar\theta^{(-1,1)}_{\dot\alpha},\ u \right\}. \quad (12.35)$$

Then one can define an analytic basis:

$$\begin{aligned} \underline{\text{AB:}} \qquad x^{\alpha\dot\alpha}_A &= x^{\alpha\dot\alpha} + 2i\left[\theta^{(0,1)\alpha}\bar\theta^{(0,-1)\dot\alpha} - \theta^{(-1,0)\alpha}\bar\theta^{(1,0)\dot\alpha}\right] \\ \theta^{(a,b)}_A &= \theta^{(a,b)}, \qquad u_A = u\,, \end{aligned} \quad (12.36)$$

in which there exists an invariant Grassmann analytic subspace:

$$\begin{aligned} \mathbb{HA}^{4+6|8} &= \left\{ x^{\alpha\dot\alpha}_A,\ \theta^{(1,-1)}_\alpha,\ \theta^{(0,1)}_\alpha,\ \bar\theta^{(1,0)}_{\dot\alpha},\ \bar\theta^{(-1,1)}_{\dot\alpha},\ u \right\} \\ &\equiv \{\zeta, u\}. \end{aligned} \quad (12.37)$$

It is closed under the full $N = 3$ super-Poincaré algebra:

$$\begin{aligned} \delta_Q x^{\alpha\dot\alpha}_A &= 4i\theta^{(0,1)\alpha}\bar\epsilon^{\dot\alpha i}u^{(0,-1)}_i + 2i\theta^{(1,-1)\alpha}\bar\epsilon^{\dot\alpha i}u^{(-1,1)}_i \\ &\quad -4i\epsilon^\alpha_i\bar\theta^{(1,0)\dot\alpha}u^{i(-1,0)} - 2i\epsilon^\alpha_i\bar\theta^{(-1,1)\dot\alpha}u^{i(1,-1)}\,, \\ \delta_Q\theta^{(a,b)}_\alpha &= \epsilon_{\alpha i}u^{i(a,b)}\,, \qquad \delta_Q\bar\theta^{(a,b)}_{\dot\alpha} = \bar\epsilon^i_{\dot\alpha}u^{(a,b)}_i\,, \qquad \delta_Q u = 0\,. \end{aligned} \quad (12.38)$$

Note that $\mathbb{HA}^{4+6|8}$ contains only 2/3 of the original 12 Grassmann variables, i.e., eight (and not 1/2 of them, as was the case in $N = 2$). Further, the analytic superspace is real under the $\widetilde{\ }$ conjugation:

$$\widetilde{x^{\alpha\dot\alpha}_A} = x^{\alpha\dot\alpha}_A, \qquad \widetilde{\theta^{(1,-1)}_\alpha} = -\bar\theta^{(-1,1)}_{\dot\alpha}, \qquad \widetilde{\theta^{(0,1)}_\alpha} = \bar\theta^{(1,0)}_{\dot\alpha}\,. \quad (12.39)$$

The superfields defined on $\mathbb{HA}^{4+6|8}$ satisfy the following Grassmann analyticity conditions:

$$\begin{aligned} D^{(1,0)}_\alpha\Phi &\equiv u^{(1,0)}_i D^i_\alpha\Phi = 0\,, \\ \bar D^{(0,1)}_{\dot\alpha}\Phi &\equiv u^{i(0,1)}\bar D_{\dot\alpha i}\Phi = 0 \end{aligned} \quad (12.40)$$

which imply

$$\Phi = \Phi(\zeta, u)\,. \quad (12.41)$$

Note also that the spinor derivatives (12.40) defining analyticity commute with the harmonic derivatives (12.23):

$$\begin{aligned} \left[D^{(2,-1)}, D^{(1,0)}_\alpha\right] &= \left[D^{(-1,2)}, D^{(1,0)}_\alpha\right] = \left[D^{(1,1)}, D^{(1,0)}_\alpha\right] = 0\,, \\ \left[D^{(2,-1)}, \bar D^{(0,1)}_{\dot\alpha}\right] &= \left[D^{(-1,2)}, \bar D^{(0,1)}_{\dot\alpha}\right] = \left[D^{(1,1)}, \bar D^{(0,1)}_{\dot\alpha}\right] = 0\,, \end{aligned} \quad (12.42)$$

i.e., they form a CR structure (in the terminology of [S4, R4, R5, R6]).

In the analytic basis (12.36) the spinor derivatives $D_\alpha^{(1,0)}$, $\bar{D}_{\dot\alpha}^{(0,1)}$ become just partial derivatives:

$$D_\alpha^{(1,0)} = \frac{\partial}{\partial\theta_A^{\alpha(-1,0)}}\,, \qquad \bar{D}_{\dot\alpha}^{(0,1)} = -\frac{\partial}{\partial\bar\theta_A^{\dot\alpha(0,-1)}}\,, \tag{12.43}$$

whereas the harmonic derivatives acquire vielbein terms:

$$\begin{aligned} D^{(2,-1)} &= \partial^{(2,-1)} + 2i\theta^{(1,-1)\alpha}\bar\theta^{(1,0)\dot\alpha}\frac{\partial}{\partial x_A^{\alpha\dot\alpha}} + \bar\theta^{(1,0)\dot\alpha}\frac{\partial}{\partial\bar\theta^{(-1,1)\dot\alpha}}\,, \\ D^{(-1,2)} &= \partial^{(-1,2)} + 2i\theta^{(0,1)\alpha}\bar\theta^{(-1,1)\dot\alpha}\frac{\partial}{\partial x_A^{\alpha\dot\alpha}} - \theta^{(0,1)\alpha}\frac{\partial}{\partial\theta^{(1,-1)\alpha}}\,, \\ D^{(1,1)} &= \partial^{(1,1)} + 4i\theta^{(0,1)\alpha}\bar\theta^{(1,0)\dot\alpha}\frac{\partial}{\partial x_A^{\alpha\dot\alpha}} \end{aligned} \tag{12.44}$$

(here we have not included terms vanishing on analytic superfields).

Note an interesting new feature of the $N = 3$ analytic harmonic superspace (12.37) as compared to its $N = 2$ prototype. In it one can single out a complex analytic subspace of Grassmann dimension 6, such that it is still closed under $N = 3$ supersymmetry [G12]:

$$\begin{aligned} \mathbb{HA}_+^{4+6|6} &= \left\{x_{A+}^{\alpha\dot\alpha} = x_A^{\alpha\dot\alpha} + 2i\theta^{(1,-1)\alpha}\bar\theta^{(-1,1)\dot\alpha},\ \theta_\alpha^{(1,-1)},\ \theta_\alpha^{(0,1)},\ \bar\theta_{\dot\alpha}^{(1,0)},\ u\right\}, \\ \delta_Q x_{A+}^{\alpha\dot\alpha} &= 4i\left(\theta^{(0,1)\alpha}u_i^{(0,-1)} + \theta^{(1,-1)\alpha}u_i^{(-1,1)}\right)\bar\epsilon^{\dot\alpha i} - 4i\epsilon_i^\alpha\bar\theta^{(1,0)\dot\alpha}u^{i(-1,0)}\,. \end{aligned} \tag{12.45}$$

Under the $\tilde{}$ conjugation this subspace goes into another analytic subspace:

$$\mathbb{HA}_-^{4+6|6} = \left\{x_{A-}^{\alpha\dot\alpha} = x_A^{\alpha\dot\alpha} - 2i\theta^{(1,-1)\alpha}\bar\theta^{(-1,1)\dot\alpha},\ \bar\theta_{\dot\alpha}^{(-1,1)},\ \bar\theta_{\dot\alpha}^{(1,0)},\ \theta_\alpha^{(0,1)},\ u\right\}. \tag{12.46}$$

These two conjugate subspaces are analogous to the left- and right-handed chiral subspaces, the full analytic subspace (12.37) being the analog of the standard real superspace. As shown in [G12], the superspace (12.37) can be treated as a real (with respect to the $\tilde{}$ conjugation) hypersurface in the complex one (12.45) with $\tilde{}$ complexified harmonics, quite similarly to the view on the real superspace as a hypersurface in the complex chiral superspace (recall Section 3.2). Actually, (12.45) or (12.46) can be understood as the analytic superspaces corresponding to a different choice of the harmonic coset, namely $SU(3)/U(2)$, with the Grassmann coordinates of the same chirality forming $SU(2)$ doublets and the remaining one being a singlet. At present, the significance of these complex superspaces is not fully understood. An important manifestation of this 'short' Grassmann analyticity in $N = 3$ SYM theory is mentioned in Section 12.7.

12.5 From covariant to manifest analyticity: An equivalent interpretation of the $N = 3$ SYM constraints

The harmonic superspace formalism developed above allowed us to interpret the $N = 3$ SYM constraints (12.1) as integrability conditions (12.17) for the existence of analytic superfields (12.18), (12.41) in an $N = 3$ SYM background. As in the $N = 2$ case, in order to make sure that $\mathcal{D}^{(1,0)}_\alpha$ and $\bar{\mathcal{D}}^{(0,1)}_{\dot\alpha}$ depend linearly on the harmonic variables, we have to impose further constraints which are the covariant version of (12.42):

$$\begin{aligned}
\left[\mathcal{D}^{(2,-1)}, \mathcal{D}^{(1,0)}_\alpha\right] &= \left[\mathcal{D}^{(-1,2)}, \mathcal{D}^{(1,0)}_\alpha\right] = \left[\mathcal{D}^{(1,1)}, \mathcal{D}^{(1,0)}_\alpha\right] = 0\,, \\
\left[\mathcal{D}^{(2,-1)}, \bar{\mathcal{D}}^{(0,1)}_{\dot\alpha}\right] &= \left[\mathcal{D}^{(-1,2)}, \bar{\mathcal{D}}^{(0,1)}_{\dot\alpha}\right] = \left[\mathcal{D}^{(1,1)}, \bar{\mathcal{D}}^{(0,1)}_{\dot\alpha}\right] = 0\,. \qquad (12.47)
\end{aligned}$$

This is obviously a necessary condition, it is also sufficient as one can verify by studying the harmonic expansion (see (12.32)). The basic reason is that the two derivatives $D^{(2,-1)}$ and $D^{(-1,2)}$ together with the decomposable one $D^{(1,1)}$ (12.26) correspond to the three complex dimensions of our harmonic space, so the spinor gauge connections satisfying (12.47) can depend on the harmonics only trivially. Of course, the covariant harmonic derivatives in (12.47) still coincide with the rigid ones, since the original gauge group has u-independent parameters $\tau(x, \theta, \bar\theta)$. For the same reason they satisfy the flat algebra

$$\left[\mathcal{D}^{(2,-1)}, \mathcal{D}^{(-1,2)}\right] = \mathcal{D}^{(1,1)}\,, \qquad \left[\mathcal{D}^{(2,-1)}, \mathcal{D}^{(1,1)}\right] = \left[\mathcal{D}^{(-1,2)}, \mathcal{D}^{(1,1)}\right] = 0\,. \qquad (12.48)$$

Following the strategy developed in the case $N = 2$, we can solve the constraints (12.17) by introducing a gauge bridge $b(x, \theta, \bar\theta, u)$ such that the covariant derivatives in (12.17) take a 'pure gauge' form

$$\mathcal{D}^{(1,0)}_\alpha = e^{-ib} D^{(1,0)}_\alpha e^{ib}\,, \qquad \bar{\mathcal{D}}^{(0,1)}_{\dot\alpha} = e^{-ib} \bar{D}^{(0,1)}_{\dot\alpha} e^{ib}\,. \qquad (12.49)$$

The bridge can be chosen real in the sense $\tilde{b} = b$, which is in agreement with the property $\widetilde{\mathcal{D}^{(1,0)}_\alpha} = \bar{\mathcal{D}}^{(0,1)}_{\dot\alpha}$. As usual, the solution (12.49) has some gauge freedom:

$$e^{ib'} = e^{i\lambda} e^{ib} e^{-i\tau}\,, \qquad D^{(1,0)}_\alpha \lambda = \bar{D}^{(0,1)}_{\dot\alpha} \lambda = 0 \qquad (12.50)$$

with analytic parameters λ and u-independent parameters τ.

Note that if a superfield $\phi(x, \theta, u)$ satisfies the covariant analyticity constraint (12.18), after a gauge-frame rotation with the bridge $b(x, \theta, \bar\theta, u)$:

$$\Phi = e^{ib} \phi\,, \qquad (12.51)$$

it becomes manifestly analytic, $\Phi = \Phi(\zeta, u)$ (12.41). On the other hand, in the old frame one could define u-independent superfields (since the gauge parameter τ is itself u-independent). After the gauge-frame rotation this manifest u independence becomes covariant. In other words, in the new gauge frame the connections of the spinor derivatives (12.49) disappear; instead, there appear gauge connections in the harmonic derivatives in (12.48):

$$\mathcal{D}^{(a,c)} \equiv D^{(a,c)} + iV^{(a,c)} = e^{ib} D^{(a,c)} e^{-ib} \, ; \tag{12.52}$$

$$V^{(a,c)} = -ie^{ib} D^{(a,c)} e^{-ib} \, . \tag{12.53}$$

Since the bridge b is real, the newly defined connections have the following properties:

$$\widetilde{V^{(2,-1)}} = V^{(-1,2)} \, , \qquad \widetilde{V^{(1,1)}} = -V^{(1,1)} \, , \tag{12.54}$$

and similarly for the complex conjugate connections. They have the usual gauge transformation law:

$$V^{(a,b)\prime} = e^{i\lambda} \left(V^{(a,b)} - iD^{(a,b)} \right) e^{-i\lambda} \, . \tag{12.55}$$

The connections (12.54) satisfy constraints following from (12.47) and (12.48). The constraints (12.47) simply mean that the connections (12.54) are analytic:

$$\begin{aligned} D_\alpha^{(1,0)} V^{(2,-1)} = D_\alpha^{(1,0)} V^{(-1,2)} = D_\alpha^{(1,0)} V^{(1,1)} &= 0 \, , \\ \bar{D}_{\dot\alpha}^{(0,1)} V^{(2,-1)} = \bar{D}_{\dot\alpha}^{(0,1)} V^{(-1,2)} = \bar{D}_{\dot\alpha}^{(0,1)} V^{(1,1)} &= 0 \, . \end{aligned} \tag{12.56}$$

These constraints can be explicitly solved in the analytic basis (12.36) yielding the form (12.41) for (12.54).

The second set of constraints (12.48) have a completely new nature. Indeed, in the case $N = 2$ we had only one harmonic covariant derivative $\mathcal{D}^{++}$ in the CR structure and, correspondingly, only one harmonic connection V^{++} which was analytic but did not satisfy any further constraints. Now we have three connections, and according to (12.48) they have to obey the *zero harmonic curvature conditions* (see (12.60) below):

$$\begin{aligned} D^{(2,-1)} V^{(-1,2)} - D^{(-1,2)} V^{(2,-1)} + i \left[V^{(2,-1)}, V^{(1,-2)} \right] &= V^{(1,1)} , \\ D^{(2,-1)} V^{(1,1)} - D^{(1,1)} V^{(2,-1)} + i \left[V^{(2,-1)}, V^{(1,1)} \right] &= 0 \, , \\ D^{(-1,2)} V^{(1,1)} - D^{(1,1)} V^{(-1,2)} + i \left[V^{(-1,2)}, V^{(1,1)} \right] &= 0 \, . \end{aligned} \tag{12.57}$$

The reason for this is that all of them originate from *the same bridge* b (12.53). In other words, the constraints (12.57) are just the integrability conditions for the existence of a common bridge for all the three connections above. Equivalently, they guarantee the existence of covariantly u-independent superfields:

$$\mathcal{D}^{(2,-1)} \Phi = \mathcal{D}^{(-1,2)} \Phi = \mathcal{D}^{(1,1)} \Phi = 0 \, . \tag{12.58}$$

The constraints (12.57) make the case $N = 3$ radically different from $N = 2$. In the case $N = 2$, given an arbitrary Grassmann analytic connection V^{++}, one could in principle always find the corresponding bridge. In the case $N = 3$

the Grassmann analytic connections must in addition satisfy the zero curvature conditions (12.57), if they are to be simultaneously expressed in the 'pure gauge' form (12.53).

This new phenomenon has another interpretation. Let us recall that the original constraints (12.1) implied the equations of motion for the theory. After the introduction of $SU(3)/U(1) \times U(1)$ harmonic variables we were able to rewrite them in the equivalent form (12.17) together with (12.47) and (12.48). Moreover, the introduction of the λ gauge frame enabled us to solve the constraints (12.17) in the 'pure gauge' form (12.49), and (12.47) in the form of the analyticity conditions (12.56). The only constraints still remaining unsolved are (12.48) or (12.57). The conclusion is that in the λ frame the on-shell equations of $N = 3$ SYM are equivalent to the constraints (12.57) on the analytic superfields $V^{(2,-1)}$, $V^{(-1,2)}$, $V^{(1,1)}$. Let us stress once again that just these equations enable one to find a common bridge (12.53) and to go from the λ frame back to the original τ frame picture.

12.6 Off-shell action

Going off shell would mean trying to relax the dynamical constraints (12.1) (τ frame) or (12.57) (λ frame) and finding an action from which they will follow as Euler–Lagrange equations. In the τ frame it is not clear whether this can be done at all, whereas in the λ frame it is very easy. Let us assume that we are given three analytic harmonic connections $V^{(2,-1)}$, $V^{(-1,2)}$, $V^{(1,1)}$ with the reality properties (12.54) *which do not obey any constraints*. Then the commutators

$$\begin{aligned} \left[\mathcal{D}^{(2,-1)}, \mathcal{D}^{(-1,2)}\right] &= \mathcal{D}^{(1,1)} + iF^{(1,1)}, \\ \left[\mathcal{D}^{(1,1)}, \mathcal{D}^{(2,-1)}\right] &= iF^{(3,0)}, \\ \left[\mathcal{D}^{(-1,2)}, \mathcal{D}^{(1,1)}\right] &= iF^{(0,3)} \end{aligned} \tag{12.59}$$

define three non-vanishing curvatures. The equations of motion (12.57) are then equivalent to

$$F^{(1,1)} = F^{(3,0)} = F^{(0,3)} = 0, \tag{12.60}$$

i.e., to the requirement that the harmonic curvatures in (12.59) vanish. The crucial point here is that we have three superfields $V^{(2,-1)}$, $V^{(-1,2)}$, $V^{(1,1)}$ and exactly three equations (12.60) for them. Matching the numbers of fields and equations is a necessary condition for finding an action from which these equations could be derived. Indeed, such an action exists:

$$\begin{aligned} S^{N=3}_{\text{SYM}} &= \int du\, d\zeta_A^{(-2,-2)}\, \text{Tr}\Big\{ V^{(2,-1)} \left(D^{(-1,2)} V^{(1,1)} - D^{(1,1)} V^{(-1,2)}\right) \\ &\quad - V^{(-1,2)} \left(D^{(2,-1)} V^{(1,1)} - D^{(1,1)} V^{(2,-1)}\right) \end{aligned}$$

$$
\begin{aligned}
&+V^{(1,1)}\left(D^{(2,-1)}V^{(-1,2)} - D^{(-1,2)}V^{(2,-1)}\right)\\
&-\left(V^{(1,1)}\right)^2 + 2i\,V^{(1,1)}\left[V^{(2,-1)}, V^{(-1,2)}\right]\Big\}\\
= &\int du\, d\zeta_A^{(-2,-2)}\, \mathrm{Tr}\Big\{V^{(2,-1)}F^{(0,3)} + V^{(-1,2)}F^{(3,0)} + V^{(1,1)}F^{(1,1)}\\
&-i\,V^{(1,1)}\left[V^{(2,-1)}, V^{(-1,2)}\right]\Big\}, \qquad (12.61)
\end{aligned}
$$

where $du\, d\zeta_A^{(-2,-2)}$ is the integration measure of the analytic superspace. Varying (12.61) with respect to the V's one easily obtains the equations of motion (12.60). The second form of the action in (12.61) clearly shows that the Lagrangian is not a tensor. It is gauge invariant only up to total harmonic derivatives. Indeed, under the λ gauge transformations (12.55) one finds, after integration by parts,

$$
\delta S = \int du\, d\zeta_A^{(-2,-2)}\, \mathrm{Tr}\left\{\lambda\left(\mathcal{D}^{(2,-1)}F^{(0,3)} + \mathcal{D}^{(-1,2)}F^{(3,0)} + \mathcal{D}^{(1,1)}F^{(1,1)}\right)\right\}, \qquad (12.62)
$$

which vanishes as a consequence of a Bianchi identity following from (12.59). Note also the exact matching of the $U(1)$ charges in (12.61).

One can say that the action (12.61) has the form of a Chern–Simons term involving the three harmonic connections V and curvatures F. It is well known that a Chern–Simons action can only exist in three dimensions. This shows once again why we had to choose our harmonic space three-dimensional.*

It should be stressed that the presence of three harmonic gauge connections with just three equations for them is only one reason why it becomes possible to construct an off-shell action for $N = 3$ SYM theory. Two other reasons are the vanishing dimension of the integration measure of the $N = 3$ analytic harmonic superspace and the charge assignment $(-2, -2)$ of this measure, which precisely match the zero dimension and the charges $(2, 2)$ of the analytic Lagrangian. The conspiracy of these three properties in the case of $N = 3$ SYM theory can be regarded as a 'miracle'. Unfortunately, such a miracle does not happen in the maximally supersymmetric $N = 4$ SYM theory.

Concluding this section we note that the first of the constraints (12.57) is purely algebraic (conventional). It serves as a definition of $V^{(1,1)}$ in terms of $V^{(2,-1)}$ and $V^{(-1,2)} = \widetilde{V}^{(2,-1)}$. Therefore, one can use it to eliminate $V^{(1,1)}$ from the action(12.61). Then the action becomes second-order in the space-time derivatives which are hidden in the analytic basis expressions for the harmonic derivatives (see, e.g., (12.44)). For possible applications it is instructive to explicitly give this form of the $N = 3$ SYM action:

$$
S^{N=3}_{\mathrm{SYM}} = \int du\, d\zeta_A^{(-2,-2)}\, \mathrm{Tr}\Big\{\left(D^{(2,-1)}V^{(-1,2)}\right)^2 + \left(D^{(-1,2)}V^{(2,-1)}\right)^2
$$

* We mean three *complex* dimensions, but the action is still real in the sense of $\widetilde{\ }$ conjugation.

$$-4\,D^{(-1,2)}V^{(2,-1)}\,D^{(2,-1)}V^{(-1,2)}+2\,D^{(2,-1)}V^{(2,-1)}\,D^{(-1,2)}V^{(-1,2)}$$
$$+2i\left[V^{(2,-1)},V^{(-1,2)}\right]\left(D^{(2,-1)}V^{(-1,2)}-D^{(-1,2)}V^{(2,-1)}\right)$$
$$-\left[V^{(2,-1)},V^{(-1,2)}\right]\left[V^{(2,-1)},V^{(-1,2)}\right]\Big\}. \qquad (12.63)$$

12.7 Components on and off shell

The off-shell $N = 3$ SYM theory described in the preceding section is highly unusual. In some respects it is similar to the off-shell theory of the q^+ hypermultiplet. Both theories have infinitely many auxiliary fields coming from the harmonic expansion of the analytic prepotentials. On shell they are eliminated by equations of motion which restrict the harmonic dependence of the prepotentials ((5.27) for q^+ and (12.57) for V). At the same time the $N = 3$ SYM theory resembles the $N = 2$ SYM one in the sense that in both cases there are infinitely many gauge degrees of freedom contained in the harmonic expansion of the analytic λ gauge parameters. So, exhibiting the finite set of on-shell physical fields of the $N = 3$ theory would mean going to a Wess–Zumino gauge (thus gauging away an infinite set of pure gauge fields) and at the same time imposing the equations of motion (thus eliminating an infinite set of auxiliary fields). Doing this explicitly is a non-trivial exercise and we do not undertake it here. Instead, we only indicate where the physical fields occur, e.g., in the harmonic connection $V^{(2,-1)}$:

$$V^{(2,-1)}(x_A,\theta^{(1,-1)},\theta^{(0,1)},\bar\theta^{(1,0)},\bar\theta^{(-1,1)},u)$$
$$=\cdots+i\theta^{(1,-1)}_{\alpha}\bar\theta^{(1,0)}_{\dot\alpha}A^{\alpha\dot\alpha}(x_A)+\bar\theta^{(1,0)}_{\dot\alpha}\bar\theta^{\dot\alpha(1,0)}u^{(0,-1)i}\phi_i(x_A) \qquad (12.64)$$
$$+\bar\theta^{(1,0)}_{\dot\alpha}\bar\theta^{\dot\alpha(-1,1)}\theta^{(1,-1)\alpha}u^{(1,-1)}_i\chi^i_\alpha(x_A)+\theta^{(1,-1)}_\alpha\theta^{(1,-1)\alpha}\theta^{(0,1)\beta}\psi_\beta(x_A)+\cdots.$$

In order to present the off-shell content of the theory we have to fully develop the differential geometry formalism and construct all the non-vanishing curvature tensors which represent the gauge-covariant degrees of freedom. Note that in the case of $N = 2$ SYM the curvature W satisfies the Bianchi identity $D^{++}W = 0$ (7.85) which makes it u-independent and guarantees the finite off-shell content of the theory. However, in $N = 3$ SYM there are no such restrictions off shell, so the curvature tensors like those in (12.59) will have a non-trivial harmonic dependence. Another essential difference between the cases $N = 2$ and $N = 3$ lies in the fact that there is no off-shell τ-frame geometry for $N = 3$ SYM. Indeed, the bridge from the λ to the τ frame only exists if the on-shell constraints are imposed. So, off shell in $N = 3$ SYM we can only develop λ-frame geometry. The starting point is the analytic harmonic connection $V^{(2,-1)}$. The connection $V^{(-1,2)}$ is its $\widetilde{\ }$ conjugate, and $V^{(1,1)}$ is obtained from the conventional algebraic constraint (12.57). The central problem now is to find the harmonic connection $V^{(-2,1)}$ as the solution of the constraint (see (12.27))

$$\left[\mathcal{D}^{(2,-1)}, \mathcal{D}^{(-2,1)}\right] = -D^0_{(1)}\,. \tag{12.65}$$

This constraint is again conventional (it defines $V^{(-2,1)}$ in terms of $V^{(2,-1)}$), but this time it is differential rather than algebraic. The situation here is similar to the case of $N = 2$ SYM where one has to solve the constraint

$$\left[\mathcal{D}^{++}, \mathcal{D}^{--}\right] = D^0 \tag{12.66}$$

for V^{--} in terms of V^{++}. Once again, the solution of (12.65) almost always exists* and is unique. To see this it is sufficient to look at the linearized (Abelian) version of (12.65):

$$D^{(2,-1)}V^{(-2,1)} - D^{(-2,1)}V^{(2,-1)} = 0\,. \tag{12.67}$$

The homogeneous equation

$$D^{(2,-1)}V_0^{(-2,1)} = 0 \tag{12.68}$$

has only the trivial solution $V_0 = 0$. The reason for this is that the derivatives $D^{(2,-1)}$, $D^{(-1,2)}$ and $D^0_{(1)}$ form the algebra of $SU(2) \subset SU(3)$ (see (12.65)). Then, by just looking at the first charges in equation (12.68), we see that it is the direct analog of the $SU(2)/U(1)$ harmonic equation $D^{++}V^{--} = 0$ which admits no non-trivial solution.

Having found $V^{(-2,1)}$, one obtains $V^{(1,-2)}$ as its $\widetilde{\ }$ conjugate. Further, the non-trivial spinor covariant derivatives $\mathcal{D}^{(-1,1)}_\alpha, \mathcal{D}^{(0,-1)}_\alpha, \bar{\mathcal{D}}^{(-1,0)}_{\dot\alpha}, \bar{\mathcal{D}}^{(1,-1)}_{\dot\alpha}$ in the λ frame can be obtained from the trivial ones $D^{(1,0)}_\alpha$ and $\bar{D}^{(0,1)}_{\dot\alpha}$ by means of conventional constraints, e.g.,

$$\mathcal{D}^{(-1,1)}_\alpha = -\left[\mathcal{D}^{(-2,1)}, D^{(1,0)}_\alpha\right], \qquad \text{etc.} \tag{12.69}$$

Commuting or anticommuting all these covariant derivatives one can derive a great number of curvature tensors which satisfy the corresponding Bianchi identities. As mentioned earlier, off shell these tensors non-trivially depend on the harmonic variables and thus define infinite sets of gauge-covariant physical and auxiliary fields. Some steps toward constructing the off-shell differential geometry for $N = 3$ SYM theory in the harmonic superspace were undertaken in [K4, K5].

It is worth mentioning [A6] that *on shell* the harmonic projections of the basic such tensor, the covariant superfield strength $\overline{W}^{ik} \equiv \epsilon^{ikl}\overline{W}_l$ defined in (12.1), reveal 'short' Grassmann harmonic analyticities (with six θ's) as a consequence

* An example of a choice of $V^{(2,-1)}$ for which there is no $V^{(-1,2)}$ is given in [R6]. It is similar to that mentioned in footnote $*$ in Section 7.2.3.

of the Bianchi identities (12.2). Indeed, at the linearized level the latter amount to the following conditions on, e.g., $\overline{W}^{(0,1)} \equiv \overline{W}^{ik} u_i^{(1,0)} u_k^{(-1,1)} = -\overline{W}_i u^{i(0,1)}$:

$$D_\alpha^{(1,0)} \overline{W}^{(0,1)} = \bar{D}_{\dot\alpha}^{(0,1)} \overline{W}^{(0,1)} = D_\alpha^{(-1,1)} \overline{W}^{(0,1)} = 0\,. \tag{12.70}$$

These are just the $N = 3$ Grassmann analyticity conditions stating that $\overline{W}^{(0,1)}$ live in the complex 'short' analytic superspace (12.46) (in the analytic basis):

$$\overline{W}^{(0,1)} = \overline{W}^{(0,1)}(x_{A-}, \bar\theta^{(-1,1)},\ \bar\theta^{(1,0)},\ \theta^{(0,1)},\ u)\,. \tag{12.71}$$

Like in the case of the q^+ hypermultiplet, the definition of $\overline{W}^{(0,1)}$ as a homogeneous harmonic function of weight 1 can be reformulated as the harmonic analyticity conditions

$$D^{(2,-1)} \overline{W}^{(0,1)} = D^{(-1,2)} \overline{W}^{(0,1)} = 0\,. \tag{12.72}$$

The set of analyticity conditions (12.70), (12.72) is equivalent to the Bianchi identities (12.2) and can be shown to yield precisely the equations of motion for the $N = 3$ gauge field strength multiplet in the linearized approximation. In the full non-linear case the harmonic analyticity conditions of course become covariant in the analytic basis.

This comment demonstrates the important rôle played by the complex analytic superspaces (12.45), (12.46) in the harmonic superspace geometry of $N = 3$ SYM theory (alongside the standard real analytic superspace (12.37)).

12.8 Conformal invariance

The four-dimensional supersymmetric Yang–Mills theories are known to be conformally invariant, both on and off shell (whenever an off-shell formulation exists). It is then natural to expect that the off-shell $N = 3$ SYM theory developed here should have the same property. To check this one needs a realization of the $N = 3$ superconformal algebra $SU(2, 2|3)$ in the analytic harmonic superspace. This can be done following the general scheme developed in Chapter 9 for the case $N = 2$. Here we only give the conformal supersymmetry transformations (their commutators with the ordinary supersymmetry (12.38) produce the whole superconformal algebra).* The harmonic variables transform as follows:

$$\begin{aligned}
\delta u_i^{(0,-1)} &= \delta u^{i(-1,0)} = 0\,,\\
\delta u_i^{(1,0)} &= \lambda^{(1,1)} u_i^{(0,-1)} + \lambda^{(2,-1)} u_i^{(-1,1)}\,,\\
\delta u^{i(0,1)} &= -\lambda^{(1,1)} u^{i(-1,0)} - \lambda^{(-1,2)} u^{i(1,-1)}\,,\\
\delta u_i^{(-1,1)} &= \lambda^{(-1,2)} u_i^{(0,-1)}\,,\\
\delta u^{i(1,-1)} &= -\lambda^{(2,-1)} u^{i(-1,0)}\,,
\end{aligned} \tag{12.73}$$

* The full set of $N = 3$ superconformal transformations can be found in [G12].

where

$$
\begin{aligned}
\lambda^{(1,1)} &= -4i\left(\bar{\eta}_{\dot{\beta}j}\bar{\theta}^{(1,0)\dot{\beta}}u^{j(0,1)} + \theta^{(0,1)\beta}\eta^j_\beta u^{(1,0)}_j\right),\\
\lambda^{(2,-1)} &= -4i\left(\bar{\eta}_{\dot{\beta}j}\bar{\theta}^{(1,0)\dot{\beta}}u^{j(1,-1)} + \theta^{(1,-1)\beta}\eta^j_\beta u^{(1,0)}_j\right),\\
\lambda^{(-1,2)} &= -4i\left(\bar{\eta}_{\dot{\beta}j}\bar{\theta}^{(-1,1)\dot{\beta}}u^{j(0,1)} + \theta^{(0,1)\beta}\eta^j_\beta u^{(-1,1)}_j\right),
\end{aligned}
\tag{12.74}
$$

and η^i_α and $\bar{\eta}_{\dot{\alpha}i}$ are the parameters of conformal supersymmetry. The parameters λ (12.74) obey the relations

$$
\begin{aligned}
&D^{(1,1)}\lambda^{(a,b)} = 0\,, \qquad D^{(2,-1)}\lambda^{(2,-1)} = 0\,,\\
&D^{(2,-1)}\lambda^{(-1,2)} = -D^{(-1,2)}\lambda^{(2,-1)} = \lambda^{(1,1)}
\end{aligned}
$$

(recall the analogous ones in the $N = 2$ case, eq. (9.35)). The remaining coordinates of the analytic superspace transform as follows:

$$
\begin{aligned}
\delta x^{\alpha\dot{\alpha}}_A &= -2i\bar{\eta}_{\dot{\beta}i}x^{\alpha\dot{\beta}}_{A-}\bar{\theta}^{(-1,1)\dot{\alpha}}u^{i(1,-1)} - 4i\bar{\eta}_{\dot{\beta}i}x^{\alpha\dot{\beta}}_A\bar{\theta}^{(1,0)\dot{\alpha}}u^{i(-1,0)}\\
&\quad - 2i\eta^i_\beta x^{\beta\dot{\alpha}}_{A+}\theta^{(1,-1)\alpha}u^{(-1,1)}_i - 4i\eta^i_\beta x^{\beta\dot{\alpha}}_A\theta^{(0,1)\alpha}u^{(0,-1)}_i\,,\\
\delta\theta^{(1,-1)\alpha} &= -x^{\alpha\dot{\beta}}_{A+}\bar{\eta}_{\dot{\beta}j}u^{j(1,-1)} - 4i\left(\theta^{(0,1)\alpha}u^{(0,-1)}_j + \theta^{(1,-1)\alpha}u^{(-1,1)}_j\right)\theta^{(1,-1)\beta}\eta^j_\beta\,,\\
\delta\theta^{(0,1)\alpha} &= x^{\alpha\dot{\beta}}_{A-}\bar{\eta}_{\dot{\beta}j}u^{j(0,1)} - 4i\theta^{(0,1)\alpha}u^{(0,-1)}_j\theta^{(0,1)\beta}\eta^j_\beta\,,\\
\delta\bar{\theta}^{(-1,1)\dot{\alpha}} &= -\widetilde{\delta\theta^{(1,-1)\alpha}}\,, \qquad \delta\bar{\theta}^{(1,0)\dot{\alpha}} = \widetilde{\delta\theta^{(0,1)\alpha}}\,, \qquad \widetilde{\delta x^{\alpha\dot{\alpha}}_A} = \delta x^{\alpha\dot{\alpha}}_A\,,
\end{aligned}
\tag{12.75}
$$

where $x^{\alpha\dot{\alpha}}_{A\pm} = x^{\alpha\dot{\alpha}}_A \pm 2i\theta^{(1,-1)\alpha}\bar{\theta}^{(-1,1)\dot{\alpha}}$.

Note that the transformation laws (12.73), (12.74), (12.75) are consistent only with the $\tilde{}$ conjugation, like the $N = 2$ ones (see Section 9.1.3). These transformations have the important property that they preserve the supervolume of the analytic superspace (recall (9.34) in the case $N = 2$):

$$
\mathrm{Ber}\left(\frac{\partial(\zeta+\delta\zeta, u+\delta u)}{\partial(\zeta, u)}\right) \simeq 1 + \frac{\partial}{\partial x_A}\delta x_A + \frac{\partial}{\partial u}\delta u - \frac{\partial}{\partial\theta}\delta\theta = 1\,. \tag{12.76}
$$

The harmonic derivatives have non-trivial conformal transformation laws (recall (9.33)):

$$
\begin{aligned}
\delta D^{(2,-1)} &= -\lambda^{(2,-1)}D^0_{(1)}\,, \qquad \delta D^{(1,-2)} = -\lambda^{(1,-2)}D^0_{(2)}\,,\\
\delta D^{(1,1)} &= \lambda^{(2,-1)}D^{(-1,2)} - \lambda^{(-1,2)}D^{(2,-1)} - \lambda^{(1,1)}(D^0_{(1)} + D^0_{(2)})\,.
\end{aligned}
\tag{12.77}
$$

Correspondingly, we postulate the following transformation laws for the harmonic connections:

$$
\delta V^{(2,-1)} = \delta V^{(-1,2)} = 0\,, \qquad \delta V^{(1,1)} = \lambda^{(2,-1)}V^{(-1,2)} - \lambda^{(-1,2)}V^{(2,-1)} \tag{12.78}
$$

and curvatures:

$$\delta F^{(1,1)} = 0\,, \qquad \delta F^{(3,0)} = -\lambda^{(2,-1)} F^{(1,1)}\,, \qquad \delta F^{(0,3)} = \lambda^{(-1,2)} F^{(1,1)}\,. \tag{12.79}$$

Given all this, it is not hard to verify that the $N = 3$ SYM off-shell action (12.61) is indeed superconformally invariant.

Note that the 'short' conjugate analytic superspaces (12.45), (12.46) can also be shown to be closed under the $N = 3$ superconformal group provided that the relevant sets of harmonic variables are assumed complex (in the $\tilde{}$ sense) and conjugate to each other [G12] (such harmonics parametrize $SL(3, \mathbb{C})$ rather than $SU(3)$). The above superconformal transformations in the real $N = 3$ analytic harmonic superspace are induced by those in the 'short' analytic subspace when the former is treated as a real hypersurface in the latter.

12.9 Final remarks

There still remain several unsolved problems in the harmonic $N = 3$ superfield approach.

The construction of off-shell $N = 3$ supergravity [C5, F8] is a challenge. The straightforward generalization of the rigid superconformal transformations in the $N = 3$ analytic superspace to diffeomorphisms and the appropriate covariantization of the $N = 3$ analyticity-preserving harmonic derivatives do not immediately give the off-shell $N = 3$ Weyl multiplet, in contradistinction to the $N = 2$ case. Some additional constraints have to be imposed on the relevant analytic vielbeins. It is likely that in the $N = 3$ case these vielbeins are not the fundamental geometric objects of conformal supergravity as opposed to the $N = 2$ case. The complex analytic superspaces (12.45), (12.46) seem to play some important, though for the time being not fully understood, rôle in the geometry of $N = 3$ conformal supergravity. If the latter is known, the Einstein $N = 3$ supergravity action could be constructed as the off-shell action of a set of $N = 3$ gauge multiplets in a conformal $N = 3$ supergravity background. Indeed, these multiplets are the only representatives of $N = 3$ 'matter', and as such they provide the unique choice for compensator superfields.

Another related problem is to reveal the possible purely bosonic implications of the harmonic $SU(3)$ analyticity similar to those of the harmonic $SU(2)$ analyticity in hyper-Kähler and quaternionic geometries and in self-dual Yang–Mills theory.

An interesting and urgent practical task is to construct the complete quantum formalism of $N = 3$ SYM theory in $N = 3$ harmonic superspace. The corresponding Feynman supergraph techniques are expected to be in a sense simpler than those of $N = 2$ SYM. This should be so because of the presence of only two self-interaction vertices in (12.63) or of only one vertex in the first-order action (12.61), in contrast to the infinite number of vertices in the off-shell $N = 2$ SYM action (7.98). The quantization of $N = 3$ SYM theory in the $N = 3$ harmonic superspace formulation was considered in [D8], but some further work is still needed to bring the quantization scheme into a form suitable for computations.

One of the merits of having a self-consistent $N = 3$ harmonic supergraph technique is the radical simplification of the proof of the absence of ultraviolet divergences in $N = 3$ SYM theory and, hence, in $N = 4$ SYM theory to which the former is reduced on shell.* As was argued by the authors of [D8], the form of the $N = 3$ gauge superfields propagators they have found is such that one can apply the standard arguments of the non-renormalization theorems [G26, G41, H16]. Namely, one can (i) remove the appropriate number of spinor derivatives from the propagators to restore the full $N = 3$ superspace Grassmann integration measure at each vertex ($d^8\theta_A \to d^{12}\theta$) and (ii) write down the propagators using spinor derivatives acting on the $N = 3$ Grassmann delta-functions $\delta^{12}(1|2)$ without any other θ dependence. Then, doing all the θ integrations except one, it is possible to reduce, schematically, any graph to a sum of terms of the following type:

$$\sim \prod_{\text{all vertices}} du_v \int d^{12}\theta \prod_{e=1}^{N} d^4 p_e \left[(D)^{n_1} J^1(p_1, u_1, \theta) \ldots (D)^{n_N} J^N(p_N, u_N, \theta)\right]$$
$$\times\, \delta^4\left(\sum_{e=1}^{N} p_e\right) \prod_{l=1}^{L} \int d^4 k_l\, F(k_l, p_e, u_v)\,. \tag{12.80}$$

Here $J^1, \ldots, J^N$ are dimensionless external sources (we do not write down their $U(1) \times U(1)$ indices) on which some spinor derivatives can act, N is the number of external vertices and L is the number of loops. The relevant momenta are denoted, respectively, by p_e and k_l. Taking into account that the total graph is dimensionless and counting dimensions in (12.80), it is easy to see that the loop integrals are superficially convergent. One can hope that this simple reasoning will not be obscured by possible harmonic divergences due to coincident singularities in the $SU(3)$ harmonic distributions. Like in the $N = 2$ case (Section 8.5), these specific divergences can be expected never to show up in the final expressions for the harmonic supergraphs.

* The ultraviolet finiteness of $N = 4$ SYM theory was shown by different methods in [B11, B12, H16, M2, S15].

13
Conclusions

We have presented the basics of the harmonic superspace approach but not the latest developments. Some of the most recent are mentioned in the Introduction as the motivation for writing this book. During the past few years this method has been advanced in several directions, some of which we briefly overview here.

Firstly, even at the early stages it was realized that higher N supersymmetries provide many more possibilities for choosing the harmonic superspaces and their analytic subspaces due to the higher rank of the relevant R symmetry groups. In [I7] an exhaustive list of the coset spaces of such groups in terms of the corresponding harmonic variables was presented. No *a priori* reasons exist for preferring any of these higher N harmonic superspaces and the right choice is imposed by the problem at hand. For instance, as we have seen, the problem of constructing an off-shell formulation of $N = 3$ super Yang–Mills theory (which coincides with $N = 4$ SYM on shell) is solved using an $N = 3$ harmonic superspace based on the coset $SU(3)/U(1) \times U(1)$ [G5, G6]. Unfortunately, the analogous problem in $N = 4$ has not yet been solved. The only recent development has been the harmonic superspace (based upon the coset $SU(4)/SU(2) \times SU(2) \times U(1)$) reformulation of the on-shell constraints of $N = 4$ SYM proposed in [H19, H21]. It has proved useful in analyzing the general properties of the correlation functions of this theory in the context of the AdS/CFT correspondence.

Harmonic superspace has been successfully used for constructing off-shell actions for a number of lower-dimensional supersymmetric theories. Among them are the $N = 3, 4$ gauge theories in three dimensions [Z6] which naturally occur as world-volume theories for intersecting branes [G29]. A large class of two-dimensional sigma models admit a formulation in harmonic superspace with a clear geometric interpretation. These include all hyper-Kähler sigma models with (4,4) supersymmetry (they are obtained by dimensional reduction of the general self-interaction of hypermultiplets), the (4,0) and (4,4) sigma models with torsion [D7, I6, I10] (in the (4,4) case a modified harmonic superspace with two sets of $SU(2)$ harmonics is required [I10]*). The corresponding analytic superspace Lagrangians can be regarded as the unconstrained potentials of the target space geometry, very much like the general Lagrangian for hypermultiplets can be viewed as the potential of hyper-Kähler geometry. As an extension of the geometric applications overviewed in Chapter 5 (Sections 5.3.2, 5.3.5), the efficiency of the harmonic superspace techniques for the explicit computations of some quaternionic metrics was recently demonstrated in [I11, I12].

Harmonic spaces and superspaces find interesting applications in integrable systems. As was discussed in Section 11.2.1, harmonics on the Euclidean Lorentz group $SO(4)$ allow one to reformulate the standard self-duality Yang–Mills equations as integrability conditions very similar to the $N = 2$ super-Yang–Mills constraints [G20]. The harmonic approach to self-duality is closely related to, although not identical with, the twistor approach. In refs. [D12, D13, D14] super-self-duality was treated in a similar way. Harmonic $d = 1$, $N = 4$ superfields have been used for constructing the $N = 4$ super-KdV hierarchy [D5].

Another line of application of the harmonic method concerns strings and superstrings, and it makes use of non-compact groups, in particular the Lorentz group in various dimensions. They are useful for separating the first- and second-class constraints for superparticles and strings in a covariant way [D4, D5, D6, G3, N3, N4, N5, S16]. These Lorentz harmonics also appear in the twistor-like approach to superstrings and branes (see [S20] and references therein). Particle mechanics in harmonic superspace was considered in [A1, K2].

It is beyond the scope of our presentation here to elaborate on these and further applications and developments of the harmonic superspace method. We leave this to the ambitious reader.

* For further generalizations see the recent paper [Z5].

Appendix
Notations, conventions and useful formulas

A.1 Two-component spinors

In this book we use the two-component spinor formalism. Here is a brief summary of the most important conventions.

The dotted and undotted spinor indices are raised and lowered as follows:

$$\psi^\alpha = \epsilon^{\alpha\beta}\psi_\beta\,, \qquad \bar\chi^{\dot\alpha} = \epsilon^{\dot\alpha\dot\beta}\bar\chi_{\dot\beta}\,; \tag{A.1}$$

$$\psi_\alpha = \epsilon_{\alpha\beta}\psi^\beta\,, \qquad \bar\chi_{\dot\alpha} = \epsilon_{\dot\alpha\dot\beta}\bar\chi^{\dot\beta}\,, \tag{A.2}$$

where the antisymmetric ϵ symbols have the properties:

$$\epsilon_{12} = \epsilon_{\dot1\dot2} = -\epsilon^{12} = -\epsilon^{\dot1\dot2} = 1\,, \qquad \epsilon_{\alpha\beta}\epsilon^{\beta\gamma} = \delta_\alpha^\gamma\,, \qquad \epsilon_{\dot\alpha\dot\beta}\epsilon^{\dot\beta\dot\gamma} = \delta_{\dot\alpha}^{\dot\gamma}\,. \tag{A.3}$$

The convention for the contraction of a pair of spinor indices is

$$\psi^\alpha\lambda_\alpha \equiv \psi\lambda\,, \qquad \bar\chi_{\dot\alpha}\bar\rho^{\dot\alpha} \equiv \bar\chi\bar\rho\,, \qquad \psi^2 \equiv \psi^\alpha\psi_\alpha\,, \qquad \bar\psi^2 \equiv \bar\psi_{\dot\alpha}\bar\psi^{\dot\alpha}\,. \tag{A.4}$$

The sigma matrices σ^a are defined as follows:

$$(\sigma^a)_{\alpha\dot\alpha} = (1, \vec\sigma)_{\alpha\dot\alpha}\,, \qquad (\tilde\sigma^a)^{\dot\alpha\alpha} = \epsilon^{\dot\alpha\dot\beta}\epsilon^{\alpha\beta}(\sigma^a)_{\beta\dot\beta} = (1, -\vec\sigma)^{\dot\alpha\alpha} \tag{A.5}$$

and have the basic properties:

$$\begin{aligned}
&\sigma^a\tilde\sigma^b = \eta^{ab} - i\sigma^{ab}\,, \qquad \tilde\sigma^a\sigma^b = \eta^{ab} - i\tilde\sigma^{ab}\,,\\
&(\sigma^a)_{\alpha\dot\alpha}(\tilde\sigma_a)^{\dot\beta\beta} = 2\delta_\alpha^\beta\delta_{\dot\alpha}^{\dot\beta}\,, \qquad (\sigma_a)_{\alpha\dot\alpha}(\tilde\sigma^b)^{\dot\alpha\alpha} = 2\delta_a^b\,,\\
&\sigma^{ab} = -\sigma^{ba}\,, \qquad \tilde\sigma^{ab} = -\tilde\sigma^{ba}\,, \qquad (\sigma^{ab})_\alpha{}^\alpha = (\tilde\sigma^{ab})_{\dot\alpha}{}^{\dot\alpha} = 0\,.
\end{aligned} \tag{A.6}$$

A Minkowski four-vector can be written down as a two-component bi-spinor:

$$x^{\dot\alpha\alpha} = x^a\tilde\sigma_a^{\dot\alpha\alpha}\,, \qquad x^a = \frac12 x^{\dot\alpha\alpha}\sigma^a_{\alpha\dot\alpha}\,, \qquad x^2 = x^a x_a = \frac12 x^{\dot\alpha\alpha}x_{\alpha\dot\alpha}\,. \tag{A.7}$$

The $SU(2)$ isospinor indices are raised and lowered with the help of the symbols ϵ_{ij}, ϵ^{ij} which have the same properties as $\epsilon_{\alpha\beta}$, $\epsilon^{\alpha\beta}$ above.

A.2 Harmonic variables and derivatives

The harmonic variables u_i^+, u_i^- are defined by the relations:

$$u^{+i}u_i^- = 1 \quad \Leftrightarrow \quad u_i^+ u_k^- - u_k^+ u_i^- = \epsilon_{ik}\,. \tag{A.8}$$

The reduction formulas for the products of symmetrized harmonic monomials are

$$\begin{aligned} u_i^+ u^+_{(j_1} \dots u^+_{j_n} u^-_{k_1} \dots u^-_{k_m)} &= u^+_{(i} u^+_{j_1} \dots u^-_{k_m)} \\ &\quad + \frac{m}{m+n+1} \epsilon_{i(k_1} u^+_{j_1} \dots u^+_{j_n} u^-_{k_2} \dots u^-_{k_m)}, \\ u_i^- u^+_{(j_1} \dots u^+_{j_n} u^-_{k_1} \dots u^-_{k_m)} &= u^-_{(i} u^+_{j_1} \dots u^-_{k_m)} \\ &\quad - \frac{n}{m+n+1} \epsilon_{i(j_1} u^+_{j_2} \dots u^-_{k_m)}\,. \end{aligned} \tag{A.9}$$

The harmonic integral is defined by the rules:

$$(i) \qquad \int du\; f^{(q)}(u) = 0 \qquad \text{if } q \neq 0\,, \tag{A.10}$$

$$(ii) \qquad \int du\; 1 = 1\,, \tag{A.11}$$

$$(iii) \qquad \int du\; u^+_{(i_1} \dots u^+_{i_n} u^-_{j_1} \dots u^-_{j_n)} = 0 \qquad \text{for } n \geq 1\,. \tag{A.12}$$

From these rules follow orthogonality relations for the symmetrized products of $u^\pm$:

$$\int du\; (u^+)^{(m}(u^-)^{n)}(u^+)_{(k}(u^-)_{l)} = \frac{(-1)^n m!n!}{(m+n+1)!} \delta^{(i_1}_{(j_1} \dots \delta^{i_{m+n})}_{j_{m+n})} \delta_{ml}\delta_{nk}\,, \tag{A.13}$$

where

$$(u^+)^{(m}(u^-)^{n)} \equiv u^{+(i_1} \dots u^{+i_m} u^{-j_1} \dots u^{-j_n)}\,. \tag{A.14}$$

The harmonic derivatives $D^\pm$ and the $U(1)$ charge operator D^0 in the central (or real) basis of $\mathbb{HR}^{4+2|8}$ are defined by

$$\begin{aligned} D^{++} &= u^{+i}\frac{\partial}{\partial u^{-i}} \equiv \partial^{++}\,, \qquad D^{--} = u^{-i}\frac{\partial}{\partial u^{+i}} \equiv \partial^{--}\,, \\ D^0 &= u^{+i}\frac{\partial}{\partial u^{+i}} - u^{-i}\frac{\partial}{\partial u^{-i}} \equiv \partial^0 \end{aligned} \tag{A.15}$$

and form the $SU(2)$ algebra

$$\left[D^{++}, D^{--}\right] = D^0\,, \qquad \left[D^0, D^{\pm\pm}\right] = \pm 2D^{\pm\pm}\,. \tag{A.16}$$

Their action on the harmonics is given by

$$D^0 u_i^\pm = \pm u_i^\pm\,, \qquad D^{\pm\pm}u_i^\mp = u_i^\pm\,, \qquad D^{\pm\pm}u_i^\pm = 0\,. \tag{A.17}$$

In the analytic basis (A.25) of $\mathbb{HR}^{4+2|8}$ these derivatives read

$$\begin{aligned} D_A^{++} &= \partial^{++} - 2i\theta^+\sigma^a\bar{\theta}^+\frac{\partial}{\partial x_A^a} + \theta^{+\alpha}\frac{\partial}{\partial\theta^{-\alpha}} + \bar{\theta}^{+\dot{\alpha}}\frac{\partial}{\partial\bar{\theta}^{-\dot{\alpha}}}\,, \\ D_A^{--} &= \partial^{--} - 2i\theta^-\sigma^a\bar{\theta}^-\frac{\partial}{\partial x_A^a} + \theta^{-\alpha}\frac{\partial}{\partial\theta^{+\alpha}} + \bar{\theta}^{-\dot{\alpha}}\frac{\partial}{\partial\bar{\theta}^{+\dot{\alpha}}}\,, \\ D_A^0 &= \partial^0 + \theta^{+\alpha}\frac{\partial}{\partial\theta^{+\alpha}} - \theta^{-\alpha}\frac{\partial}{\partial\theta^{-\alpha}} + \bar{\theta}^{+\dot{\alpha}}\frac{\partial}{\partial\bar{\theta}^{+\dot{\alpha}}} - \bar{\theta}^{-\dot{\alpha}}\frac{\partial}{\partial\bar{\theta}^{-\dot{\alpha}}} \end{aligned} \quad (A.18)$$

and obey the same algebra (A.16). The commutation relations of the harmonic derivatives with the spinor ones are (in any basis)

$$[D^{\pm\pm}, D^{\mp}_{\alpha,\dot{\alpha}}] = D^{\pm}_{\alpha,\dot{\alpha}}\,, \qquad [D^{\pm\pm}, D^{\pm}_{\alpha,\dot{\alpha}}] = 0\,. \quad (A.19)$$

The harmonic derivative D^{++} acts on the harmonic distributions according to the rule:

$$D_1^{++}\frac{1}{(u_1^+u_2^+)^n} = \frac{1}{(n-1)!}(D_1^{--})^{n-1}\delta^{(n,-n)}(u_1, u_2)\,. \quad (A.20)$$

A.3 Spinor derivatives

The supersymmetric covariant derivatives in the real basis of the $N = 2$ superspace $\mathbb{R}^{4|8}$ have the explicit form:

$$\begin{aligned} D_a &= \frac{\partial}{\partial x^a}\,, \\ D_\alpha^i &= \frac{\partial}{\partial\theta_i^\alpha} + i\bar{\theta}^{\dot{\alpha}i}(\sigma^a)_{\alpha\dot{\alpha}}\frac{\partial}{\partial x^a}\,, \\ \bar{D}_{\dot{\alpha}i} &= -\frac{\partial}{\partial\bar{\theta}^{\dot{\alpha}i}} - i\theta_i^\alpha(\sigma^a)_{\alpha\dot{\alpha}}\frac{\partial}{\partial x^a}\,. \end{aligned} \quad (A.21)$$

and obey the following algebra:

$$\{D_\alpha^i, D_\beta^j\} = \{\bar{D}_{\dot{\alpha}i}, \bar{D}_{\dot{\beta}j}\} = 0\,, \qquad \{D_\alpha^i, \bar{D}_{\dot{\alpha}j}\} = -2i\delta_j^i(\sigma^a)_{\alpha\dot{\alpha}}D_a \equiv -2i\delta_j^i\partial\!\!\!/_{\alpha\dot{\alpha}}\,, \quad (A.22)$$

where $\partial\!\!\!/_{\alpha\dot{\alpha}} = (\sigma^a)_{\alpha\dot{\alpha}}\partial/\partial x^a$. In the harmonic extension $\mathbb{HR}^{4+2|8}$ of $\mathbb{R}^{4|8}$ one considers harmonic-projected spinor derivatives:

$$D_\alpha^\pm = u_i^\pm D_\alpha^i, \qquad \bar{D}_{\dot{\alpha}}^\pm = u_i^\pm \bar{D}_{\dot{\alpha}}^i \quad (A.23)$$

satisfying the algebra:

$$\begin{aligned} &\{D_\alpha^+, D_\beta^+\} = \{D_\alpha^+, \bar{D}_{\dot{\alpha}}^+\} = \{\bar{D}_{\dot{\alpha}}^+, \bar{D}_{\dot{\beta}}^+\} = 0\,, \\ &\{D_\alpha^-, D_\beta^-\} = \{D_\alpha^-, \bar{D}_{\dot{\alpha}}^-\} = \{\bar{D}_{\dot{\alpha}}^-, \bar{D}_{\dot{\beta}}^-\} = 0\,, \\ &\{D_\alpha^+, \bar{D}_{\dot{\alpha}}^-\} = -\{D_\alpha^-, \bar{D}_{\dot{\alpha}}^+\} = -2i\partial\!\!\!/_{\alpha\dot{\alpha}}\,. \end{aligned} \quad (A.24)$$

In the analytic basis in $\mathbb{HR}^{4+2|8}$,

$$
\begin{aligned}
x_A^a &= x^a - 2i\theta^{(i}\sigma^a\bar\theta^{j)}u_i^+u_j^- , \\
\theta^\pm_{A\alpha} &= u_i^\pm\theta^i_\alpha , \qquad \bar\theta^\pm_{A\dot\alpha} = u_i^\pm\bar\theta^i_{\dot\alpha} ,
\end{aligned}
\tag{A.25}
$$

the spinor derivatives D^+ become particularly simple:

$$
\begin{aligned}
D^+_\alpha &= \frac{\partial}{\partial\theta^{-\alpha}} , \qquad \bar D^+_{\dot\alpha} = \frac{\partial}{\partial\bar\theta^{-\dot\alpha}} , \\
D^-_\alpha &= -\frac{\partial}{\partial\theta^{+\alpha}} + 2i\bar\theta^{-\dot\alpha}\not\partial_{\alpha\dot\alpha} , \qquad \bar D^-_{\dot\alpha} = -\frac{\partial}{\partial\bar\theta^{+\dot\alpha}} - 2i\theta^{-\alpha}\not\partial_{\alpha\dot\alpha} .
\end{aligned}
\tag{A.26}
$$

A.4 Conjugation rules

Complex conjugation rules:

$$
\begin{aligned}
\overline{\theta_{\alpha i}} &= \bar\theta^i_{\dot\alpha} , \qquad \overline{\theta^i_\alpha} = -\bar\theta_{\dot\alpha i} ; \\
\overline{u^{+i}} &= u_i^- , \qquad \overline{u_i^+} = -u^{-i} .
\end{aligned}
\tag{A.27}
$$

$\star$ conjugation rules:

$$
\begin{aligned}
&(u^{+i})^\star = u^{-i} , \qquad (u_i^+)^\star = u_i^- , \\
&(u^{-i})^\star = -u^{+i} , \qquad (u_i^-)^\star = -u_i^+ , \\
&(u_i^\pm)^{\star\star} = -u_i^\pm .
\end{aligned}
\tag{A.28}
$$

$\sim$ conjugation rules:

$$
\widetilde{(u_i^\pm)} = u^{\pm i} , \qquad \widetilde{(u^{\pm i})} = -u_i^\pm .
\tag{A.29}
$$

Conjugation rules for $\theta^\pm$:

(A.30)

	θ^+	θ^-	$\bar\theta^+$	$\bar\theta^-$
$\overline{}$	$-\bar\theta^-$	$\bar\theta^+$	θ^-	$-\theta^+$
$\star$	θ^-	$-\theta^+$	$\bar\theta^-$	$-\bar\theta^+$
$\sim$	$\bar\theta^+$	$\bar\theta^-$	$-\theta^+$	$-\theta^-$

Conjugation rules for $D^\pm$:

(A.31)

	D^+	D^-	$\bar D^+$	$\bar D^-$
$\overline{}$	$\bar D^-$	$-\bar D^+$	$-D^-$	D^+
$\star$	D^-	$-D^+$	$\bar D^-$	$-\bar D^+$
$\sim$	$-\bar D^+$	$-\bar D^-$	D^+	D^-

Conjugation rules for D^{++}, D^{--}, D^0:

	D^{++}	D^{--}	D^0
$\bar{\ }$	$-D^{--}$	$-D^{++}$	$-D^0$
$\star$	$-D^{--}$	$-D^{++}$	$-D^0$
$\sim$	D^{++}	D^{--}	D^0

(A.32)

A.5 Superspace integration measures

The integral over the full harmonic superspace $\mathbb{HR}^{4+2|8}$ is defined as

$$\begin{aligned}\int du\, d^{12}X &\equiv \int du\, d^4x\, d^8\theta = \int du\, d^4x_A\, d^4\theta^+\, d^4\theta^- \\ &= \int du\, d^4x_A\, (D^-)^4(D^+)^4\,,\end{aligned} \tag{A.33}$$

where

$$(D^{\pm})^4 = \frac{1}{16}(D^{\pm})^2(\bar{D}^{\pm})^2\,, \tag{A.34}$$

and

$$(D^{\pm})^2 = D^{\pm\alpha} D^{\pm}_{\alpha}\,, \qquad (\bar{D}^{\pm})^2 = \bar{D}^{\pm}_{\dot{\alpha}} \bar{D}^{\pm\dot{\alpha}}\,. \tag{A.35}$$

The integral over the analytic subspace $\mathbb{HA}^{4+2|4}$ is defined by

$$\int du\, d\zeta^{(-4)} \equiv \int du\, d^4x_A\, d^4\theta^+ = \int du\, d^4x_A\, (D^-)^4\,. \tag{A.36}$$

The integrals over the right and left mutually conjugated chiral subspaces of $\mathbb{HR}^{4+2|8}$, up to surface terms originating from the nilpotent shifts between $x^{\dot{\alpha}\beta}$ and $x_R^{\dot{\alpha}\beta}$, $x_L^{\dot{\alpha}\beta}$, are defined by

$$\int du\, d\zeta_R \equiv \int du\, d^4x_R\, d^4\bar{\theta} = \int du\, d^4x\, (\bar{D})^4\,, \tag{A.37}$$

$$\int du\, d\zeta_L \equiv \int du\, d^4x_L\, d^4\theta = \int du\, d^4x\, (D)^4\,, \tag{A.38}$$

where

$$\begin{aligned}(D)^4 &\equiv \frac{1}{48} D^{\alpha i} D^j_{\alpha} D^{\beta}_i D_{\beta j} = \frac{1}{16}\,(D^+)^2(D^-)^2\,, \\ (\bar{D})^4 &\equiv \frac{1}{48} \bar{D}^i_{\dot{\alpha}} \bar{D}^{\dot{\alpha} j} \bar{D}_{\dot{\beta} i} D^{\dot{\beta}}_j = \frac{1}{16}\,(\bar{D}^+)^2(\bar{D}^-)^2\,.\end{aligned} \tag{A.39}$$

The relations among all four measures above, up to total derivatives of the integrand, are given by

$$\int du\, d^{12}X = \int du\, d\zeta^{(-4)}\, (D^+)^4 = \int du\, d\zeta_R\, (D)^4 = \int du\, d\zeta_L\, (\bar{D})^4 . \tag{A.40}$$

The Grassmann integration measures are normalized so that

$$\int d^8\theta\, \theta^8 = \int d^4\theta^+\, (\theta^+)^4 = \int d^4\theta\, (\theta)^4 = \int d^4\bar{\theta}\, (\bar{\theta})^4 = 1 , \tag{A.41}$$

where

$$\begin{aligned} &\theta^8 = (\theta^+)^4(\theta^-)^4 = (\theta)^4(\bar{\theta})^4 , \qquad (\theta^\pm)^4 = (\theta^\pm)^2(\bar{\theta}^\pm)^2 , \\ &(\theta)^4 = (\theta^+)^2(\theta^-)^2 , \qquad (\bar{\theta})^4 = (\bar{\theta}^+)^2(\bar{\theta}^-)^2 . \end{aligned} \tag{A.42}$$

References

[A1] V. P. Akulov, D. P. Sorokin, I. A. Bandos, Particle mechanics in harmonic superspace, *Mod. Phys. Lett.* **A 3** (1988) 1633–1645.

[A2] L. Alvarez-Gaumé, D. Z. Freedman, Ricci-flat Kähler manifolds and supersymmetry, *Phys. Lett.* **B 94** (1980) 171–173.

[A3] L. Alvarez-Gaumé, D. Z. Freedman, Geometrical structure and ultraviolet finiteness in the supersymmetric σ model, *Commun. Math. Phys.* **80** (1981) 443–451.

[A4] L. Alvarez-Gaumé, D. Z. Freedman, Potentials for the supersymmetric non-linear sigma model, *Commun. Math. Phys.* **91** (1983) 97–101.

[A5] L. Alvarez-Gaumé, P. Ginsparg, Finiteness of Ricci flat supersymmetric non-linear sigma models, *Commun. Math. Phys.* **102** (1985) 311–326.

[A6] L. Andrianopoli, S. Ferrara, E. Sokatchev, B. M. Zupnik, Shortening of primary operators in N-extended $SCFT_4$ and harmonic-superspace analyticity, *Adv. Theor. Math. Phys.* **4** (2000) 1149–1179.

[A7] M. F. Atiyah, *Geometry of Yang–Mills Fields*, Academia Nazionale del Lincei Scuola Normale Superiore, Pisa, 1979.

[A8] M. F. Atiyah, N. J. Hitchin, Low-energy scattering of non-abelian monopoles, *Phys. Lett.* **A 107** (1985) 21–25.

[A9] M. F. Atiyah, N. J. Hitchin, V. G. Drinfeld, Yu. I. Manin, Construction of instantons, *Phys. Lett.* **A 65** (1978) 185–187.

[B1] J. Bagger, Supersymmetric sigma models, In *Proceedings of the Bonn-NATO Advanced Study Institute on Supersymmetry*, New York, Plenum, 1985, p. 213–257.

[B2] J. Bagger, A. S. Galperin, E. A. Ivanov, V. I. Ogievetsky, Gauging $N = 2$ sigma models in harmonic superspace, *Nucl. Phys.* **B 303** (1988) 522–542.

[B3] J. Bagger, E. Witten, The gauge invariant supersymmetric non-linear sigma model, *Phys. Lett.* **B 118** (1983) 103–106.

[B4] J. Bagger, E. Witten, Matter couplings in $N = 2$ supergravity, *Nucl. Phys.* **B 222** (1983) 1–10.

[B5] A. A. Belavin, A. M. Polyakov, A. S. Schwarz, Yu. S. Tyupkin, Pseudoparticle solutions of the Yang–Mills equations, *Phys. Lett.* **B 59** (1975) 85–87.

[B6] A. Belavin, V. Zacharov, Yang–Mills equations and inverse scattering problem, *Phys. Lett.* **B 73** (1977) 53–57.

[B7] F. Berezin, *The Method of Second Quantization*, Academic Press, New York, 1966.

[B8] E. Bergshoeff, M. de Roo, B. de Wit, Extended conformal supergravity, *Nucl. Phys.* **B 182** (1981) 173–204.

[B9] L. Biedenharn, J. D. Louck, *Angular Momentum in Quantum Physics*, Addison-Wesley, Reading, 1981.

[B10] P. Breitenlohner, M. F. Sohnius, An almost simple off-shell version of SU(2) Poincaré supergravity, *Nucl. Phys.* **B 178** (1981) 152–176.

[B11] L. Brink, O. Lindgren, B. Nilsson, The ultra-violet finiteness of the $N = 4$ Yang–Mills theory, *Phys. Lett.* **B 123** (1983) 323–328.

[B12] L. Brink, O. Lindgren, B. Nilsson, $N = 4$ Yang–Mills theory on the light cone, *Nucl. Phys.* **B 212** (1983) 401–412.

[B13] L. Brink, J. H. Schwarz, J. Scherk, Supersymmetric Yang–Mills theories, *Nucl. Phys.* **B 121** (1977) 77–92.

[B14] E. I. Buchbinder, I. L. Buchbinder, E. A. Ivanov, S. M. Kuzenko, Central charge as the origin of holomorphic effective action in $N = 2$ gauge theory, *Mod. Phys. Lett.* **A 13** (1998) 1071–1082.

[B15] I. L. Buchbinder, E. I. Buchbinder, E. A. Ivanov, S. M. Kuzenko, B. A. Ovrut, Effective action of the $N = 2$ Maxwell multiplet in harmonic superspace, *Phys. Lett.* **B 412** (1997) 309–319.

[B16] I. L. Buchbinder, E. I. Buchbinder, S. M. Kuzenko, Non-holomorphic effective potential in $N = 4$ $SU(n)$ SYM, *Phys. Lett.* **B 446** (1999) 216–233.

[B17] I. L. Buchbinder, S. M. Kuzenko, *Ideas and Methods of Supersymmetry and Supergravity*, IOP Publishing, Bristol and Philadelphia, 1995.

[C1] E. Calabi, Métriques Kähleriennes et fibrés holomorphes, *Ann. Scient. Ecole Norm. Sup.* **12** (1979) 269–294.

[C2] W. B. Campbell, Tensor and spin spherical functions ${}_sY_{lm}(\theta, \phi)$, *J. Math. Phys.* **12** (1971) 1763–1770.

[C3] R. Camporesi, Harmonic analysis and propagators on homogeneous spaces, *Phys. Rep.* **196** (1990) N 1–2, 1–133.

[C4] E. Cartan, *Leçons sur la Géométrie des Espaces de Riemann*, Gauthier-Villars, Paris, 1946.

[C5] L. Castellani, A. Ceresole, R. D'Auria, S. Ferrara, P. Fré, E. Maina, *Nucl. Phys.* **B 268** (1986) 317–348.

[C6] L. Castellani, S. J. Gates, P. van Nieuwenhuizen, Constraints for $N = 2$ superspace from extended supergravity in ordinary space, *Phys. Rev.* **D 22** (1980) 2364–2370.

[C7] S. Cecotti, Homogeneous Kähler manifolds and T-algebras in $N = 2$ supergravity and superstrings, *Commun. Math. Phys.* **124** (1989) 23–55.

[C8] S. Cecotti, Singularity-theory and $N = 2$ supersymmetry, *Int. J. Mod. Phys.* **A 6** (1991) 2427–2496.

[C9] S. Cecotti, S. Ferrara, L. Girardello, Geometry of type II superstrings and the moduli of superconformal field theories, *Int. J. Mod. Phys.* **A 4** (1989) 2475–2529.

[C10] S. Cecotti, S. Ferrara, L. Girardello, M. Porrati, Super-Kähler geometry in supergravity and superstrings, *Phys. Lett.* **B 185** (1987) 345–350.

[C11] S. Coleman, J. Wess, B. Zumino, Structure of phenomenological Lagrangians I, *Phys. Rev.* **177** (1969) 2239–2247; C.Callan, S. Coleman, J. Wess, B. Zumino, Structure of phenomenological Lagrangians II, *Phys. Rev.* **177** (1969) 2247–2268.

[C12] E. Cremmer, C. Kounnas, A. Van Proeyen, J. P. Derendinger, S. Ferrara, B. de Wit, L. Girardello, Vector multiplets coupled to N = 2 supergravity: super-Higgs effect, flat potentials and geometric structure, *Nucl. Phys.* **B 250** (1985) 385–426.

[C13] T. L. Curtright, D. Z. Freedman, Nonlinear σ-model with extended supersymmetry in four dimensions, *Phys. Lett.* **B 90** (1980) 71–74.

[D1] B. Delamotte, F. Delduc, Renormalization properties of the self-interactions of a massive vector superfield, *Nucl. Phys.* **B 267** (1986) 473–481.

[D2] B. Delamotte, F. Delduc, P. Fayet, $N = 2$ massive gauge superfields in harmonic superspace, *Phys. Lett.* **B 176** (1986) 409–415.

[D3] B. Delamotte, P. Fayet, Spontaneous electroweak breaking and massive gauge superfields in six dimensions, *Phys. Lett.* **B 195** (1987) 563–568.

[D4] F. Delduc, A. Galperin, E. Sokatchev, Lorentz-harmonic (super)fields and (super)particles, *Nucl. Phys.* **B 368** (1992) 143–174.

[D5] F. Delduc, E. Ivanov, $N = 4$ super KdV equation, *Phys. Lett.* **B 309** (1993) 312–319.

[D6] F. Delduc, S. Kalitzin, E. Sokatchev, Learning the ABC of light-cone harmonic space, *Class. Quantum Grav.* **6** (1989) 1561–1575.

[D7] F. Delduc, S. Kalitzin, E. Sokatchev, Geometry of sigma models with heterotic supersymmetry, *Class. Quantum Grav.* **7** (1990) 1567–1582.

[D8] F. Delduc, J. McCabe, The quantization of $N = 3$ super-Yang–Mills off-shell in superspace, *Class. Quantum Grav.* **6** (1989) 233–254.

[D9] F. Delduc, E. Sokatchev, Superparticle with extended world-line supersymmetry, *Class. Quantum Grav.* **9** (1992) 361–376.

[D10] J. P. Derendinger, S. Ferrara, A. Masiero, Exceptional ultraviolet finite Yang–Mills theories, *Phys. Lett.* **B 143** (1984) 133–136.

[D11] S. Deser, B. Zumino, Consistent supergravity, *Phys. Lett.* **B 62** (1976) 335–337.

[D12] Ch. Devchand, V. Ogievetsky, Superselfduality as analyticity in harmonic superspace, *Phys. Lett.* **B 297** (1992) 93–98.

[D13] Ch. Devchand, V. Ogievetsky, The structure of all extended supersymmetric self-dual gauge theories, *Nucl. Phys.* **B 414** (1994) 763–782.

[D14] Ch. Devchand, V. Ogievetsky, Selfdual supergravities, *Nucl. Phys.* **B 444** (1995) 381–400.

[D15] B. de Wit, P. G. Lauwers, R. Philippe, S.-Q. Su, A. Van Proeyen, Gauge and matter fields coupled to $N = 2$ supergravity, *Phys. Lett.* **B 134** (1984) 37–43.

[D16] B. de Wit, P. G. Lauwers, A. Van Proeyen, Lagrangians of $N = 2$ supergravity matter systems, *Nucl. Phys.* **B 255** (1985) 569–608.

[D17] B. de Wit, R. Philippe, A. Van Proeyen, The improved tensor multiplet in $N = 2$ supergravity, *Nucl. Phys.* **B 219** (1983) 143–166.

[D18] B. de Wit, M. Roček, Improved tensor multiplets, *Phys. Lett.* **B 109** (1982) 439–443.

[D19] B. de Wit, J. W. van Holten, Multiplets of linearized SO(2) supergravity, *Nucl. Phys.* **B 155** (1979) 530–542.

[D20] B. de Wit, J. W. van Holten, A. Van Proeyen, Transformation rules of $N = 2$ supergravity multiplets, *Nucl. Phys.* **B 167** (1980) 186–204.

[D21] B. de Wit, J.W. van Holten, A. Van Proeyen, Structure of supergravity, *Nucl. Phys.* **B 184** (1981) 77–109; **B 222** (1983) 516 E.

[D22] B. de Wit, A. Van Proeyen, Potentials and symmetries of general gauged $N = 2$ supergravity-Yang–Mills models, *Nucl. Phys.* **B 245** (1984) 89–117.

[D23] N. Dragon, Torsion and curvature in extended supergravity, *Z. Phys.* **C 2** (1979) 29–32.

[D24] N. Dragon, E. Ivanov, S. M. Kuzenko, E. Sokatchev, U. Theis, N = 2 rigid supersymmetry with gauged central charge, *Nucl. Phys.* **B 538** (1998) 411–450.

[D25] N. Dragon, S. M. Kuzenko, The Higgs mechanism in $N = 2$ superspace, *Nucl. Phys.* **B 508** (1997) 229–244.

[D26] N. Dragon, S. M. Kuzenko, U. Theis, The vector-tensor multiplet in harmonic superspace, *Eur. Phys. J.* **C 4** (1997) 717–721.

[D27] B. Dubrovin, S. Novikov, A. Fomenko, *Modern Geometry: Methods and Applications*, part 1, New York, Springer, 1984.

[E1] B. Eden, P. S. Howe, C. Schubert, E. Sokatchev, P. C. West, Four-point functions in $N = 4$ supersymmetric Yang–Mills theory at two loops, *Nucl. Phys.* **B 557** (1999) 355–379.

[E2] B. Eden, P. S. Howe, C. Schubert, E. Sokatchev, P. C. West, Simplifications of four-point functions in $N = 4$ supersymmetric Yang–Mills theory at two loops, *Phys. Lett.* **B 466** (1999) 20–26.

[E3] B. Eden, P. S. Howe, C. Schubert, E. Sokatchev, P. C. West, Extremal correlators in four-dimensional SCFT, *Phys. Lett.* **B 472** (2000) 323–331.

[E4] B. Eden, C. Schubert, E. Sokatchev, Three-loop four-point correlator in $N = 4$ SYM, *Phys. Lett.* **B 482** (2000) 309–314.

[E5] B. Eden, A. C. Petkou, C. Schubert, E. Sokatchev, Partial non-renormalisation of the stress-tensor four-point function in $N = 4$ SYM and the AdS/CFT correspondence, `hep-th/0009106`.

[E6] T. Eguchi, P. Gilkey, A. Hanson, Gravitation, gauge theories and differential geometry, *Phys. Rep.* **66** (1980) 213–393.

[E7] T. Eguchi, A. Hanson, Self-dual solutions to Euclidean gravity, *Ann. Phys.* **120** (1979) 82–106.

[E8] T. Eguchi, A. Hanson, Gravitational instantons, *Gen. Relat. Grav.* **11** (1979) 315–320.

[E9] S. Eremin, E. Ivanov, Holomorphic effective action of $N = 2$ SYM theory from harmonic superspace with central charges, *Mod. Phys. Lett.* **A 15** (2000) 1859–1878.

[E10] M. Evans, F. Gürsey, V. Ogievetsky, From 2-D conformal to 4-D selfdual theories: quaternionic analyticity, *Phys. Rev.* **D 47** (1993) 3496–3508.

[F1] P. Fayet, Fermi–Bose hypersymmetry, *Nucl. Phys.* **B 113** (1976) 135–155.

[F2] P. Fayet, Supersymmetric theories of particles and interactions, *Phys. Rep.* **105** (1984) 21–51.

[F3] P. Fayet, Supersymmetric theories of particles and interactions, *Phys. Scr.* **T 15** (1987) 46–60.

[F4] P. Fayet, S. Ferrara, Supersymmetry, *Phys. Rep.* **32** (1977) 249–334.

[F5] P. Fayet, J. Iliopoulos, Spontaneously broken supergauge symmetries and Goldstone fermions, *Phys. Lett.* **B 51** (1974) 461–464.

[F6] A. Ferber, Supertwistors and conformal supersymmetry, *Nucl. Phys.* **B 132** (1978) 55–64.

[F7] S. Ferrara, Supersymmetry and fundamental particle interactions, *Phys. Rep.* **105** (1984) 5–19.

[F8] S. Ferrara, P. Fré, L. Girardello, Spontaneously broken $N = 3$ supergravity, *Nucl. Phys.* **B 274** (1986) 600–618.

[F9] S. Ferrara, C. A. Savoy, L. Girardello, Spin sum rules in extended supersymmetry, *Phys. Lett.* **B 105** (1981) 363–368.

[F10] S. Ferrara, P. van Nieuwenhuizen, Consistent supergravity with complex spin 3/2 gauge fields, *Phys. Rev. Lett.* **37** (1976) 1669–1671.

[F11] S. Ferrara, P. van Nieuwenhuizen, The auxiliary fields of supergravity, *Phys. Lett.* **B 74** (1978) 333–335.

[F12] S. Ferrara, B. Zumino, Supergauge invariant Yang–Mills theories, *Nucl. Phys.* **B 79** (1974) 413–421.

[F13] S. Ferrara, B. Zumino, J. Wess, Supergauge multiplets and superfields, *Phys. Lett.* **B 51** (1974) 239–241.

[F14] E. S. Fradkin, A. A. Tseytlin, Conformal supergravity, *Phys. Rep.* **119** (1985) 233–362.

[F15] E. S. Fradkin, M. A. Vasiliev, Minimal set of auxiliary fields in $SO(2)$-extended supergravity, *Phys. Lett.* **B 85** (1979) 47–51.

[F16] D. Z. Freedman, P. van Nieuwenhuizen, S. Ferrara, Progress toward a theory of supergravity, *Phys. Rev.* **B 13** (1976) 3214–3218.

[G1] K. Galicki, Quaternionic Kähler and hyperkähler non-linear sigma models, *Nucl. Phys.* **B 271** (1986) 402–416.

[G2] K. Galicki, New matter couplings in $N = 2$ supergravity, *Nucl. Phys.* **B 289** (1987) 573–588.

[G3] A. Galperin, P. S. Howe, K. S. Stelle, The superparticle and the Lorentz group, *Nucl. Phys.* **B 368** (1992) 248–280.

[G4] A. Galperin, E. Ivanov, S. Kalitzin, V. Ogievetsky, E. Sokatchev, Unconstrained $N = 2$ matter, Yang–Mills and supergravity theories in harmonic superspace, *Class. Quantum Grav.* **1** (1984) 469–498.

[G5] A. Galperin, E. Ivanov, S. Kalitzin, V. Ogievetsky, E. Sokatchev, $N = 3$ supersymmetric gauge theory, *Phys. Lett.* **B 151** (1985) 215–218.

[G6] A. Galperin, E. Ivanov, S. Kalitzin, V. Ogievetsky, E. Sokatchev, Unconstrained off-shell $N = 3$ supersymmetric Yang–Mills theory, *Class. Quantum Grav.* **2** (1985) 155–166.

[G7] A. Galperin, E. Ivanov, O. Ogievetsky, Harmonic space and quaternionic manifolds, *Ann. Phys.* **230** (1994) 201–249.

[G8] A. Galperin, E. Ivanov, V. Ogievetsky, Grassmann analyticity and extended supersymmetries, *JETP Lett.* **33** (1981) 176–181.

[G9] A. Galperin, E. Ivanov, V. Ogievetsky, Superfield anatomy of the Fayet-Sohnius multiplet, *Sov. J. Nucl. Phys.* **35** (1982) 458–463.

[G10] A. Galperin, E. Ivanov, V. Ogievetsky, Superspace actions and duality transformations for $N = 2$ tensor multiplets, *Phys. Scr.* **T 15** (1987) 176–183.

[G11] A. Galperin, E. Ivanov, V. Ogievetsky, Duality transformations and most general matter self-couplings in $N = 2$ supersymmetry, *Nucl. Phys.* **B 282** (1987) 74–102.

[G12] A. S. Galperin, E. A. Ivanov, V. I. Ogievetsky, Superspaces for $N = 3$ supersymmetry, *Sov. J. Nucl. Phys.* **46** (1987) 543–556.

[G13] A. Galperin, E. Ivanov, V. Ogievetsky, E. Sokatchev, Harmonic superspace as a key to supersymmetric theories, *JETP Lett.* **40** (1984) 155–158.

[G14] A. Galperin, E. Ivanov, V. Ogievetsky, E. Sokatchev, Harmonic supergraphs: Green functions, *Class. Quantum Grav.* **2** (1985) 601–616.

[G15] A. Galperin, E. Ivanov, V. Ogievetsky, E. Sokatchev, Harmonic supergraphs: Feynman rules and examples, *Class. Quantum Grav.* **2** (1985) 617–630.

[G16] A. Galperin, E. Ivanov, V. Ogievetsky, E. Sokatchev, Conformal invariance in harmonic superspace, prepr. JINR E2-85-363, (1985), publ. in *Quantum Field theory and Quantum Statistics*, eds. I. Batalin, C. J. Isham, G. Vilkovisky, vol.2, Adam Hilger, Bristol, 1987, p. 233–248.

[G17] A. Galperin, E. Ivanov, V. Ogievetsky, E. Sokatchev, Hyper-Kähler metrics and harmonic superspace, *Commun. Math. Phys.* **103** (1986) 515–526.

[G18] A. Galperin, E. Ivanov, V. Ogievetsky, E. Sokatchev, Harmonic superspace in action: General $N = 2$ matter self-couplings, in *Supersymmetry, Supergravity, Superstrings*, eds. B. de Wit, P. Fayet, M. Grisaru, Singapore, World Scientific, 1986, p. 511–565.

[G19] A. Galperin, E. Ivanov, V. Ogievetsky, E. Sokatchev, $N = 2$ supergravity in superspace: Different versions and matter couplings, *Class. Quantum Grav.* **4** (1987) 1255–1265.

[G20] A. Galperin, E. Ivanov, V. Ogievetsky, E. Sokatchev, Gauge field geometry from complex and harmonic analyticities. I. Kähler and self-dual Yang–Mills cases, *Ann. Phys.* **185** (1988) 1–21.

[G21] A. Galperin, E. Ivanov, V. Ogievetsky, E. Sokatchev, Gauge field geometry from complex and harmonic analyticities. II. Hyper-Kähler case, *Ann. Phys.* **185** (1988) 22–45.

[G22] A. Galperin, E. Ivanov, V. Ogievetsky, P. Townsend, Eguchi–Hanson type metrics from harmonic superspace, *Class. Quantum Grav.* **3** (1986) 625–633.

[G23] A. Galperin, Nguyen Anh Ky, E. Sokatchev, $N = 2$ supergravity in superspace: Solution to the constraints and the invariant action, *Class. Quantum Grav.* **4** (1987) 1235–1254.

[G24] A. Galperin, Anh Ky Nguyen, E. Sokatchev, Coinciding harmonic singularities in harmonic supergraphs, *Mod. Phys. Lett.* **A 2** (1987) 33–39.

[G25] A. Galperin, V. Ogievetsky, $N = 2$ $D = 4$ supersymmetric sigma models and Hamiltonian mechanics, *Class. Quantum Grav.* **8** (1991) 1757–1764.

[G26] S. J. Gates, M. T. Grisaru, M. Roček, W. Siegel, *Superspace*, Benjamin/Cummings, Reading, Mass., 1983.

[G27] S. J. Gates, W. Siegel, Linearized $N = 2$ superfield supergravity, *Nucl. Phys.* **B 195** (1982) 39–60.

[G28] S. J. Gates, K. S. Stelle, P. C. West, Algebraic origins of superspace constraints in supergravity, *Nucl. Phys.* **B 169** (1980) 347–364.

[G29] J. P. Gauntlett, G. W. Gibbons, G. Papadopoulos, P. K. Townsend, Hyper-Kähler manifolds and multiply intersecting branes, *Nucl. Phys.* **B 500** (1997) 133–162.

[G30] I. M. Gelfand, M. I. Graev, N. Y. Vilenkin, *Generalized Functions*, vol.5: *Integral Geometry and Representation Theory*, New York, Academic Press, 1966.

[G31] I. M. Gelfand, R. A. Minlos, Z. Ya. Shapiro, *Representations of the Rotation and Lorentz Groups and Their Applications*, London, Pergamon, 1963.

[G32] G. W. Gibbons, S. W. Hawking, Gravitational multi-instantons, *Phys. Lett.* **B 78** (1978) 430–432.

[G33] G. W. Gibbons, S. W. Hawking, Classification of gravitational instanton symmetries, *Commun. Math. Phys.* **66** (1979) 291–310.

[G34] G. W. Gibbons, D. Olivier, P. J. Ruback, G. Valent, Multicentre metrics and harmonic superspace, *Nucl. Phys.* **B 296** (1988) 679–696.

[G35] G. W. Gibbons, P. J. Ruback, The hidden symmetries of multicentre metrics, *Commun. Math. Phys.* **115** (1988) 267–300.

[G36] G. W. Gibbons, P. Rychenkova, R. Goto, Hyper-Kähler quotient construction of BPS monopole moduli spaces, *Commun. Math. Phys.* **186** (1997) 581–599.

[G37] J. N. Goldberg, A. J. Macfarlane, E. T. Newman, F. Rohrlich, E. C. G. Sudarshan, Spin-s spherical harmonics and edth, *J. Math. Phys.* **8** (1967) 2155–2161.

[G38] Yu. A. Golfand, E. P. Lichtman, Extension of the algebra of Poincaré group generators and breakdown of P-invariance, *JETP Lett.* **13** (1971) 452–455.

[G39] Yu. A. Golfand, E. P. Lichtman, On extensions of the Poincaré algebra by bispinor generators, in *Problems of Theoretical Physics*, Nauka, Moskwa, 1972, p. 37–45.

[G40] R. Grimm, M. Sohnius, J. Wess, Extended supersymmetry and gauge theories, *Nucl. Phys.* **B 133** (1978) 275–284.

[G41] M. T. Grisaru, M. Roček, W. Siegel, Improved methods for supergraphs, *Nucl. Phys.* **B 159** (1979) 429–450.

[G42] S. S. Gubser, I. R. Klebanov, A. M. Polyakov, Gauge theory correlators from non-critical string theory, *Phys. Lett.* **B 428** (1998) 105–114.

[G43] M. Günaydin, G. Sierra, P. K. Townsend, Exceptional supergravity theories and the magic square, *Phys. Lett.* **B 133** (1983) 72–76.

[G44] F. Gürsey, H. C. Tze, Complex and quaternionic analyticity in chiral and gauge theories. I, *Ann. Phys.* **128** (1980) 29–130.

[H1] R. Haag, J.T. Lopuszansky, M. Sohnius, All possible generators of supersymmetries of the S-matrix, *Nucl. Phys.* **B 88** (1975) 257–274.

[H2] R. Hansen, E. T. Newman, P. Tod, R. Penrose, The metric and curvature of H space, *Proc. Roy. Soc.* **A 363** (1978) 445–468.

[H3] G. G. Hartwell, P. S. Howe, (N, P, Q) harmonic superspace, *Int. J. Mod. Phys.* **A 10** (1995) 3901–3920.

[H4] S. W. Hawking, G. F. R. Ellis, *The Large Scale Structure of Space-Time*, Cambridge University Press, 1973.

[H5] S. Helgason, *Differential Geometry and Symmetric Spaces*, Academic Press, New York and London, 1978.

[H6] S. Helgason, *Groups and Geometric Analysis. Integral Geometry, Invariant Differential Operators and Spherical Functions*, Academic Press, New York, 1984.

[H7] N. J. Hitchin, *Monopoles, Minimal Surfaces and Algebraic Curves*, SMS 105, Montreal, 1987.

[H8] N. J. Hitchin, A. Karlhede, U. Lindström, M. Roček, Hyperkähler metrics and supersymmetry, *Commun. Math. Phys.* **108** (1987) 535–589.

[H9] P. S. Howe, Supergravity in superspace, *Nucl. Phys.* **B 199** (1984) 309–364.

[H10] P. S. Howe, G. G. Hartwell, A superspace survey, *Class. Quantum Grav.* **12** (1995) 1823–1880.

[H11] P. S. Howe, A. Karlhede, U. Lindström, M. Roček, The geometry of duality, *Phys. Lett.* **B 168** (1986) 89–92.

[H12] P. S. Howe, M. I. Leeming, Harmonic superspaces in low dimensions, *Class. Quantum Grav.* **11** (1994) 2843–2852.

[H13] P. S. Howe, G. Sierra, P. K. Townsend, Supersymmetry in six dimensions, *Nucl. Phys.* **B 221** (1983) 331–348.

[H14] P. S. Howe, E. Sokatchev, P. C. West, Three point functions in $N = 4$ Yang–Mills, *Phys. Lett.* **B 444** (1998) 341–351.

[H15] P. S. Howe, K. S. Stelle, P. K. Townsend, The relaxed hypermultiplet: the unconstrained $N = 2$ superfield theory, *Nucl. Phys.* **B 214** (1983) 519–531.

[H16] P. S. Howe, K. S. Stelle, P. K. Townsend, Miraculous ultraviolet cancellations in supersymmetry made manifest, *Nucl. Phys.* **B 236** (1984) 125–166.

[H17] P. S. Howe, K. S. Stelle, P. C. West, A class of finite four-dimensional supersymmetric field theories, *Phys. Lett.* **B 124** (1983) 55–58.

[H18] P. S. Howe, K. S. Stelle, P. C. West, $N = 1, d = 6$ harmonic superspace, *Class. Quantum Grav.* **2** (1985) 815–821.

[H19] P. S. Howe, P. C. West, Operator product expansions in four-dimensional superconformal field theories, *Phys. Lett.* **B 389** (1996) 273–279.

[H20] P. S. Howe, P. C. West, Superconformal invariants and extended supersymmetry, *Phys. Lett.* **B 400** (1997) 307–313.

[H21] P. S. Howe, P. C. West, Non-perturbative Green's functions in theories with extended superconformal symmetry, *Int. J. Mod. Phys.* **A 14** (1999) 2659–2674.

[H22] S. H. Hsieh, S. L. Kent, E. T. Newman, Special solutions of the Sparling equation, *J. Math. Phys.* **27** (1986) 2043–2046.

[H23] C. Hull, A. Karlhede, U. Lindström, M. Roček, Nonlinear sigma models and their gauging in and out of superspace, *Nucl. Phys.* **B 266** (1986) 1–44.

[I1] C. Itzykson, J.-B. Zuber, *Quantum Field Theory*, McGraw-Hill, New York, 1980.

[I2] J. Ivancovich, C. Kozameh, E. Newman, Green's functions of the Edth operator, *J. Math. Phys.* **30** (1989) 45–52.

[I3] E. Ivanov, On the geometric meaning of the $N = 1$ Yang–Mills prepotential, *Phys. Lett.* **B 117** (1982) 59–63.

[I4] E. A. Ivanov, Intrinsic geometry of the $N = 1$ supersymmetric Yang–Mills theory, *J. Phys. A: Math. Gen.* **16** (1983) 2571–2586.

[I5] E. Ivanov, Composite gauge fields and higher derivative couplings in $N = 2$ supersymmetry, *Phys. Lett.* **B 205** (1988) 499–503.

[I6] E. A. Ivanov, Off-shell (4,4) supersymmetric sigma models with torsion as gauge theories in harmonic superspace, *Phys. Lett.* **B 356** (1995) 239–248.

[I7] E. Ivanov, S. Kalitzin, Nguyen Ai Viet, V. Ogievetsky, Harmonic superspaces of extended supersymmetry. The calculus of harmonic variables, *J. Phys. A: Math. Gen.* **18** (1985) 3433–3443.

[I8] E. A. Ivanov, S. V. Ketov, B. M. Zupnik, Induced hypermultiplet self-interactions in $N = 2$ gauge theories, *Nucl. Phys.* **B 509** (1998) 53–82.

[I9] E. A. Ivanov, J. Niederle, $N = 1$ supergravity as a non-linear realization, *Phys. Rev.* **D 45** (1992) 4545–4554.

[I10] E. Ivanov, A. Sutulin, Sigma models in (4,4) harmonic superspace, *Nucl. Phys.* **B 432** (1994) 246–280; **B 483** (1997) 531E.

[I11] E. Ivanov, G. Valent, Quaternionic Taub-NUT from the harmonic space approach, *Phys. Lett.* **B 445** (1998) 60–68.

[I12] E. Ivanov, G. Valent, Quaternionic metrics from harmonic superspace: Lagrangian approach and quotient construction, *Nucl. Phys.* **B 576** (2000) 543–577.

[K1] M. Kaku, P. K. Townsend, Poincaré supergravity as broken superconformal gravity, *Phys. Lett.* **B 76** (1978) 54–58.

[K2] S. Kalitzin, E. Nissimov, S. Pacheva, $N = 1$ superfields and $N = 2$ harmonic superfields in four-dimensions as second quantized superparticle, *Mod. Phys. Lett.* **A 2** (1987) 651–661.

[K3] S. Kalitzin, E. Sokatchev, Multi-instanton solutions in the harmonic approach to self-dual Yang–Mills equations, *Class. Quantum Grav.* **4** (1987) L173-L17.

[K4] R. Kallosh, $N = 3$ harmonic superspace, *Pis'ma ZhETF* **41** (1985) 172–174.

[K5] R. Kallosh, Superstrings and harmonic superspace, in *Quantum Field Theory and Quantum Statistics*, vol.2, eds. I. Batalin, C. J. Isham, G. Vilkovisky, Adam Hilger, Bristol, 1987, p. 485–505.

[K6] A. Karlhede, U. Lindström, M. Roček, Selfinteracting tensor multiplets in $N = 2$ superspace, *Phys. Lett.* **B 147** (1984) 297–300.

[K7] A. Karlhede, U. Lindström, M. Roček, Hyperkahler manifolds and non-linear supermultiplets, *Commun. Math. Phys.* **108** (1987) 529–534.

[K8] M. Ko, M. Ludvigsen, E. T. Newman, P. Tod, The theory of H space, *Phys. Rep.* **71** (1981) 53–139.

[K9] S. Kobayashi, K. Nomizu, *Foundations of Differential Geometry*, Wiley-Interscience, 1969.

[K10] J. Koller, Unconstrained prepotentials in extended superspace, *Phys. Lett.* **B 124** (1983) 324–328.

[K11] J. Koller, A six dimensional superspace approach to extended superfields, *Nucl. Phys.* **B 222** (1983) 319–337.

[K12] T. Kugo, P. Townsend, Supersymmetry and division algebras, *Nucl. Phys.* **B 221** (1983) 357–380.

[L1] U. Lindström, M. Roček, Scalar-tensor duality and $N = 1, 2$ non-linear sigma models, *Nucl. Phys.* **B 222** (1983) 285–308.

[L2] U. Lindström, M. Roček, New hyperkahler metrics and new supermultiplets, *Commun. Math. Phys.* **115** (1988) 21–29.

[L3] U. Lindström, M. Roček, $N = 2$ super Yang–Mills theory in projective superspace, *Commun. Math. Phys.* **128** (1990) 191–196.

[M1] J. Maldacena, The large N limit of superconformal field theories and supergravity, *Adv. Theor. Math. Phys.* **2** (1998) 231–252.

[M2] S. Mandelstam, Light cone superspace and ultraviolet finiteness of the $N = 4$ model, *Nucl. Phys.* **B 213** (1983) 149–168.

[M3] L. Mezincescu, On superfield formulation of $O(2)$-supersymmetry, Dubna preprint JINR-P2-12572 (1979).

[M4] K. J. Muck, Finiteness of (4,4) non-linear sigma models in light cone harmonic superspace, *Phys. Lett.* **B 221** (1989) 314–318.

[N1] E. T. Newman, Gauge theories, the holonomy operator and the Riemann–Hilbert problem, *J. Math. Phys.* **27** (1986) 2797–2802.

[N2] H.-P. Nilles, Supersymmetry, supergravity and particle physics, *Phys. Rep.* **110** (1984) 1–162.

[N3] E. Nissimov, S. Pacheva, S. Solomon, Covariant first and second quantization of the $N = 2$, $D = 10$ Brink–Schwarz superparticle, *Nucl. Phys.* **B 296** (1988) 462–492.

[N4] E. Nissimov, S. Pacheva, S. Solomon, Covariant canonical quantization of the Green–Schwarz superstring, *Nucl. Phys.* **B 297** (1988) 349–373.

[N5] E. Nissimov, S. Pacheva, S. Solomon, Covariant unconstrained superfield action for the linearized D=10 super-Yang–Mills theory, *Nucl. Phys.* **B 299** (1988) 183–205.

[O1] O. Ogievetsky, Harmonic representatives of instantons and self-dual monopoles, in *Proc. of Conf. on Group Theor. Methods in Physics* (Varna, 1987), Springer, 1988, p. 548–554.

[O2] O. V. Ogievetsky, Instantons and topological field theories, PhD Thesis, Physical Lebedev Institute, Moscow, 1989, p. 1–115.

[O3] V. I. Ogievetsky, in *Proc. of X-th Winter School of Theor. Physics in Karpach* (1974), vol.1, Wroclaw, 1974, p. 117–132.

[O4] V. I. Ogievetsky, L. Mezincescu, Symmetries between bosons and fermions and superfields, *Usp. Fiz. Nauk* **117** (1975) 637–683.

[O5] V. I. Ogievetsky, I. V. Polubarinov, Notoph and its possible interactions, *Sov. J. Nucl. Phys.* **4** (1967) 156–161.

[O6] V. Ogievetsky, E. Sokatchev, Structure of supergravity group, *Phys. Lett.* **B 79** (1978) 222–224.

[O7] V. I. Ogievetsky, E. S. Sokatchev, The gravitational axial superfield and the formalism of differential geometry, *Sov. J. Nucl. Phys.* **31** (1980) 424–433.

[O8] N. Ohta, H. Sugata, H. Yamaguchi, $N = 2$ harmonic superspace with central charges and its application to self-interacting massive hypermultiplets, *Ann. Phys.* **172** (1986) 26–39.

[O9] N. Ohta, H. Yamaguchi, Superfield perturbation theory in harmonic superspace, *Phys. Rev.* **D 32** (1985) 1954–1967.

[O10] D. Olivier, G. Valent, Hyperkähler generalization of Taub-NUT from harmonic superspace, *Phys. Lett.* **B 189** (1987) 79–82.

[P1] H. Pedersen, Y.-S. Poon, Hyper-Kähler metrics and a generalization of the Bogomolny equations, *Commun. Math. Phys.* **117** (1988) 569–580.

[P2] R. Penrose, Twistor algebra, *J. Math. Phys.* **8** (1967) 345–366.

[P3] R. Penrose, Nonlinear gravitons and curved twistor theory, *Gen. Relat. Grav.* **7** (1976) 31–52.

[P4] R. Penrose, R. S. Ward, in *General Relativity and Gravitation*, vol.2, ed. A. Held, Plenum, New York/London, 1980, p. 283–297.

[R1] V. Rittenberg, E. Sokatchev, Decomposition of extended superfields into irreducible representations of supersymmetry, *Nucl. Phys.* **B 193** (1981) 477–501.

[R2] V. O. Rivelles, J. G. Taylor, Off-shell no-go theorems for higher dimensional supersymmetries and supergravities, *Phys. Lett.* **B 121** (1983) 37–42.

[R3] M. Roček, Supersymmetry and non-linear sigma models, *Physica* **D 15** (1985) 75–82.

[R4] A. A. Rosly, Constraints in supersymmetric Yang–Mills theory as integrability conditions, in *Theory Group Methods in Physics*, eds. M. Markov, V. Man'ko, Nauka, Moskwa, 1983, p. 263–268.

[R5] A. A. Rosly, Gauge fields in superspace and twistors, *Class. Quantum Grav.* **2** (1985) 693–699.

[R6] A. A. Rosly, A. S. Schwarz, Supersymmetry in a space with auxiliary dimensions, *Commun. Math. Phys.* **105** (1986) 645–668.

[S1] A. Salam, J. Strathdee, Supergauge transformations, *Nucl. Phys.* **B 76** (1974) 477–482.

[S2] A. Salam, J. Strathdee, Supersymmetry and nonabelian gauges, *Phys. Lett.* **B 51** (1974) 353–355.

[S3] J. Scherk, J. Schwarz, How to get masses from extra dimensions, *Nucl. Phys.* **B 153** (1979) 61–88.

[S4] A. Schwarz, Supergravity, complex geometry and G-structures, *Commun. Math. Phys.* **87** (1982) 37–64.

[S5] N. Seiberg, E. Witten, Electric-magnetic duality, monopole condensation, and confinement in $N = 2$ supersymmetric Yang–Mills theory, *Nucl. Phys.* **B 426** (1994) 19–52; Erratum, *ibid*, **B 430** (1994) 485–486.

[S6] W. Siegel, Chiral actions for $N = 2$ supersymmetric tensor multiplets, *Phys. Lett.* **B 153** (1985) 51–54.

[S7] W. Siegel, S. J. Gates, Jr., Superfield supergravity, *Nucl. Phys.* **B 147** (1979) 77–104.

[S8] W. Siegel, M. Roček, On off-shell supermultiplets, *Phys. Lett.* **B 105** (1981) 278–289.

[S9] G. Sierra, P. K. Townsend, The hyperkähler supersymmetric σ-model in six dimensions, *Phys. Lett.* **B 124** (1983) 497–500.

[S10] G. Sierra, P. K. Townsend, The gauge invariant $N = 2$ supersymmetric sigma model with general scalar potential, *Nucl. Phys.* **B 233** (1984) 289–306.

[S11] M. F. Sohnius, Bianchi identities for supersymmetric gauge theories, *Nucl. Phys.* **B 136** (1978) 461–464.

[S12] M. F. Sohnius, Supersymmetry and central charges, *Nucl. Phys.* **B 138** (1978) 109–121.

[S13] M. F. Sohnius, Introducing supersymmetry, *Phys. Rep.* **128** (1985) 39–204.

[S14] M. F. Sohnius, K. S. Stelle, P. C. West, Representations of extended supersymmetry, in *Supergravity and Superspace*, eds. S. Hawking, M. Roček, Cambridge University Press, 1981, p. 283–330.

[S15] M. F. Sohnius, P. C. West, Conformal invariance in $N = 4$ supersymmetric Yang–Mills theory, *Phys. Lett.* **B 100** (1981) 245–250.

[S16] E. Sokatchev, Light-cone harmonic superspace and its applications, *Phys. Lett.* **B 169** (1986) 209–214.

[S17] E. Sokatchev, Off-shell six-dimensional supergravity in harmonic superspace, *Class. Quantum Grav.* **5** (1988) 1459–1471.

[S18] E. Sokatchev, An off-shell formulation of $N = 4$ supersymmetric Yang–Mills theory in twistor harmonic superspace, *Phys. Lett.* **B 217** (1989) 489–493.

[S19] E. Sokatchev, K. S. Stelle, Finiteness of (4,0) supersymmetric sigma models, *Class. Quantum Grav.* **4** (1987) 501–508.

[S20] D. Sorokin, Superbranes and superembeddings, *Phys. Rep.* **329** (2000) 1–101.

[S21] K. S. Stelle, Manifest realizations of extended supersymmetry, Santa Barbara preprint NSF-ITP-85-001.

[S22] K. S. Stelle, P. C. West, Minimal auxiliary fields for supergravity, *Phys. Lett.* **B 74** (1978) 330–332.

[S23] K. S. Stelle, P. C. West, Algebraic formulation of $N = 2$ supergravity constraints, *Phys. Lett.* **B 90** (1980) 393–397.

[U1] N. I. Usyukina, A. I. Davydychev, An approach to the evaluation of three and four point ladder diagrams, *Phys. Lett.* **B 298** (1993) 363–370.

[V1] P. van Nieuwenhuizen, Supergravity, *Phys. Rep.* **68** (1981) 189–398.

[V2] N. Ya. Vilenkin, *Special Functions and Theory of Representations of Groups*, Nauka, Moscow, 1965; *Fonctions Speciales et Théorie des Représentations des Groupes*, Paris, Dunod, 1969.

[V3] D. V. Volkov, Phenomenological Lagrangians, *Elem. Chast. Atom. Yadra* **4** (1973) 3–41.

[V4] D. V. Volkov, V. P. Akulov, On a possible universal neutrino interaction, *JETP Lett.* **16** (1972) 621–624.

[V5] D. V. Volkov, V. P. Akulov, Is the neutrino a Goldstone particle?, *Phys. Lett.* **B 46** (1973) 109–110.

[V6] D. V. Volkov, V. P. Akulov, The Goldstone fields of spin one half, *Teor. Mat. Fiz.* **18** (1973) 39–50.

[W1] R. S. Ward, On self-dual gauge fields, *Phys. Lett.* **A 61** (1977) 81–82.

[W2] R. S. Ward, A class of self-dual solutions of Einstein equations, *Proc. Roy. Soc. (London)* **A 363** (1978) 289–295.

[W3] R. S. Ward, Completely solvable gauge-field equations in dimension greater than four, *Nucl. Phys.* **B 236** (1984) 381–396.

[W4] R. S. Ward, R. O. Wells, *Twistor Geometry and Field Theory*, Cambridge University Press, 1989.

[W5] S. Weinberg, Nonlinear realizations of chiral symmetry, *Phys. Rev.* **166** (1968) 1568–1577.

[W6] J. Wess, Supersymmetry and internal symmetry, *Acta Phys. Austriaca* **41** (1975) 409–414.

[W7] J. Wess, J. Bagger, *Supersymmetry and Supergravity*, Princeton University Press, 1983.

[W8] J. Wess, B. Zumino, Supergauge transformations in four dimensions, *Nucl. Phys.* **B 70** (1974) 39–49.

[W9] J. Wess, B. Zumino, A Lagrangian model invariant under supergauge transformations, *Phys. Lett.* **B 49** (1974) 52–54.

[W10] J. Wess, B. Zumino, Supergauge invariant extension of quantum electrodynamics, *Nucl. Phys.* **B 78** (1974) 1–13.

[W11] J. Wess, B. Zumino, Superspace formulation of supergravity, *Phys. Lett.* **B 66** (1977) 361–364.

[W12] P. C. West, *Introduction to Supersymmetry and Supergravity*, World Scientific, 1990.

[W13] E. T. Whittaker, G. N. Watson, *A Course of Modern Analysis*, Fourth edition, Cambridge University Press, 1927.

[W14] E. Witten, An interpretation of classical Yang–Mills theory, *Phys. Lett.* **B 77** (1978) 394–398.

[W15] E. Witten, Topological quantum field theory, *Commun. Math. Phys.* **117** (1988) 353–386.

[W16] E. Witten, Anti-de Sitter space and holography, *Adv. Theor. Math. Phys.* **2** (1998) 253–291.

[W17] E. Witten, D. Olive, Supersymmetry algebras that include topological charges, *Phys. Lett.* **B 78** (1978) 97–101.

[Y1] J. P. Yamron, W. Siegel, Unified description of the $N = 2$ scalar multiplet *Nucl. Phys.* **B 263** (1986) 70–92.

[Y2] C. N. Yang, Condition of self-duality for SU(2) gauge fields, *Phys. Rev. Lett.* **38** (1977) 1377–1379.

[Z1] B. Zumino, Supersymmetry and Kähler manifolds, *Phys. Lett.* **B 87** (1979) 203–206.

[Z2] B. M. Zupnik, Six-dimensional supergauge theories in harmonic superspace, *Yad. Fiz.* **48** (1986) 794–802.

[Z3] B. M. Zupnik, Solution of constraints of the supergauge theory in $SU(2)/U(1)$ harmonic superspace, *Theor. Mat. Fiz.* **69** (1986) 207–213.

[Z4] B. M. Zupnik, The action of the supersymmetric $N = 2$ gauge theory in harmonic superspace, *Phys. Lett.* **B 183** (1987) 175–176.

[Z5] B. M. Zupnik, Harmonic superpotentials and symmetries in gauge theories with eight supercharges, *Nucl. Phys.* **B 554** (1999) 365–390.

[Z6] B. M. Zupnik, D. V. Hetselius, Three-dimensional extended supersymmetry in the harmonic superspace, *Sov. J. Nucl. Phys.* **48** (1988) 730–735.

Index